Trends in Partial Differential Equations of Mathematical Physics

José F. Rodrigues
Gregory Seregin
José Miguel Urbano
Editors

Birkhäuser
Basel · Boston · Berlin

Progress in Nonlinear Differential Equations and Their Applications
Volume 61

Editors:

José Francisco Rodrigues
Universidade de Lisboa / CMAF
Av. Prof. Gama Pinto, 2
1649-003 Lisboa
Portugal
rodrigue@ptmat.fc.ul.pt

Gregory Seregin
Steklov Institute of Mathematics
Russian Academy of Science
27, Fontanka
St. Petersburg 191023
Russia
seregin@pdmi.ras.ru

José Miguel Urbano
Departamento de Matemática
Universidade de Coimbra / FCT
Apartado 3008
3001-454 Coimbra
Portugal
jmurb@mat.uc.pt

2000 Mathematics Subject Classification 35-06; 35Q30, 76D05

A CIP catalogue record for this book is available from the Library of Congress, Washington D.C., USA

Bibliographic information published by Die Deutsche Bibliothek
Die Deutsche Bibliothek lists this publication in the Deutsche Nationalbibliografie; detailed bibliographic data is available in the Internet at <http://dnb.ddb.de>.

ISBN 3-7643-7165-X Birkhäuser Verlag, Basel – Boston – Berlin

Part of Springer Science+Business Media
Printed on acid-free paper produced of chlorine-free pulp. TCF ∞
Printed in Germany
ISBN-10: 3-7643-7165-X
ISBN-13: 978-3-7643-7165-4

9 8 7 6 5 4 3 2 1 www.birkhauser.ch

Table of Contents

Preface

Vsevolod Alekseevich Solonnikov is known as one of the outstanding mathematicians from the St. Petersburg Mathematical School. His remarkable results on exact estimates of solutions to boundary and initial-boundary value problems for linear elliptic, parabolic, and Stokes systems, his methods and contributions to the investigation of free boundary problems, in particular in fluid mechanics, are well known to specialists all over the world.

The International Conference on "Trends in Partial Differential Equations of Mathematical Physics" was held on the occasion of his 70th birthday in Óbidos (Portugal), from June 7 to 10, 2003. It was an organization of the "Centro de Matemática e Aplicações Fundamentais da Universidade Lisboa", in collaboration with the "Centro de Matemática da Universidade de Coimbra", the "Centro de Matemática Aplicada do IST/Universidade Técnica de Lisboa", the "Centro de Matemática da Universidade da Beira Interior", from Portugal, and with the Laboratory of Mathematical Physics of the St. Petersburg Department of the Steklov Institute of Mathematics from Russia.

The conference consisted of thirty eight invited and contributed lectures and gathered, in the charming and unique medieval town of Óbidos, about sixty participants from fifteen countries, namely USA, Switzerland, Spain, Russia, Portugal, Poland, Lithuania, Korea, Japan, Italy, Germany, France, Canada, Australia and Argentina. Several colleagues gave us a helping hand in the organization of the conference. We are thankful to all of them, and in particular to Stanislav Antontsev, Anvarbek Meirmanov and Adélia Sequeira, that integrated also the Organizing Committee. A special acknowledgement is due to Elena Frolova that helped us in compiling the short and necessarily incomplete bio-bibliographical notes below.

This book contains twenty original contributions, selected from the invited talks of the Óbidos conference and complements the special issue "Boundary-Value Problems of Mathematical Physics and Related Problems of Function Theory. Part 34" published as Vol. 306 of "Zapiski Nauchnyh Seminarov POMI" (2003), which is also dedicated to Professor Solonnikov.

Support from various institutions, in addition to the participating research centers already mentioned, has been essential for the realization of the conference and this book. We acknowledge the financial support from the Portuguese Fundação para a Ciência e Tecnologia and the Calouste Gulbenkian Foundation and the Town of Óbidos for the generous support and hospitality that provided a nice atmosphere for a very friendly meeting.

José Francisco Rodrigues, Gregory Seregin and José Miguel Urbano

Lisbon, St. Petersburg and Coimbra, September 2004

On Vsevolod Alekseevich Solonnikov and his 70th birthday

Vsevolod Alekseevich Solonnikov (V.A.) was born on the 8th of June, 1933 in Leningrad, and was brought up by his mother and nurse. His mother died in Leningrad during the Second World War, and V.A. spent several years in a children's house after the war. In his childhood, V.A. was keen on music and, in 1947, he entered the Music School at Leningrad Conservatoire where he studied violoncello. Later on, in 1951 he entered the Leningrad Conservatoire. Due to his talent, V.A. finished the first year at the Conservatoire with excellent marks, but, nevertheless, he decided to change his future profession. His interest in physics and mathematics prevailed, and in 1952 V.A. entered the physics department of Leningrad State University, starting a new period of his life that would lead him to mathematics.

At the physics department, V.A. attended lectures and seminars organized by Olga A. Ladyzhenskaya and defended a diploma thesis under her supervision. In 1957, V.A. graduated from the University with an excellent diploma and started to work at the Leningrad Branch of the Steklov Mathematical Institute, in the scientific group headed by O.A. Ladyzhenskaya.

In his candidate thesis (1961) V.A. proved the a priory estimates for solutions of boundary value problems for elliptic and parabolic partial differential equations, and also for the stationary Stokes system in Sobolev spaces. In 1965, V.A. defended his doctoral thesis. In [1,2], he constructed the solvability theory for parabolic systems of general type in Sobolev and Hölder spaces. Many interesting results for parabolic problems are presented in chapters 4 and 7 of the classical monograph [3]. In [4], he proved his famous coercive estimates for solutions to the three-dimensional non-stationary Stokes system under Dirichlet boundary conditions. With the help of these estimates, in [5] V.A. investigated the differentiability properties of generalized solutions to the non-stationary Navier–Stokes system. Later on, in 1973, V.A. gave another proof of those estimates which is nowadays classical, see [5]. In the mid-1990s, in connection with the linearization of modified Navier–Stokes equations suggested by O.A. Ladyzhenskaya, he obtained coercive estimates for their solutions in anisotropic Sobolev spaces.

Since the 1970s, V.A. concentrated his scientific interests on various mathematical problems of fluid dynamics. In the mid-70s, he began his studies of boundary value problems for Stokes and Navier–Stokes equations in domains with noncompact boundaries. Here, the important step forward was the right choice of the main functional spaces made by him, together with O.A. Ladyzhenskaya. The most significant results were the content of their fundamental papers [6,7]. In [8], they also analyzed problems for domains with arbitrary "outlets" to infinity without assumptions on boundedness of the Dirichlet integral. The review on problems for noncompact boundaries can be found in [9,10].

One of the favorite topics of V.A. is free boundary problems for viscous incompressible fluids. He started to think about them in the 1970s. One of the remarkable results in this direction was about the properties of solutions in the vicinity of the contact between a free surface and a rigid wall, see [11,12]. At the beginning of the 1980s, V.A. considered the problem of describing the motion of a drop. The series of works on this theme started in 1984 with the small paper [13], where he formulated a problem on the evolution of an isolated volume of viscous incompressible capillary fluid and gave the plan of the proof of the local unique solvability to this problem in anisotropic Sobolev–Slobodetskii spaces. To prove the solvability of the corresponded linear problem, V.A. used the method of construction of regularizers introduced by him earlier in the monograph [3].

The complete proof of the existence theorems to the problem of the motion of a viscous drop was finished by the beginning of the 1990s [14,15]. In the case of sufficiently small initial velocity and in the case the initial shape of the drop is close to a ball, V.A. proved the existence of a global solution for the problem of the motion of a finite volume of self-gravitating capillary incompressible [16] and compressible [17] fluids. He showed that in a coordinate system connected with the center of mass of the drop, for large moments of time, the solution approaches the rotation of the liquid as a rigid body around a certain axis, and the domain occupied by the fluid tends to an equilibrium figure.

In the 1990's, V.A. introduced also new ideas and obtained sharp results on some classical free boundary problems for parabolic equations. Based on new coercive estimates for solutions of linearized problems, obtained after the transformation of the free boundary problems in highly nonlinear and nonstandard evolution problems, he obtained, with some of his co-workers, interesting local solvability of the Stefan problem in Sobolev spaces [18], deep estimates for certain systems, for instance in [19], and the classical solvability of the Stefan and Muskat–Veregin two-phase problems in [20], obtaining the sharpest regularity in Hölder spaces (at least equal regularity for $t = 0$ and for $t > 0$, at the lowest possible order). A detailed presentation of this results and methods of analyzing free boundary problems where given by V.A. in the lecture notes [21].

In the last years, V.A. analyzed the stability of axially symmetric and non-symmetric equilibrium figures of rotating viscous incompressible liquid. This is a famous classic problem that had attracted the attention of great mathematicians like, for instance, Newton, Jacobi, Liouville, Poincaré and Lyapunov. Recently, he succeed to prove that the positiveness of the second variation of the energy functional on a certain subspace of functions is the sufficient (and correct) condition of stability of non-symmetric equilibrium figures [22]. This deep and remarkable result can be considered as an extension of classical results on the stability of ellipsoidal equilibrium forms of rotating fluid without taking into account the surface tension, and the rigorous justification of the principle of minimum of the energy functional was obtained after the recent analysis and progresses he made on a certain evolution free boundary problem for the Navier–Stokes equations.

After 1969, V.A. has worked also at Leningrad State University (Department of Mathematics and Mechanics). He gave basic and special courses of lectures, and wrote lecture notes together with N.N. Uraltseva. Many students obtained a diploma under his supervision, and he had also several post-graduate students. V.A. always took care of his students, attentively looking after their work and collaborating with them. Owing to the hospitality of his wife Tatiana Fedorovna, his home became native for many of his students and friends.

V.A. Solonnikov continues to be a leading researcher at the St. Petersburg Branch of the Steklov Mathematical Institute and pursues intensively and successfully his work, corresponding to multiple invitations, participations in international conferences and visits in Europe and in Asia. Among many other distinctions, in 1997 he obtained a senior research fellowship at the University of Lisbon, Portugal, and in 2002 he was awarded the prize of Humboldt Research Award, in Germany. In recent years he is being also lecturing as Professor of Rational Mechanics at the University of Ferrara, in Italy, where a second International Conference ("Directions on Partial Differential Equations", 6–9 November 2003) on the occasion of his 70th birthday took also place, following the one of Óbidos, in Portugal.

References

[1] V.A. Solonnikov, *A priory estimates for solutions of second order equations of parabolic type*, Trudy Mat. Inst. Steklov, **70** (1964), 133–212; English transl. in Amer. Mat. Soc. Transl., **75**(2) (1968).

[2] V.A. Solonnikov, *On boundary value problems for linear parabolic systems of differential equations of general form*, Trudy Mat. Inst. Steklov, **83** (1965), 3–162.

[3] O.A. Ladyzhenskaya, V.A. Solonnikov, N.N. Uraltseva, *Linear and quasi-linear equations of parabolic type*, M., Nauka 1967; English transl.: Transl. Math. Monographs, **23**, Amer. Math. Soc., Providence, RI, 1968.

[4] V.A. Solonnikov, *Estimates of the solution of a time-depended linearized system of Navier–Stokes equations*, Trudy Mat. Inst. Steklov, **70** (1964), 213–317; English transl. in Amer. Math. Soc. Transl., **75**(2) (1968).

[5] V.A. Solonnikov, *On differential properties of solutions of the first boundary value problem for the non-stationary system of Navier–Stokes equations*, Trudy Mat. Inst. Steklov, **73** (1964), 221–291.

[6] O.A. Ladyzhenskaya and V.A. Solonnikov, *On the solvability of boundary and initial-boundary value problems for the Navier–Stokes equations in regions with noncompact boundaries*, Vestnik Leningrad. Univ. (1977), **13**, 39–47; English transl. in Vestnik Leningrad Univ. Math., **10** (1982), 271–279.

[7] O.A. Ladyzhenskaya and V.A. Solonnikov, *Some problems of vector analysis, and generalized formulations of boundary value problems for the Navier–Stokes equations*, Zap. Nauchn. Sem. LOMI, **59** (1976), 81–116; English transl. in J. Soviet. Math., **10**(2) (1978).

[8] O.A. Ladyzhenskaya and V.A. Solonnikov, *Determination of solutions of boundary value problems for steady-state Stokes and Navier–Stokes equations having an unbounded Dirichlet integral*, Zap. Nauchn. Sem. LOMI, **96** (1979), 117–160; English transl. in J. Soviet Math., **21** (1983), 728–761.

[9] V.A. Solonnikov, *On the problems of a viscous incompressible liquid in domains with noncompact boundaries*, Algebra i Analyz, **4**(6) (1992), 28–53; English transl. in St. Petersburg Math. J., **4**(6) (1993), 1081–1102.

[10] V.A. Solonnikov, *Stokes and Navier–Stokes equations in domains with noncompact boundaries*, Nonlinear Partial Differential Equations and there Applications (College de France Sem., **4** (1981/82); Res. Notes in Math., **84**, Pitman (1983), 240–349.

[11] V.A. Solonnikov, *Solvability of the problem of the plane motion of a heavy viscous incompressible capillary liquid partially filling a container*, Izv. Acad. Nauk SSSR Ser. Mat., **43** (1979), 203–236; English transl. in Math. USSR Izv., **14** (1980), 193–221.

[12] V.A. Solonnikov, *Solvability of three-dimensional problems with a free boundary for a system of steady-state Navier–Stokes equations*, Zap. Nauchn. Sem. LOMI, **84** (1979), 252–285; English transl. in J. Soviet Math., **21** (1983), 427–450.

[13] V.A. Solonnikov, *Solvability of the problem of evolution of an isolated volume of a viscous incompressible capillary fluid*, Zap. Nauchn. Sem. LOMI, **140** (1984), 179–186; English transl. in J. Soviet. Mat., **32** (1986), 223–228.

[14] V.A. Solonnikov, *An initial-boundary value problem for a Stokes system that arises in the study of a problem with a free boundary*, Trudy Mat. Inst. Steklov, **188** (1990), 150–188; English transl. in Proc. Steklov Inst. Math., **3** (1991), 191–239.

[15] V.A. Solonnikov, *Solvability of a problem on the evolution of a viscous incompressible fluid bounded by a free surface on a finite time interval*, Algebra i Analiz, **3**(1) (1991), 222–257; English transl. in St. Petersburg Math. J., **3**(1) (1992), 189–220.

[16] V.A. Solonnikov, *Unsteady motion of a finite isolated mass of a self-gravitating fluid*, Algebra i Analiz, **1**(1) (1989), 207–249; English transl. in Leningrad Math. J., **1**(1) (1990), 227–276.

[17] V.A. Solonnikov, *Evolution free boundary problem for equations of motion of viscous compressible barotropic self-gravitating fluid*, Stability Appl. Anal. Contin. Media, **3**(4) (1993), 257–275.

[18] V.A. Solonnikov, E. Frolova, *L_p-theory of the Stefan problem*, Zap. Nauchn. Sem. S.-Peterburg. Otdel. Mat. Inst. Steklov (POMI), **249** (1997) 299–323.

[19] J.F. Rodrigues, V.A. Solonnikov, F. Yi, *On a parabolic system with time derivative in the boundary conditions and related free boundary problems*, Math. Ann., **315** (1999), 61–95.

[20] G. I. Bizhanova, V.A. Solonnikov, *On problems with free boundaries for second-order parabolic equations* (Russian), Algebra i Analiz, **12**(6) (2000), 98–139; translation in St. Petersburg Math. J., **12**(6) (2001), 949–981.

[21] V.A. Solonnikov, *Lectures on evolution free boundary problems: classical solutions*, Lecture Notes in Math., Springer, **1812** (2003), 123–175.

[22] V.A. Solonnikov, *On stability of non-symmetric equilibrium figures of rotating viscous incompressible fluid*, Interfaces and Free Boundaries, **6**(4) (2004).

List of Participants

Carlos J.S. Alves, Instituto Superior Técnico, Portugal, calves@math.ist.utl.pt
Herbert Amann, Univ. of Zürich, Switzerland, amann@math.unizh.ch
Stanislav Antontsev, Univ. da Beira Interior, Portugal, anton@ubi.pt
Assis Azevedo, Univ. do Minho, Portugal, assis@math.uminho.pt
Michael Bildhauer, Saarland Univ., Germany, bibi@math.uni-sb.de
Lorenzo Brandolese, Univ. Paris XII, Belgium, brandole@cmla.ens-cachan.fr
Luis Caffarelli, Univ. of Texas at Austin, USA, caffarel@math.utexas.edu
João Paulo Carvalho Dias, Univ. de Lisboa, Portugal, dias@ptmat.fc.ul.pt
Irina Denisova, Institute of Mechanical Eng. Problems, Russia, ira@wave.ipme.ru
Emmanuele DiBenedetto, Vanderbilt Univ., USA, em.diben@vanderbilt.edu
Carsten Ebmeyer, Univ. Bonn, Germany, ebmeyer@uni.bonn.de
Antonio Fasano, Univ. Firenze, Italy, fasano@math.unifi.it
Mário Figueira, Univ. de Lisboa, Portugal, figueira@ptmat.fc.ul.pt
Elena Frolova, St. Petersburg Electrotehnical Univ., Russia, elenafr@mail.ru
Diogo Gomes, Instituto Superior Técnico, Portugal, dgomes@math.ist.utl.pt
César Rafael Gonçalves, IPG-ESTG, Portugal, crg@netvisao.pt
Gabriele Grillo, Politecnico di Torino, Italy, gabriele.grillo@polito.it
Giovanna Guidoboni, Univ. of Ferrara, Italy, gio@dm.unife.it
Eurica Henriques, UTAD, Portugal, eurica@utad.pt
John G. Heywood, Univ. of British Columbia, Canada, heywood@math.ubc.ca
Bum Ja Jin, Seoul National Univ., Republic of Korea, bumjajin@hanmail.net
Olga Ladyzhenskaya, POMI, St. Petersburg, Russia
Anvarbek Meirmanov, Univ. Hanty-Mansiisk, Russia
Fernando Miranda, Univ. do Minho, Portugal, fmiranda@math.uminho.pt
Ilia Moguilevski, Tver State Univ., Russia, ilia.mogilevski@tversu.ru
Piotr Mucha, Warsaw Univ., Poland, mucha@hydra.mimuw.edu.pl
Serguei Nazarov, Instituto Superior Técnico, Portugal, nazarov@math.ist.utl.pt
Basil Nicolaenko, Arizona State Univ., USA, byn@stokes.la.asu.edu
Mariarosaria Padula, Univ. of Ferrara, Italy, pad@unife.it

Carlo Domenico Pagani, Politecnico di Milano, Italy, carpag@mate.polimi.it
Dário Pierotti, Politecnico di Milano, Italy, dario.pierotti@polimi.it
Konstantin Pileckas, Inst. of Math. and Informatics, Lithuania, pileckas@ktl.mii.lt
Vladislav Pukhnachov, Lavrentyev Inst. of Hydrodynamics, Russia, pukh@hydro.nsc.ru
Diego Rapoport, Univ. of Buenos Aires and Quilmes, Argentina, drapo@unq.edu.ar
Reimund Rautmann, Univ. of Paderborn, Germany, rautmann@math.uni-paderborn.de
Ludmila Rivkind, Univ. Dortmund, Germany, rivkind@mathematik.uni-dortmund.de
José Francisco Rodrigues, Univ. de Lisboa, Portugal, rodrigue@ptmat.fc.ul.pt
Jorge Salazar, Univ. de Lisboa, Portugal, salazar@ptmat.fc.ul.pt
Lisa Santos, Univ. do Minho, Portugal, lisa@math.uminho.pt
Luís Saraiva, Univ. de Lisboa, Portugal, mmff5@ptmat.fc.ul.pt
Adélia Sequeira, Inst. Superior Técnico, Portugal, adelia.Sequeira.Silva@math.ist.utl.pt
Gregory Seregin, Steklov Mathematical Institute, St. Petersburg, Russia, seregin@pdmi.ras.ru
Timofey Shilkin, Steklov Mathematical Institute, St. Petersburg, Russia, shilkin@pdmi.ras.ru
Sergey Shmarev, Univ. of Oviedo, Spain, shmarev@orion.ciencias.uniovi.es
Jorge Drumond Silva, Inst. Superior Técnico, Portugal, jsilva@math.ist.utl.pt
Ana Leonor Silvestre, Inst. Superior Técnico, Portugal, Ana.Silvestre@math.ist.utl.pt
Vsevolod Solonnikov, Steklov Mathematical Institute, St. Petersburg, Russia, solonnik@pdmi.ras.ru
Maria Specovius-Neugebauer, Univ. of Kassel, Germany, specovi@mathematik.uni-kassel.de
Victor Starovoitov, CAESAR, Germany, starovoitov@caesar.de
Atusi Tani, Keio Univ., Japan, tani@math.keio.ac.jp
Gérard Tronel, Univ. Pierre et Marie Curie, France, g.tronel@wanadoo.fr
Neil S. Trudinger, ANU, Canberra, Australia, neil.trudinger@anu.edu.au
José Miguel Urbano, Univ. de Coimbra, Portugal, jmurb@mat.uc.pt
Juha Videman, Instituto Superior Técnico, Portugal, videman@math.ist.utl.pt
Vadim Yurinsky, Univ. da Beira Interior, Portugal, yurinsky@ubi.pt

Progress in Nonlinear Differential Equations
and Their Applications, Vol. 61, 1–14

Stopping a Viscous Fluid by a Feedback Dissipative Field: Thermal Effects without Phase Changing

S.N. Antontsev, J.I. Díaz and H.B. de Oliveira

Dedicated to Professor V.A. Solonnikov on the occasion of his 70th birthday.

Abstract. We show how the action on two simultaneous effects (a suitable coupling about velocity and temperature and a low range of temperature but upper that the phase changing one) may be responsible of stopping a viscous fluid without any changing phase. Our model involves a system, on an unbounded pipe, given by the planar stationary Navier-Stokes equation perturbed with a sublinear term $\mathbf{f}(\mathbf{x}, \theta, \mathbf{u})$ coupled with a stationary (and possibly nonlinear) advection diffusion equation for the temperature θ.

After proving some results on the existence and uniqueness of weak solutions we apply an energy method to show that the velocity $\mathbf{u}$ vanishes for $\mathbf{x}$ large enough.

Mathematics Subject Classification (2000). 76A05 ,76D07, 76E30, 35G15.

Keywords. Non-Newtonian fluids, nonlinear thermal diffusion equations, feedback dissipative field, energy method, heat and mass transfer, localization effect.

1. Introduction

It is well known (see, for instance, [6, 8, 14]) that in phase changing flows (as the Stefan problem) usually the solid region is assumed to remain static and so we can understand the final situation in the following way: the thermal effect are able to stop a viscous fluid.

The main contribution of this paper is to show how the action on two simultaneous effects (a suitable coupling about velocity and temperature and a low range of temperature but upper the phase changing one) may be responsible of stopping a viscous fluid without any changing phase. This philosophy could be useful in the monitoring of many flows problems, specially in metallurgy.

We shall consider a, non-standard, Boussinesq coupling among the temperature θ and the velocity $\mathbf{u}$. Motivated by our previous works (see [1, 2, 3, 4]), we assume the body force field is given in a non-linear feedback form, $\mathbf{f} : \Omega \times \mathbb{R} \times \mathbb{R}^2 \rightarrow \mathbb{R}^2$, $\mathbf{f} = (f_1(\mathbf{x}, \theta, \mathbf{u}), f_2(\mathbf{x}, \theta, \mathbf{u}))$, where $\mathbf{f}$ is a Carathéodory function (*i.e.*, continuous on θ and $\mathbf{u}$ and measurable in $\mathbf{x}$) such that, for every $\mathbf{u} \in \mathbb{R}^2$, $\mathbf{u} = (u, v)$, for any $\theta \in [m, M]$, and for almost all $\mathbf{x} \in \Omega$

$$-\mathbf{f}(\mathbf{x}, \theta, \mathbf{u}) \cdot \mathbf{u} \geq \delta\, \chi_{\mathbf{f}}(\mathbf{x})\, |u|^{1+\sigma(\theta)} - g(\mathbf{x}, \theta) \tag{1.1}$$

for some $\delta > 0$, σ a Lipschitz continuous function such that

$$0 < \sigma^- \leq \sigma(\theta) \leq \sigma^+ < 1, \quad \theta \in [m, M], \tag{1.2}$$

and

$$g \in \mathrm{L}^1\left(\Omega^{x_g} \times \mathbb{R}\right), \quad g \geq 0,\ g(\mathbf{x}, \theta) = 0 \text{ a.e. in } \Omega^{x_g} \text{ for any } \theta \in [m, M], \tag{1.3}$$

for some $x_{\mathbf{f}}$, x_g, with $0 \leq x_g < x_{\mathbf{f}} \leq \infty$ and $x_{\mathbf{f}}$ large enough, where $\Omega^{x_g} = (0, x_g) \times (0, L)$ and $\Omega_{x_g} = (x_g, \infty) \times (0, L)$. The function $\chi_{\mathbf{f}}$ denotes the characteristic function of the interval $(0, x_{\mathbf{f}})$, *i.e.*, $\chi_{\mathbf{f}}(\mathbf{x}) = 1$, if $x \in (0, x_{\mathbf{f}})$ and $\chi_{\mathbf{f}}(\mathbf{x}) = 0$, if $x \notin (0, x_{\mathbf{f}})$. We shall not need any monotone dependence assumption on the function $\sigma(\theta)$.

It seems interesting to notice that the term $\mathbf{f}(\mathbf{x}, \theta, \mathbf{u})$ plays a similar role to the one in the penalized changing phase problems (see equation (3.13) of [14]), although our formulation and our methods of proof are entirely different. We shall prove that the fluid is stopped at a finite distance of the semi-infinite strip entrance by reducing the nonlinear system to a fourth order non-linear scalar equation for which the localization of solutions is obtained by means of a suitable energy method (see [5]).

2. Statement of the problem

In the domain $\Omega = (0, \infty) \times (0, L)$, $L > 0$, we consider a planar stationary thermal flow of a fluid governed by the following system

$$(\mathbf{u} \cdot \nabla)\mathbf{u} = \nu \triangle \mathbf{u} - \nabla p + \mathbf{f}(\mathbf{x}, \theta, \mathbf{u}), \tag{2.4}$$

$$\mathrm{div}\mathbf{u} = 0, \tag{2.5}$$

$$\mathbf{u} \cdot \nabla \mathcal{C}(\theta) = \triangle \varphi(\theta), \tag{2.6}$$

where $\mathbf{u} = (u, v)$ is the vector velocity of the fluid, θ its absolute temperature, p is the hydrostatic pressure, ν is the kinematics viscosity coefficient,

$$\mathcal{C}(\theta) := \int_{\theta_0}^{\theta} C(s)\, ds \quad \text{and} \quad \varphi(\theta) := \int_{\theta_0}^{\theta} \kappa(s)\, ds,$$

with $C(\theta)$ and $\kappa(\theta)$ being the specific heat and the conductivity, respectively. Assuming $\kappa > 0$ then φ is invertible and so $\theta = \varphi^{-1}(\overline{\theta})$ for some real argument $\overline{\theta}$. Then we can define functions

$$\overline{\mathcal{C}}(\overline{\theta}) := \mathcal{C} \circ \varphi^{-1}(\overline{\theta}), \quad \overline{\mathbf{f}}(\mathbf{x}, \overline{\theta}, \mathbf{u}) := \mathbf{f} \circ \varphi^{-1}(\overline{\theta}), \quad \overline{\mu}(\overline{\theta}) := \mu \circ \varphi^{-1}(\overline{\theta}).$$

We point out that functions $\overline{\mathcal{C}}, \overline{\mathbf{f}}$ and $\overline{\mu}$ are Lipschitz continuous functions of $\overline{\theta}$. Substituting these expressions in (2.4)–(2.6), we get, omitting the bars,

$$(\mathbf{u} \cdot \nabla)\mathbf{u} = \nu \triangle \mathbf{u} - \nabla p + \mathbf{f}(\mathbf{x}, \theta, \mathbf{u}), \tag{2.7}$$

$$\mathrm{div}\mathbf{u} = 0, \tag{2.8}$$

$$\mathbf{u} \cdot \nabla \mathcal{C}(\theta) = \triangle \theta. \tag{2.9}$$

To these equations we add the following boundary conditions on $\mathbf{u}$

$$\mathbf{u} = \mathbf{u}_*, \quad \text{on } x = 0, \tag{2.10}$$

$$\mathbf{u} = \mathbf{0}, \quad \text{on } y = 0,\, L, \tag{2.11}$$

$$\mathbf{u} \to \mathbf{0}, \quad \text{when } x \to \infty, \tag{2.12}$$

and on θ

$$\theta = \theta_*, \quad \text{on} \;\; x = 0,\; y = 0,\, L, \tag{2.13}$$

$$\theta \to 0, \quad \text{when} \;\; x \to \infty, \tag{2.14}$$

where $\mathbf{u}_*$ and θ_* are given functions with a suitable regularity to be indicated later on and

$$0 \leq m \leq \theta_*(\mathbf{x}) \leq M < \infty. \tag{2.15}$$

We assume the possible non-zero velocity $\mathbf{u}_*$ and temperature θ_* satisfy the compatibility conditions

$$\mathbf{u}_*(0) = \mathbf{u}_*(L) = \mathbf{0}, \quad \int_0^L u_*(s)ds = 0, \tag{2.16}$$

$$\theta_*(x,y) \to 0, \quad \text{when } x \to \infty \;\; \text{for any} \;\; y \in [0, L]. \tag{2.17}$$

3. Existence theorem

As in [1, 2, 3, 4], we introduce the functional spaces

$$\widetilde{\mathbf{H}}(\Omega) = \left\{ \mathbf{u} \in \mathbf{H}(\Omega) : \mathbf{u}(0,.) = \mathbf{u}_*(.),\; \mathbf{u}(.,0) = \mathbf{u}(.,L) = \mathbf{0},\; \lim_{x \to \infty} |\mathbf{u}| = 0 \right\},$$

$$\widetilde{\mathbf{H}}_0(\Omega) = \left\{ \mathbf{u} \in \mathbf{H}(\Omega) : \mathbf{u}(0,.) = \mathbf{u}(.,0) = \mathbf{u}(.,L) = \mathbf{0},\; \lim_{x \to \infty} |\mathbf{u}| = 0 \right\},$$

where $\mathbf{H}(\Omega) = \left\{ \mathbf{u} \in \mathbf{H}^1(\Omega) : \mathrm{div}\mathbf{u} = 0 \right\}$, and assume that

$$\mathbf{u}_* \in \mathbf{H}^{\frac{1}{2}}(0, L). \tag{3.18}$$

We shall search solutions $(\theta, \mathbf{u})$ such that, additionally to assumptions (2.14) and (2.12), satisfy

$$\int_\Omega |\nabla \theta|^2 d\mathbf{x} < \infty \quad \text{and} \quad \int_\Omega |\nabla \mathbf{u}|^2 d\mathbf{x} < \infty.$$

Moreover, due to the fact that the Poincaré inequality

$$\int_\Omega |w|^p \, d\mathbf{x} \leq \left(\frac{L}{\pi} \right)^p \int_\Omega |\nabla w|^p \, d\mathbf{x}, \tag{3.19}$$

holds for every $w \in \mathrm{W}_0^{1,p}(\Omega)$ and $1 \leq p < \infty$ (see, *e.g.*, [10]), our searched solutions $(\theta, \mathbf{u})$ will be elements of the Sobolev space $\mathrm{H}^1(\Omega) \times \mathbf{H}^1(\Omega)$.

Let us still denote by $\mathbf{u}_*$ and θ_* the extensions of the boundary data to the whole domain Ω in a way such that

$$\mathbf{u}_* \in \widetilde{\mathbf{H}}(\Omega) \quad \text{and} \quad \theta_* \in \mathrm{W}^{1,q}(\Omega) \cap \mathrm{C}^\alpha(\overline{\Omega}), \quad 2 < q < \infty,\ \alpha > 0. \tag{3.20}$$

Definition 3.1. *The pair* $(\theta, \mathbf{u})$ *is said to be a* weak solution *of* (2.7)–(2.14) *if:*

(i) $\theta - \theta_* \in \mathrm{W}_0^{1,q}(\Omega) \cap \mathrm{C}^\alpha(\overline{\Omega})$, $\alpha > 0$, $2 < q < \infty$, $m \leq \theta \leq M$ *and for any test function* $\zeta \in \mathrm{W}_0^{1,q'}(\Omega)$ $(1/q + 1/q' = 1)$

$$\int_\Omega (\nabla\theta - \mathcal{C}(\theta)\mathbf{u}) \cdot \nabla\zeta \, d\mathbf{x} = 0.$$

(ii) $\mathbf{u} \in \widetilde{\mathbf{H}}(\Omega)$, $\mathbf{u} - \mathbf{u}_* \in \widetilde{\mathbf{H}}_0(\Omega)$, $\mathbf{f}(\mathbf{x}, \theta(\mathbf{x}), \mathbf{u}(x)) \in \mathbf{L}^1_{\mathrm{loc}}(\Omega)$ *and for every* $\varphi \in \widetilde{\mathbf{H}}_0(\Omega) \cap \mathbf{L}^\infty(\Omega)$ *with compact support,*

$$\nu \int_\Omega \nabla\mathbf{u} : \nabla\varphi \, d\mathbf{x} + \int_\Omega \mathbf{u} \cdot \nabla\mathbf{u} \cdot \varphi \, d\mathbf{x} = \int_\Omega \mathbf{f} \cdot \varphi \, d\mathbf{x}. \tag{3.21}$$

In this section, we shall assume that $\mathbf{f} : \Omega \times \mathbb{R} \times \mathbb{R}^2 \to \mathbb{R}^2$ is given by the structural condition

$$\mathbf{f}(\mathbf{x}, \theta, \mathbf{u}) = -\delta\, \chi_{\mathbf{f}}(\mathbf{x})(|u|^{\sigma(\theta)-1}u, 0) - \mathbf{h}(\mathbf{x}, \theta, \mathbf{u}), \tag{3.22}$$

for any $\mathbf{u} = (u, v)$, any $\theta \in [m, M]$ and almost every $\mathbf{x} \in \Omega$, for some $\delta > 0$, $0 \leq x_{\mathbf{f}} \leq \infty$ and $\sigma(\theta)$ satisfies (1.2). Here, $\mathbf{h}(\mathbf{x}, \theta, \mathbf{u})$ is a Carathéodory function such that

$$\mathbf{h}(\mathbf{x}, \theta, \mathbf{u}) \cdot \mathbf{u} \geq -g(\mathbf{x}, \theta), \tag{3.23}$$

for every $\mathbf{u} \in \mathbb{R}^2$, for any $\theta \in [m, M]$ and almost all $\mathbf{x} \in \Omega$, for some function g satisfying (1.3), and we assume

$$H_K \in \mathrm{L}^1(\Omega^{x_{\mathbf{f}}}) \text{ for all } K > 0, \quad H_K(\mathbf{x}) = \sup_{|\mathbf{u}| \leq K,\ \theta \in [m,M]} |\mathbf{h}(\mathbf{x}, \theta, \mathbf{u})|. \tag{3.24}$$

Theorem 3.1. *Under conditions* (1.2), (2.15)–(2.17), (3.18), (3.20) *and* (3.22)–(3.24), *the problem* (2.7)–(2.14) *has, at least, one weak solution* $(\theta, \mathbf{u})$.

Proof. We will prove this theorem in several steps.
First step: an auxiliary problem for the temperature θ. Let

$$\mathbf{w} \in \mathbf{L}^2(\Omega) \cap \mathbf{L}^q(\Omega), \quad \text{with } 2 < q < \infty, \tag{3.25}$$

be a given function and let us consider the following problem for the temperature

$$\mathbf{w} \cdot \nabla\mathcal{C}(\theta) = \triangle\theta \tag{3.26}$$

completed with the boundary conditions (2.13)–(2.14). Since $\mathcal{C}$ is Lipschitz continuous we know (see, *e.g.*, [7, 10, 11]) that problem (3.26), (2.13)–(2.14), assuming (3.25), has a unique weak solution θ such that

$$\|\theta\|_{\mathrm{W}^{1,q}(\Omega)}, \ \|\theta\|_{\mathrm{C}^\alpha(\overline{\Omega})} \leq C\left(q, \|\mathbf{w}\|_{\mathbf{L}^q(\Omega)}, \|\theta_*\|_{\mathrm{W}^{1,q}(\Omega)}\right) \tag{3.27}$$

where $\alpha = 1 + [2/q] - 2/q > 0$. Moreover, from the Maximum Principle,

$$m \leq \theta(x) \leq M.$$

Then we can define the non-linear operator

$$\Lambda : \mathbf{L}^2(\Omega) \cap \mathbf{L}^q(\Omega) \to \mathrm{W}^{1,q}(\Omega) \cap \mathrm{C}^{\alpha}(\overline{\Omega}), \quad \Lambda(\mathbf{w}) = \theta, \tag{3.28}$$

with $\alpha = 1 + [2/q] - 2/q > 0$, $2 < q < \infty$. The operator Λ is continuous, because from [10, 11], we get that given a sequence $\mathbf{w}_n$ such that

$$\|\mathbf{w}_n - \mathbf{w}\|_{\mathbf{L}^2(\Omega)} + \|\mathbf{w}_n - \mathbf{w}\|_{\mathbf{L}^q(\Omega)} \to 0, \quad \text{as } n \to \infty,$$

then

$$\|\Lambda(\mathbf{w}_n) - \Lambda(\mathbf{w})\|_{\mathrm{W}^{1,q}(\Omega)} + \|\Lambda(\mathbf{w}_n) - \Lambda(\mathbf{w})\|_{\mathrm{C}^{\alpha}(\overline{\Omega})} \to 0, \quad \text{as } n \to \infty.$$

Second step: an auxiliary problem for the velocity $\mathbf{u}$. Let ω be a given function such that

$$\omega \in \mathrm{W}^{1,q}(\Omega) \cap \mathrm{C}^{\alpha}(\overline{\Omega}), \quad 2 < q < \infty, \quad \alpha > 0, \quad m \leq \omega \leq M \tag{3.29}$$

and let us consider the problem for the velocity constituted by the following equation of motion

$$(\mathbf{u} \cdot \nabla)\mathbf{u} = \nu \triangle \mathbf{u} - \nabla p + \mathbf{f}(\mathbf{x}, \omega, \mathbf{u}), \tag{3.30}$$

the equation of continuity (2.8) and the boundary conditions (2.10)–(2.12). Applying the results of [4] (which is possible due to the assumptions (3.22)–(3.24) and (3.29)), the problem (3.30), (2.8), (2.10)–(2.12) has, at least, one weak solution $\mathbf{u} \in \mathbf{H}^1(\Omega)$ which satisfies

$$\int_{\Omega} \left(|\nabla \mathbf{u}|^2 + \chi_{\mathbf{f}} |u|^{1+\sigma(\theta)} + |\mathbf{h}(\mathbf{x}, \omega, \mathbf{u}) \cdot \mathbf{u}| \right) d\mathbf{x} \leq C, \tag{3.31}$$

where

$$C = C\left(L, m, M, \delta, \nu, \|\mathbf{u}_*\|_{\mathbf{H}^{\frac{1}{2}}(0,L)}, \|g\|_{\mathrm{L}^1(\Omega^{x_g} \times \mathbb{R})}\right)$$

and, in fact,

$$C = C_0\left(L, m, M, \delta, \nu \; \|\mathbf{u}_*\|^2_{\mathbf{H}^{\frac{1}{2}}(0,L)}\right),$$

if $g = 0$. Then we can define the non-linear operator

$$\Pi : \mathrm{W}^{1,q}(\Omega) \cap \mathrm{C}^{\alpha}(\overline{\Omega}) \to \mathbf{L}^q(\Omega), \quad \Pi(\omega) = \mathbf{u}, \tag{3.32}$$

with $2 < q < \infty$ and $\alpha > 0$, which is continuous.

Third step: application of Schauder's theorem. Given $q > 2$, formulas (3.28) and (3.32) allow to define the composition non-linear operator

$$\Upsilon = \Pi\Lambda : \mathbf{L}^2(\Omega) \cap \mathbf{L}^q(\Omega) \to \mathbf{L}^q(\Omega). \tag{3.33}$$

From (3.31) we get that Υ transforms $\mathbf{L}^2(\Omega) \cap \mathbf{L}^q(\Omega)$ into a bounded subset of $\mathbf{H}^1(\Omega)$ and, from the Sobolev compact embedding $\mathbf{H}^1(\Omega) \to \mathbf{L}^q(\Omega)$, $2 < q < \infty$, it is completely continuous. Then, according to Schauder's theorem, (3.33) has, at least, a fixed point. This proves the existence of a weak solution $(\theta, \mathbf{u})$ to problem (2.7)–(2.14). □

Remark 3.1. *Questions about the solvability of boundary value problems for the Navier-Stokes system in domains with noncompact boundaries were discussed by many authors amongst whom Solonnikov* [13].

4. Uniqueness of weak solution

For the sake of simplicity in the exposition we will assume in this section that the coupling thermal force obeys to the special form

$$\mathbf{f}(\mathbf{x}, \theta, \mathbf{u}) = -\delta\ \chi_{\mathbf{f}}(\mathbf{x})(|u(\mathbf{x})|^{\sigma(\theta)-1}u(\mathbf{x}), 0). \tag{4.34}$$

The main result of this section, concerning the uniqueness of solutions, is the following.

Theorem 4.1. *Assume* (1.2) *and* (2.15)–(2.17). *We additionally suppose that*

$$|\mathcal{C}'(\theta)| \leq \lambda \quad \textit{for every } \theta \in [m, M], \tag{4.35}$$

and

$$|\sigma'(\theta)| \leq \lambda \quad \textit{for every } \theta \in [m, M], \tag{4.36}$$

for $\lambda \leq \lambda^*$ *and for some small enough positive constant* $\lambda^* > 0$. *Then, if* $\|\mathbf{u}_*\|_{\mathbf{H}^{\frac{1}{2}}(0,L)} \leq \varepsilon^*$ *for some small enough positive constant* $\varepsilon^* > 0$, *the problem* (2.7)–(2.14), (4.34) *has a unique weak solution* $(\mathbf{u}, \theta)$.

Proof. Let $(\theta_1, \mathbf{u}_1)$, $\mathbf{u}_1 = (u_1, v_1)$, and $(\theta_2, \mathbf{u}_2)$, $\mathbf{u}_2 = (u_2, v_2)$, be two weak solutions to problem (2.7)–(2.14) and let us set $\theta = \theta_1 - \theta_2$, $\mathbf{u} = \mathbf{u}_1 - \mathbf{u}_2$. According to Definition 3.1, $\mathbf{u} = \mathbf{u_1} - \mathbf{u_2} \in \mathbf{H}^1_0(\Omega)$ and $\theta = \theta_1 - \theta_2 \in \mathrm{H}^1_0(\Omega)$. Moreover, functions θ, $\mathbf{u}$ satisfy to

$$\int_\Omega \left[\nabla\theta - (\mathcal{C}(\theta_1)\mathbf{u}_1 - \mathcal{C}(\theta_2)\mathbf{u}_2)\,\mathbf{u}\right] \cdot \nabla\zeta\, d\mathbf{x} = 0,$$

$$\nu \int_\Omega \nabla\mathbf{u} : \nabla\varphi\, d\mathbf{x} + \int_\Omega [(\mathbf{u}_1 \cdot \nabla)\mathbf{u}_1 - (\mathbf{u}_2 \cdot \nabla)\mathbf{u}_2] \cdot \varphi\, d\mathbf{x}$$

$$= -\delta \int_\Omega \chi_{\mathbf{f}}(\mathbf{x}) \left(|u_1(\mathbf{x})|^{\sigma(\theta_1)-1}u_1(\mathbf{x}) - |u_2(\mathbf{x})|^{\sigma(\theta_2)-1}u_2(\mathbf{x}), 0\right) \cdot \varphi\, d\mathbf{x}.$$

Setting $\zeta = \theta$ and $\varphi = \mathbf{u}$, we came to the relations

$$\int_\Omega |\nabla\theta|^2 d\mathbf{x} = \int_\Omega (\mathcal{C}(\theta_1) - \mathcal{C}(\theta_2))\,\mathbf{u}_1 \cdot \nabla\theta\, d\mathbf{x} + \int_\Omega \mathcal{C}(\theta_2)\mathbf{u} \cdot \nabla\theta\, d\mathbf{x} := J_1 + J_2, \tag{4.37}$$

$$\nu \int_\Omega |\nabla\mathbf{u}|^2\, d\mathbf{x} + I_1 = I_2 + I_3, \tag{4.38}$$

where

$$I_1 := \delta \int_\Omega \chi_{\mathbf{f}}(\mathbf{x}) \left(|u_1(\mathbf{x})|^{\sigma(\theta_1)-1} u_1(\mathbf{x}) - |u_2(\mathbf{x})|^{\sigma(\theta_1)-1} u_2(\mathbf{x}), 0 \right) \cdot \mathbf{u}\, d\mathbf{x},$$

$$I_2 := -\delta \int_\Omega \chi_{\mathbf{f}}(\mathbf{x}) \left(|u_2(\mathbf{x})|^{\sigma(\theta_1)-1} u_2(\mathbf{x}) - |u_2(\mathbf{x})|^{\sigma(\theta_2)-1} u_2(\mathbf{x}), 0 \right) \cdot \mathbf{u}\, d\mathbf{x}$$

$$I_3 := -\int_\Omega (\mathbf{u} \cdot \nabla)\mathbf{u}_2 \cdot \mathbf{u}\, d\mathbf{x}.$$

Estimate for the temperature. Using (4.35) and Cauchy's inequality, we get

$$|J_1| \leq \lambda \int_\Omega |\theta|\, |\mathbf{u}_1|\, |\nabla\theta| d\mathbf{x} \leq \frac{1}{4} \int_\Omega |\nabla\theta|^2\, d\mathbf{x} + \lambda^2 \int_\Omega |\theta|^2 |\mathbf{u}_1|^2\, d\mathbf{x} \tag{4.39}$$

and

$$|J_2| \leq C \int_\Omega |\mathbf{u}|\, |\nabla\theta| d\mathbf{x} \leq \frac{1}{4} \int_\Omega |\nabla\theta|^2\, d\mathbf{x} + C^2 \int_\Omega |\mathbf{u}|^2\, d\mathbf{x}, \tag{4.40}$$

with $C = C(m, M) = \max_{m \leq \theta \leq M} |\mathcal{C}(\theta)|$. In the sequel the letter C will be used for different constants depending on L, m, M, δ and ν. We use the Poincaré inequalities

$$\int_0^L |\mathbf{u}|^2 dy \leq C \int_\Omega |\nabla\mathbf{u}|^2\, d\mathbf{x} \tag{4.41}$$

and

$$|\theta(x, y)|^2 \leq L \int_0^L |\theta_y(x, s)|^2\, ds, \tag{4.42}$$

to obtain, from (3.31), that

$$\int_\Omega |\theta|^2 |\mathbf{u}_1|^2 d\mathbf{x} \leq \tag{4.43}$$

$$C \int_0^\infty \left(\int_0^L |\theta_y(x, s)|^2\, ds \right) \left(\int_\Omega |\nabla\mathbf{u}_1|^2\, d\mathbf{x} \right)\, dx \leq C \int_\Omega |\nabla\theta|^2\, d\mathbf{x}.$$

Joining (4.37), (4.39), (4.40) and (4.43), we arrive to

$$\frac{1}{2} \int_\Omega |\nabla\theta|^2 d\mathbf{x} \leq C\lambda^2 \int_\Omega |\nabla\theta|^2\, d\mathbf{x} + C \int_\Omega |\nabla\mathbf{u}|^2\, d\mathbf{x}.$$

Choosing λ such that

$$2C\lambda^2 < 1, \tag{4.44}$$

it results

$$\int_\Omega |\nabla\theta|^2 d\mathbf{x} \leq C \int_\Omega |\nabla\mathbf{u}|^2 d\mathbf{x}. \tag{4.45}$$

Estimate for the velocity. Applying the inequality

$$\sigma\, |\xi - \eta|^{\sigma+1} \leq \left(|\xi|^{\sigma-1} \xi - |\eta|^{\sigma-1} \eta \right) (\xi - \eta) \left(|\xi|^{\sigma+1} + |\eta|^{\sigma+1} \right)^{\frac{1-\sigma}{1+\sigma}},$$

with $0 < \sigma < 1$ and using (1.2), we can write

$$0 < \delta\,\sigma^- \int_\Omega \chi_{\mathbf{f}}(\mathbf{x})|u_1 - u_2|^{\sigma(\theta_1)+1}\left(|u_1|^{\sigma(\theta_1)+1} + |u_2|^{\sigma(\theta_1)+1}\right)^{\frac{\sigma(\theta_1)-1}{\sigma(\theta_1)+1}} d\mathbf{x}$$
$$\leq I_1. \tag{4.46}$$

By Lagrange's theorem,

$$|u_2|^{\sigma(\theta_1)-1}u_2 - |u_2|^{\sigma(\theta_2)-1}u_2 = \sigma'(\theta_*)|u_2|^{\sigma(\theta_*)}\ln|u_2|\,\theta,$$

for every θ_* in the interval with extremities θ_1 and θ_2. Then we conclude

$$|I_2| \leq \delta \int_\Omega |\sigma'||u_2|^{\sigma(\theta_*)}\ln|u_2||\,\theta||\mathbf{u}|\,d\mathbf{x}.$$

By (4.36), Cauchy's inequality and (4.41) we obtain

$$|I_2| \leq \frac{\nu}{2}\int_\Omega |\nabla\mathbf{u}|^2\,d\mathbf{x} + \lambda^2 C I_{21},\ I_{21} = \int_\Omega |\theta|^2\,|u_2|^{2\sigma(\theta_*)}\left(\ln|u_2|\right)^2 d\mathbf{x}.$$

Using (4.42) we get

$$I_{21} \leq L\int_0^\infty \left(\int_0^L |\nabla\theta(x,s)|^2\,ds\right)\left(\int_0^L |u_2(x,y)|^{2\sigma(\theta_*)}\left(\ln|u_2|\right)^2 dy\right) dx.$$

Now we recall the following elementary inequalities

$$|u_2|^{2\sigma(\theta_*)}\left(\ln|u_2|\right)^2 \leq C \quad \text{for } |u_2| \leq 1, \quad C = C(\sigma^-,\,\sigma^+)$$
$$|u_2|^{2\sigma(\theta_*)}\left(\ln|u_2|\right)^2 \leq \frac{1}{\varepsilon^2}|u_2|^2 \quad \text{for } |u_2| \geq 1,\ \varepsilon = 1-\sigma^+ > 0.$$

Then, separating in two integrals for $|u_2| < 1$ and for $|u_2| \geq 1$, we obtain

$$\int_0^L |u_2(x,y)|^{2\sigma(\theta_*)}\left(\ln|u_2|\right)^2 dy \leq C\left(1 + \int_0^L |u_2(x,y)|^2 dy\right).$$

Using (3.31) and (4.41)

$$\int_0^L |u_2(x,y)|^{2\sigma(\theta_*)}\left(\ln|u_2|\right)^2 dy \leq C.$$

Finally we obtain

$$I_{21} \leq C\int_\Omega |\nabla\theta|^2 d\mathbf{x}$$

and consequently

$$|I_2| \leq \frac{\nu}{2}\int_\Omega |\nabla\mathbf{u}|^2\,d\mathbf{x} + \lambda^2 C\int_\Omega |\nabla\theta|^2 d\mathbf{x}$$

Using (4.45)

$$|I_2| \leq \frac{\nu}{2}\int_\Omega |\nabla\mathbf{u}|^2\,d\mathbf{x} + \lambda^2 C\int_\Omega |\nabla\mathbf{u}|^2 d\mathbf{x}.$$

Last inequality, (4.38) and (4.46), give us

$$\nu \int_{\Omega} |\nabla \mathbf{u}|^2 \, d\mathbf{x} + \delta \int_{\Omega} \chi_{\mathbf{f}}(\mathbf{x}) |u_1 - u_2|^{\sigma(\theta_1)-1} \left(|u_1|^{\sigma(\theta_1)+1} + |u_2|^{\sigma(\theta_1)+1} \right)^{\frac{\sigma(\theta_1)-1}{\sigma(\theta_1)+1}} d\mathbf{x}$$

$$\leq |I_3| + \left(\frac{\nu}{2} + \lambda^2 C \right) \int_{\Omega} |\nabla \mathbf{u}|^2 \, d\mathbf{x}. \tag{4.47}$$

Choosing λ in (4.47) such that

$$2C\lambda^2 < \nu,$$

we get that

$$\int_{\Omega} |\nabla \mathbf{u}|^2 \, d\mathbf{x} + \delta \int_{\Omega} \chi_{\mathbf{f}}(\mathbf{x}) |u_1 - u_2|^{\sigma(\theta_1)-1} \left(|u_1|^{\sigma(\theta_1)+1} + |u_2|^{\sigma(\theta_1)+1} \right)^{\frac{\sigma(\theta_1)-1}{\sigma(\theta_1)+1}} d\mathbf{x}$$

$$\leq C|I_3|. \tag{4.48}$$

By using (3.31) and some well-known estimates (see, *e.g.*, [9, 12]) we can estimate $|I_3|$ in the following way

$$|I_3| \leq \int_{\Omega} |\nabla \mathbf{u}_2| \, |\mathbf{u}|^2 \, d\mathbf{x} \tag{4.49}$$

$$\leq \|\nabla \mathbf{u_2}\|_{\mathbf{L}^2(\Omega)} \|\mathbf{u}\|^2_{\mathbf{L}^4(\Omega)} \leq C \|\mathbf{u}_*\|_{\mathbf{H}^{\frac{1}{2}}(0,L)} \|\nabla \mathbf{u}\|^2_{\mathbf{L}^2(\Omega)}.$$

Thus, by assuming that $C\|\mathbf{u}_*\|_{\mathbf{H}^{\frac{1}{2}}(0,L)} < 1$, using (4.49) and Poincaré's inequality (3.19), we obtain, from (4.48), that $\mathbf{u}_1 = \mathbf{u}_2$ and, as consequence of (4.45), $\theta_1 = \theta_2$. □

Remark 4.1. *The conditions* (4.35) *and* (4.36) *may be replaced by the condition*

$$M - m = \lambda$$

for some λ *small enough, where* m *and* M *are given in* (2.15). *Here* σ, $\mathcal{C} \in \mathrm{C}^2(m, M)$ *and according to Lagrange's theorem,* $\mathcal{C}' = \mathcal{C}''\theta$ *and* $\sigma' = \sigma''\theta$. *Then* $|\mathcal{C}'| \leq \max_{\theta \in [m,M]} |\mathcal{C}''| \, |M - m| \leq C\lambda$ *and* $|\sigma'| \leq \max_{\theta \in [m,M]} |\sigma''| \, |M - m| \leq C\lambda$.

Remark 4.2. *It seems possible to prove the uniqueness of solutions for the problem* (2.7)–(2.14) *with the body forces field given by* (3.22) *by proceeding as in* [4] *once we assume the following non-increasing condition*

$$(\mathbf{f}(\mathbf{x}, \theta, \mathbf{u}_1) - \mathbf{f}(\mathbf{x}, \theta, \mathbf{u}_2)) \cdot (\mathbf{u}_1 - \mathbf{u}_2) \leq 0 \tag{4.50}$$

for every $\mathbf{u}_1, \mathbf{u}_2 \in \mathbb{R}^2$, *for any* $\theta \in [m, M]$ *and almost all* $\mathbf{x} \in \Omega$.

5. Localization effect

In this section we study the localization effect for the velocity $\mathbf{u}$ associated to the problem (2.7)–(2.14). It turns out that the qualitative property of the spatial localization of $\mathbf{u}$ is independent of the temperature component θ. So, if we are not interested to know how big is the support of $\mathbf{u}$ but merely in knowing that support of $\mathbf{u}$ is a compact subset of Ω we can assume θ be given. In this way, our problem becomes simpler than before (since there is none PDE for θ) and so, given θ such that

$$\theta \in \mathrm{L}^{\infty}(\Omega), \quad \theta(\mathbf{x}) \in [m, M] \text{ for a.e. } \mathbf{x} \in \Omega. \tag{5.51}$$

we consider the following auxiliary problem

$$(\mathbf{u} \cdot \nabla)\mathbf{u} = \nu \triangle \mathbf{u} - \nabla p + \mathbf{f}(\mathbf{x}, \theta, \mathbf{u}), \tag{5.52}$$

$$\text{div } \mathbf{u} = 0, \tag{5.53}$$

$$\mathbf{u} = \mathbf{u}_*, \quad \text{on } x = 0, \tag{5.54}$$

$$\mathbf{u} = \mathbf{0}, \quad \text{on } y = 0,\, L, \tag{5.55}$$

$$\mathbf{u} \to \mathbf{0}, \quad \text{when } x \to \infty, \tag{5.56}$$

where the forces field satisfy (1.1)–(1.3). In Section 3 (see (3.31)) has been established the existence of a weak solution $\mathbf{u}$ having a finite global energy

$$E := \int_{\Omega} \left(|\nabla \mathbf{u}|^2 + \chi_{\mathbf{f}} |u|^{1+\sigma(\theta)} \right) d\mathbf{x} \tag{5.57}$$

and consequently, from (1.2) and assuming that $|u| \leq 1$,

$$\mathcal{E} := \int_{\Omega} \left(|\nabla \mathbf{u}|^2 + \chi_{\mathbf{f}} |u|^{1+\sigma^+} \right) d\mathbf{x} < \infty. \tag{5.58}$$

As in [3, 4] we introduce the associated stream function ψ

$$u = \psi_y \quad \text{and} \quad v = -\psi_x \quad \text{in } \Omega \tag{5.59}$$

and we reduce the study of problem (5.52)–(5.56), to the consideration of the following fourth order problem where the pressure term does not appear anymore,

$$\nu \triangle^2 \psi + \tfrac{\partial f_1}{\partial y} - \tfrac{\partial f_2}{\partial x} = \psi_y \triangle \psi_x - \psi_x \triangle \psi_y \quad \text{in } \Omega, \tag{5.60}$$

$$\psi(x,0) = \psi(x,L) = \tfrac{\partial \psi}{\partial n}(x,0) = \tfrac{\partial \psi}{\partial n}(x,L) = 0 \quad \text{for } x \in (0,\infty), \tag{5.61}$$

$$\psi(0,y) = \textstyle\int_0^y u_*(s)ds, \; \tfrac{\partial \psi}{\partial n}(0,y) = v_*(y) \quad \text{for } y \in (0,L), \tag{5.62}$$

$$\psi(x,y), \; |\nabla \psi(x,y)| \to 0, \text{ as } x \to \infty \quad \text{and for } y \in (0,L). \tag{5.63}$$

Here $\mathbf{f} = (f_1, f_2) = (f_1(\mathbf{x}, \theta, \psi_y, -\psi_x), f_2(\mathbf{x}, \theta, \psi_y, -\psi_x))$ and we recall that θ is assumed to be given. The notion of weak solution is adapted again to the information we have on the function $\mathbf{f}$.

Definition 5.1. *Given* θ *satisfying* (5.51), *a function* ψ *is a weak solution of problem* (5.60)–(5.63), *if:*

(i) $\psi \in \mathrm{H}^2(\Omega)$, $\mathbf{f}(\mathbf{x},\theta,\psi_y,-\psi_x) \in \mathbf{L}^1_{\mathrm{loc}}(\Omega)$;

(ii) $\psi(0,y) = \int_0^y u_*(s)ds$, $\frac{\partial\psi}{\partial n}(0,y) = v_*(y)$, $\psi(x,0) = \psi(x,L) = \frac{\partial\psi}{\partial n}(x,0) = \frac{\partial\psi}{\partial n}(x,L) = \psi(0,L) = 0$, *and* ψ, $|\nabla\psi| \to 0$, *when* $x \to \infty$;

(iii) *For every* $\phi \in \mathrm{H}^2_0(\Omega) \cap \mathrm{W}^{1,\infty}(\Omega)$ *with compact support,*

$$\nu \int_\Omega \triangle\psi\triangle\phi \, d\mathbf{x} - \int_\Omega (f_1\phi_y - f_2\phi_x)\, d\mathbf{x} = \int_\Omega \triangle\psi\,(\psi_x\phi_y - \psi_y\phi_x)\, d\mathbf{x}. \tag{5.64}$$

To establish the localization effect, we proceed as in [3, 4] and we prove the followings lemmas.

Lemma 5.1. *Given* θ *satisfying* (5.51), *if* $\mathbf{u}$ *is a weak solution of* {(5.52)–(5.56), (1.1)–(1.3)} *in the sense of (ii) of Definition* 3.1, *then* ψ, *given by* (5.59), *is a weak solution of* (5.60)–(5.63) *in the sense of Definition* 5.1.

Lemma 5.2. *Given* θ *satisfying (*5.51*), let* ψ *be a weak solution of (*5.60*)-(*5.63*) with* E *finite. Assume that* $\mathbf{f}$ *satisfies (*1.1*)-(*1.3*) with* $x_{\mathbf{f}} = \infty$. *Then, for every* $a > x_g$, *and every positive integer* $m \geq 2$

$$\int_\Omega \left(\nu|\mathrm{D}^2\psi|^2 + \delta|\psi_y|^{1+\sigma^+}\right)(x-a)_+^m d\mathbf{x}$$

$$\leq 2m\nu \int_\Omega |\triangle\psi||\psi_x|(x-a)_+^{m-1}d\mathbf{x} + 2m\nu \int_\Omega |\psi_y||\psi_{xy}|(x-a)_+^{m-1}d\mathbf{x} \tag{5.65}$$

$$+m(m-1)\nu \int_\Omega |\triangle\psi||\psi|(x-a)_+^{m-2}d\mathbf{x} + m \int_\Omega |\triangle\psi||\psi_y||\psi|(x-a)_+^{m-1}d\mathbf{x},$$

where $|\mathrm{D}^2\psi|^2 = \psi_{xx}^2 + 2\psi_{xy}^2 + \psi_{yy}^2$.

From the left-hand side of (5.65), it will arise the energy type term which depends on a

$$\mathcal{E}_m(a) = \int_\Omega \left(|\mathrm{D}^2\psi|^2 + |\psi_y|^{1+\sigma^+}\right)(x-a)_+^m d\mathbf{x}$$

and we observe that

$$\mathcal{E}_0(0) = \mathcal{E}, \quad (\mathcal{E}_m(a))^{(k)} = (-1)^k \frac{m!}{(m-k)!}\mathcal{E}_{m-k}(a), \; 0 \leq k \leq m.$$

Then, the following lemma is proved as in [4], where now σ depends on the temperature θ and satisfies (1.2).

Lemma 5.3. *Let* ψ *be a weak solution of* (5.60)–(5.63) *and let us assume* $\mathbf{f}$ *satisfies* (1.1)–(1.3) *with* $x_{\mathbf{f}} = \infty$. *Then, the following differential inequality holds for* $a \geq x_g$ *(*x_g *is given in* (1.3)*) :*

$$\mathcal{E}_m(a) \leq C\,(\mathcal{E}_{m-2}(a))^{\mu_1} + C\,(\mathcal{E}_{m-2}(a))^{\mu_2}, \quad \textit{for any } \theta \in [m, M],$$

for every integer $m > 3$, where $C = C(L, m, \delta, \nu, \sigma^{\pm})$ are different positive constants and $\mu_j = \mu_j(m, \sigma^+) > 1$, $j = 1, 2$. Moreover, $\mathcal{E}_2(a) < \infty$ for any $a \geq x_g$. In fact,

$$\mathcal{E}_2(a) \leq C\left(\mathcal{E}_0(a)\right)^{\mu_1} + C\left(\mathcal{E}_0(a)\right)^{\mu_2}, \quad \text{for any } \theta \in [m, M],$$

where C are different positive constants, the first an absolute constant and the others such that $C = C(L, \delta, \nu, \sigma^{\pm})$, and $\mu_j = \mu_j(\sigma^+) > 1$, $j = 1, 2$.

Starting with the case $x_{\mathbf{f}} = \infty$, we take $m = 4$ in Lemma 5.3 and then we have the fractional differential inequality

$$\mathcal{E}_4(a) \leq C\left(\mathcal{E}_2(a)\right)^{\mu_1} + C\left(\mathcal{E}_2(a)\right)^{\mu_2},$$

where, according to what we have done in [4], $\mu_j = \mu_j(\sigma^+) > 1$, $j = 1, 2$ and $C = C(L, m, \delta, \nu, \sigma^{\pm})$ means two different positive constants. Using Lemma 5.3 with $m = 2$ and because of the finiteness of $\mathcal{E}$ (see (5.58)), we can easily see that $\mathcal{E}_2(a)$ is finite. Then, using Lemma 5.1 of [4] and proceeding as in this last reference, we prove the support of $\mathcal{E}_0(a)$ is a bounded interval $[0, a^*]$ with $a^* \leq a'$, where a' is an upper limit to a^* and given by

$$a' = \frac{C}{1 - \sigma^+} \mathcal{E}^{\frac{1}{2(7+\sigma^+)}}, \quad C = C(E, L, \delta, \nu, \sigma^{\pm}).$$

Then $\mathcal{E}_0(a) = 0$ for $a > a'$, which implies $\mathbf{u} = \mathbf{0}$ almost everywhere for $x > a'$.

For the case $x_{\mathbf{f}} < \infty$, the proof follows exactly as in [3].

Remark 5.1. *We obtain the same localization effect if we consider the non-constant semi-infinite strip $\Omega = (0, \infty) \times (L_1(x), L_2(x))$, with L_1, $L_2 \in \mathrm{C}^2(0, \infty)$, $k_1 \leq |L_2(x) - L_1(x)| \leq k_2$, $|L_1'(x)|$, $|L_2'(x)| \leq k_3$, and $|L_1''(x)|$, $|L_2''(x)| \leq k_4$ for all $x \geq 0$, where k_i, $i = 1, \ldots, 4$, are positive constants.*

6. Case of a temperature depending viscosity

A harder, but very interesting, problem arises when the viscosity depends also on the temperature (which is very often the case in many concrete applications). In this case, the equation of motion (5.52) must be replaced by

$$(\mathbf{u} \cdot \nabla)\mathbf{u} = \operatorname{div}(2\nu(\theta)\mathbf{D}) - \nabla p + \mathbf{f}(\mathbf{x}, \theta, \mathbf{u}), \tag{6.66}$$

where $\mathbf{D} = \left(\nabla \mathbf{u} + \nabla \mathbf{u}^T\right)/2$ is the rate of strain tensor. We assume that

$$0 < \nu^- \leq \nu(\theta) \leq \nu^+ < \infty, \tag{6.67}$$

for some constants ν^- and ν^+, and the equation (3.21) of (ii) of Definition 3.1 is replaced by

$$2\int_\Omega \nu(\theta)\mathbf{D} : \nabla\varphi \, d\mathbf{x} + \int_\Omega \mathbf{u} \cdot \nabla \mathbf{u} \cdot \varphi \, d\mathbf{x} = \int_\Omega \mathbf{f} \cdot \varphi \, d\mathbf{x}. \tag{6.68}$$

The main goal of this section is to indicate how the localization effect can be proved still in this case. We assume the existence of, at least, one weak solution

$(\theta, \mathbf{u})$, in the sense of Definition 3.1 with (3.21) replaced by (6.68), to problem (6.66), (2.5)–(2.14) having a finite global energy (5.57).

To establish the localization effect, we proceed as in Section 5 by introducing the stream function (5.59) associated with the vector velocity and we reduce the problem {(6.66), (5.53)–(5.56)} to the following one,

$$[\nu(\theta)(\psi_{xx} - \psi_{yy})]_{xx} + [\nu(\theta)(\psi_{yy} - \psi_{xx})]_{yy} + 4[\nu(\theta)\psi_{xy}]_{xy} \quad (6.69)$$

$$+\frac{\partial f_1}{\partial y} - \frac{\partial f_2}{\partial x} = \psi_y \triangle \psi_x - \psi_x \triangle \psi_y$$

$$\psi(x,0) = \psi(x,L) = \tfrac{\partial \psi}{\partial n}(x,0) = \tfrac{\partial \psi}{\partial n}(x,L) = 0 \quad \text{for } x \in (0,\infty), \quad (6.70)$$

$$\psi(0,y) = \textstyle\int_0^y u_*(s)ds, \ \tfrac{\partial \psi}{\partial n}(0,y) = v_*(y) \quad \text{for } y \in (0,L), \quad (6.71)$$

$$\psi(x,y), \ |\nabla\psi(x,y)| \to 0, \ \text{as } x \to \infty \quad \text{and for } y \in (0,L), \quad (6.72)$$

where again $\mathbf{f} = (f_1, f_2) = (f_1(\mathbf{x}, \theta, \psi_y, -\psi_x), f_2(\mathbf{x}, \theta, \psi_y, -\psi_x))$ and the notion of weak solution to problem (6.69)–(6.72) is adapted, from Definition 5.1, by replacing (5.64) by

$$\int_\Omega \nu(\theta)\left[(\psi_{xx} - \psi_{yy})(\phi_{xx} - \phi_{yy}) + 4\psi_{xy}\phi_{xy}\right] d\mathbf{x}$$

$$-\int_\Omega (f_1\phi_y - f_2\phi_x)\, d\mathbf{x} = \int_\Omega \triangle\psi\,(\psi_x\phi_y - \psi_y\phi_x)\, d\mathbf{x}.$$

In this case, the counterpart of (5.65) is

$$\int_\Omega \left(\nu^- |\mathrm{D}^2\psi|^2 + \delta|\psi_y|^{1+\sigma^+}\right)(x-a)_+^m d\mathbf{x}$$

$$\le 2m\nu^+ \int_\Omega (|\psi_{xx}| + |\psi_{yy}|)|\psi_x|(x-a)_+^{m-1} d\mathbf{x} + 2m\nu^+ \int_\Omega |\psi_{xy}||\psi_y|(x-a)_+^{m-1} d\mathbf{x}$$

$$+m(m-1)\nu^+ \int_\Omega (|\psi_{xx}| + |\psi_{yy}|)|\psi|(x-a)_+^{m-2} d\mathbf{x}$$

$$+m \int_\Omega |\triangle\psi||\psi_y||\psi|(x-a)_+^{m-1} d\mathbf{x}.$$

Proceeding as in Section 5 and using the assumptions (1.2) and (6.67), we obtain the same localization effect mentioned in the precedent section.

References

[1] S.N. Antontsev, J.I. Díaz, H.B. de Oliveira. Stopping a viscous fluid by a feedback dissipative external field: I. The stationary Stokes equations. *Book of abstracts of NSEC8*, Euler International Mathematical Institute, St. Petersburg, 2002.

[2] S.N. Antontsev, J.I. Díaz, H.B. de Oliveira. On the confinement of a viscous fluid by means of a feedback external field. *C.R. Mécanique* **330** (2002), 797–802.

[3] S.N. Antontsev, J.I. Díaz, H.B. de Oliveira. Stopping a viscous fluid by a feedback dissipative field: I. The stationary Stokes problem. To appear in *J. Math. Fluid Mech.*

[4] S.N. Antontsev, J.I. Díaz, H.B. de Oliveira. Stopping a viscous fluid by a feedback dissipative field: I. The stationary Navier-Stokes problem. To appear in *Rend. Lincei Mat. Appl.*

[5] S.N. Antontsev, J.I. Díaz, S.I. Shmarev. *Energy Methods for Free Boundary Problems: Applications to Non-linear PDEs and Fluid Mechanics.* Progress in Nonlinear Differential Equations and Their Applications, Vol. 48, Birkhäuser, Boston, 2002.

[6] J.R. Canon, E. DiBenedetto, G.H. Knightly. The bidimensional Stefan problem with convection: the time-dependent case. *Comm. Partial Differential Equations*, **8** (1983), 1549–1604.

[7] J. Carrillo, M. Chipot, On some nonlinear elliptic equations involving derivatives of the nonlinearity, *Proc. Roy. Soc. Edinburgh Sect. A* **100** (3–4) (1985), 281–294.

[8] E. DiBenedetto, M. O'Leary. Three-dimensional conduction-convection problems with change of phase. *Arch. Rational Mech. Anal.* **123** (1993), 99–117.

[9] G.P. Galdi. *An Introduction to the Mathematical Theory of the Navier-Stokes Equations: Nonlinear Steady Problems.* Springer-Verlag, New York, 1994.

[10] D. Gilbarg, N.S. Trudinger. *Elliptic Partial Differential Equations of Second Order.* Springer-Verlag, Berlin Heidelberg, 1998.

[11] O.A. Ladyzhenskaya, N.N. Ural'tseva. *Linear and Quasilinear Elliptic Equations.* Academic Press, New York, 1968.

[12] O.A. Ladyzhenskaya. *The Mathematical Theory of Viscous Incompressible Fluids.* Gordon and Breach Science Publishers Inc., New York, 1969.

[13] V.A. Solonnikov. On the solvability of boundary and initial boundary value problems for the Navier-Stokes systems in domains with noncompact boundaries. *Pacific J. Math.* **93** (2) (1981), 443–458.

[14] X. Xu, M. Shillor. The Stefan problem with convection and Joule's heating. *Adv. Differential Equations* **2** (1997), 667–691.

S.N. Antontsev
Universidade da Beira Interior
Covilhã, Portugal
e-mail: `anton@ubi.pt`

J.I. Díaz
Universidad Complutense
Madrid, Spain
e-mail: `ji_diaz@mat.ucm.es`

H.B. de Oliveira
Universidade do Algarve
Faro, Portugal
e-mail: `holivei@ualg.pt`

Progress in Nonlinear Differential Equations
and Their Applications, Vol. 61, 15–26

Ultracontractive bounds for nonlinear evolution equations governed by the subcritical p-Laplacian

Matteo Bonforte and Gabriele Grillo

Abstract. We consider the equation $\dot{u} = \triangle_p(u)$ with $2 \leq p < d$ on a compact Riemannian manifold. We prove that any solution $u(t)$ approaches its (time independent) mean $\overline{u}$ with quantitative bounds on the rate of convergence $\|u(t) - \overline{u}\|_\infty \leq C\|u_0 - \overline{u}\|_r^\gamma / t^\beta$ for any $q \in [2, +\infty]$ and $t > 0$. The proof is based upon the connection between logarithmic Sobolev inequalities and decay propertiesof nonlinear semigroups.

Mathematics Sucject Classification (2000). Primary 47H20; Secondary 35K55, 58D07, 35K65.

Keywords. Contractivity properties, asymptotics of nonlinear evolutions, p–Laplacian on manifolds.

1. Introduction

Let (M, g) be a smooth, connected and compact Riemannian manifold without boundary, whose dimension is denoted by d with $d \geq 3$. Let ∇ be the Riemannian gradient and $\mathrm{d}x$ the Riemannian measure and consider, for $2 \leq p < d$ (the *subcritical case*), the following functional:

$$\mathcal{E}_p(u) = \int_M |\nabla u|^p \, \mathrm{d}x \tag{1.1}$$

for any $u \in \mathrm{L}^2(M)$, where we adopt the convention that $\mathcal{E}_p(u)$ equals $+\infty$ if the distributional gradient of u does not belong to $\mathrm{L}^p(M)$. It is well–known that $\mathcal{E}_p$ is a convex, lower semicontinuous functional. The subgradient of the functional $\mathcal{E}_p/p$, denoted by $\triangle_p$, generates a (nonlinear) strongly continuous nonexpansive semigroup $\{T_t : t \geq 0\}$ on $\mathrm{L}^2(M)$. On smooth functions, the operator $\triangle_p$ reads

$$\triangle_p u = \operatorname{div}\left(|\nabla u|^{p-2} \nabla u\right),$$

$|\cdot| = |\cdot|_x$ indicating the norm in the tangent space at x. We refer to [16] as a complete general reference for parabolic equations driven by operators of p–Laplace type in the Euclidean setting, there can be found also existence results (for weak solutions) in Euclidean setting as well as other properties of the solution. It should be remarked here that the function $u(t,x) := (T_t u)(x)$ is also a weak solution in the sense of [16] of the equation $\dot{u} = \triangle_p u$, so we will speak equivalently of weak solution or time evolution associated to the semigroup at hand. To be more precise, by *weak solution* to equation

$$\begin{cases} \dot{u} = \triangle_p(u), \text{ on } (0,+\infty) \times M \\ u(0,\cdot) = u_0 \in \mathrm{L}^2(M) \end{cases} \tag{1.2}$$

corresponding to the initial datum $u_0 \in \mathrm{L}^2(M)$ we mean that

$$u \in \mathrm{L}^p((0,T); W^{1,p}(M)) \cap C([0,T]; \mathrm{L}^2(M))$$

for any $T > 0$ and that, for any positive and bounded test function

$$\varphi \in W^{1,2}(0,T; \mathrm{L}^2(M)) \cap \mathrm{L}^p((0,T); W^{1,p}(M)), \quad \varphi(T) = 0,$$

one has:

$$\begin{aligned} \int_M u_0(x)\varphi(0,x)\,\mathrm{d}x = &- \int_0^T \int_M u(t)\varphi'(t,x)\,\mathrm{d}x\,\mathrm{d}t \\ &+ \int_0^T \int_M |\nabla u(t,x)|^{p-2}\,\nabla u(t,x) \cdot \nabla\varphi(t,x)\,\mathrm{d}x\,\mathrm{d}t. \end{aligned}$$

Let us denote by $\overline{u}$ the mean of an integrable function u:

$$\overline{u} = \frac{1}{\mathrm{vol}(M)} \int_M u\ \mathrm{d}x.$$

Let finally $u(t) := T_t u$ be the time evolution associated to the semigroup at hand and to the initial datum $u(0) = u \in \mathrm{L}^1(M)$ (or the weak solution to problem (1.2), as well). Then $\overline{u(t)}$ does not depend upon time, so that it equals $\overline{u}$: we prove this fact by means of abstract semigroup theory in Lemma 3.1.

Theorem 1.1. *Let (M,g) be a smooth, connected and compact Riemannian manifold without boundary and with dimension $d > 2$. Consider, for any $t > 0$, the solution $u(t)$ to the problem (1.2) with $u(0) \in L^q(M)$ with $q \geq 1$. Then the following ultracontractive bound holds true for all $t \in (0,1]$:*

$$\|u(t) - \overline{u}\|_\infty \leq \frac{C(p,q,d,\overline{A},\mathrm{Vol}(M))}{t^\beta} \|u(0) - \overline{u}\|_q^\gamma \tag{1.3}$$

with:

$$\beta = \frac{d}{pq + d(p-2)}, \qquad \gamma = \frac{pq}{pq + d(p-2)} \tag{1.4}$$

where $\overline{A}$ are the constants appearing in the Sobolev inequality

$$\|u - \overline{u}\|_{pd/(d-p)} \leq \overline{A}\,\|\nabla u\|_p$$

If $t > 1$ one has instead, for all data belonging to $L^2(M)$:

$$\|u(t) - \overline{u}\|_\infty \leq \frac{C(p,2,d,\overline{A},\mathrm{Vol}(M))}{\left(Bt + \|u(0) - \overline{u}\|_2^{2-p}\right)^{\frac{2p}{(2p+d(p-2))(p-2)}}} \tag{1.5}$$

and in particular, for any $\varepsilon \in [0,1]$*:*

$$\|u(t)-\overline{u}\|_\infty \leq \frac{C(p,2,d,\overline{A},\mathrm{Vol}(M))\|u(0)-\overline{u}\|_2^{\frac{2p(1-\varepsilon)}{(2p+d(p-2))}}}{(Bt)^{\frac{2p\varepsilon}{(2p+d(p-2))(p-2)}}} \tag{1.6}$$

where

$$B = \frac{(p-2)}{\overline{A}^p \mathrm{Vol}(M)^{p\frac{2p+d(p-2)}{2(p-d)}}}.$$

The proof will show that identical conclusions hold for the solutions to the equation $\dot{u} = \triangle_p u$ in bounded Euclidean domains, or in compact manifolds with smooth boundary, with homogeneous Neumann boundary conditions.

Corollary 1.2 (absolute bound). *For all* $t > 2$*, all* $\varepsilon \in (0,1)$ *and all initial data* u_0 *in* $\mathrm{L}^1(M)$ *there exists* $c_\varepsilon > 0$ *such that*

$$\|u(t)-\overline{u}\|_\infty \leq c_\varepsilon t^{-(1-\varepsilon)/(p-2)}. \tag{1.7}$$

independently of the initial datum u_0*. Moreover, if the initial datum belongs to* $L^r(M)$ *with* $\|u(0)\|_r < 1$ *then*

$$\|u(t)-\overline{u}\|_\infty \leq c_\varepsilon \|u(0)-\overline{u}\|_r^\varepsilon t^{-(1-\varepsilon)/(p-2)}$$

for all $t \geq 2\|u(0)-\overline{u}\|_r^{2-p}$.

The proof of this corollary is identical to the proof of the corollary (1.2) of [5] since the proof presented there is independent on the range of p.

A few comments on the *sharpness of the bound* are now given:

- (compact manifold or Neumann cases). It is known from the results of [1] that a lower bound of the form

$$\|u(t)-\overline{u}\|_2 \geq \frac{C}{t^{1/(p-2)}}$$

holds for any L^2 data and all t sufficiently large. A similar bound for the L^∞ norm thus holds as well. Hence the bounds in Corollary 1.2 are close to the optimal ones for large time. For small times a comparison with the Barenblatt solutions ([16]) shows that the power of time is the correct one for data belonging to L^1, while for data in L^q with $q > 1$ the L^∞ our result is better in the sense that norm diverges at a *slower* rate depending on q, a property which is familiar in the theory of linear ultracontractive semigroups but which seems to have not been explicitly stated so far in the nonlinear context.

- (Dirichlet case). A similar result can be shown on compact manifolds with smooth boundary, homogeneous Dirichlet boundary conditions being assumed. The main difference is in the fact that the solutions approach zero when t tends to infinity. The proof stems from the appropriate Sobolev inequality for functions in $W_0^{1,p}(M)$ and is easier than in the previous case. For short time remarks similar to the Neumann case hold. By using the optimal logarithmic Sobolev inequality of

[14] for the p–energy functional, bounds which are sharp also for general L^1 data and small times can be proved easily by the present methods in the Euclidean case.

A comparison with some previous results is now given. While a discussion of similar problems *in the whole* $\mathbb{R}^n$ has been given long ago in [19] by entirely different methods, and recently improved in [12], nothing seem to have appeared, apart of some estimates of a somewhat similar nature given in [16] (in any case the Neumann case and the compact manifold case are not discussed there) concerning asymptotics of evolution equations driven by the p–Laplacian in bounded domains before the recent work [8]. In this paper a similar discussion is given for the *Euclidean* p–Laplacian with Dirichlet boundary conditions on a bounded Euclidean domain: the solution approaches zero, instead of $\overline{u}$, in the course of time. In [10] a generalization of such results to a much larger class of operators is given, but Dirichlet boundary conditions are still assumed. The Dirichlet boundary conditions determine the form of the Sobolev inequalities on which our work relies and thus the situation is different from the very beginning. We shall also comment later on the case of Dirichlet boundary conditions is much easier and can be dealt with in the present case as well, and that the case of Neumann boundary conditions displays exactly the same properties discussed in Theorem (1.1). In the case of the present type of evolutions it seems that even the fact that $u(t)$ approaches $\overline{u}$ in the course of time is new. Similar results have been proved in [5] in the case $p > d$.

2. Entropy and Logarithmic Sobolev Inequalities

In this section we will prove a family of logarithmic Sobolev inequalities, which will be an essential tool in the rest of the paper. They involve the entropy or Young functional below:

$$J(r,u) = \int_M \log\left(\frac{|u|}{\|u\|_r}\right)\frac{|u|^r}{\|u\|_r^r}\,dx \tag{2.1}$$

well defined for any $r \geq 1$ and $u \in X = \bigcap_{p=1}^{+\infty} L^p(M)$.

Proposition 2.1. *The logartithmic Sobolev inequality*

$$pJ(p,u) \leq \frac{d}{p}\left[\varepsilon A\frac{\|\nabla f\|_p^p}{\|f\|_p^p} + \varepsilon \mathrm{Vol}(M)^{p/p^*}\frac{|\overline{f}|^p}{\|f\|_p^p} - \log\varepsilon\right] \tag{2.2}$$

holds true for any $\varepsilon > 0$, for all $f \in W^{1,p}(M)$, $1 \leq p < d$, $d \geq 2$. Here

$$A = 2^p\overline{A}$$

and $\overline{A}$ is the constant appearing in the classical Sobolev inequality:

$$\|u-\overline{u}\|_{p^*} \leq \overline{A}\,\|\nabla u\|_p\,, \qquad p^* = \frac{p\,d}{d-p} \tag{2.3}$$

Proof. First we notice that

$$\begin{aligned}\|u\|_{p^*} - |\overline{u}|\mathrm{Vol}(M)^{1/p^*} &= \|u\|_{p^*} - \|\overline{u}\|_{p^*}\\ &\leq \|u-\overline{u}\|_{p*} \leq \overline{A}\|\nabla u\|_p.\end{aligned}$$

Thus

$$\begin{aligned}\|u\|_{p^*}^p &\le \left(\overline{A}\|\nabla u\|_p + |\overline{u}|\mathrm{Vol}(M)^{1/p^*}\right)^p \\ &\le 2^{p-1}\left(\overline{A}\|\nabla u\|_p^p + |\overline{u}|^p\mathrm{Vol}(M)^{p/p^*}\right)\end{aligned} \tag{2.4}$$

where we have used the numerical Young inequality $(a+b)^p \le 2^{p-1}(a^p+b^p)$. Now we prove the LSI (2.2):

$$\begin{aligned}pJ(p,u) &= \int_M \log\left(\frac{|u(x)|^p}{\|u\|_p^p}\right)\frac{|u(x)|^p}{\|u\|_p^p}\,\mathrm{d}x = \frac{d-p}{p}\int_M \log\left(\frac{|u(x)|^{\frac{p^2}{d-p}}}{\|u\|_p^{\frac{p^2}{d-p}}}\right)\frac{|u(x)|^p}{\|u\|_p^p}\,\mathrm{d}x \\ &\le \frac{d-p}{p}\log\left(\int_M \frac{|u(x)|^{\frac{p^2}{d-p}+p}}{\|u\|_p^{\frac{p^2}{d-p}+p}}\,\mathrm{d}x\right) \\ &= \frac{d-p}{p}\log\left(\frac{\|u\|_{\frac{pd}{d-p}}^{\frac{pd}{d-p}}}{\|u\|_p^{\frac{pd}{d-p}}}\right) = \frac{d}{p}\log\frac{\|u\|_{p^*}^p}{\|u\|_p^p} \\ &\le \frac{d}{p}\log\left(\frac{2^{p-1}\overline{A}\|\nabla u\|_p^p + 2^{p-1}\mathrm{Vol}(M)^{p/p^*}|\overline{u}|^p}{\|u\|_p^p}\right) \\ &\le \frac{d}{p}\varepsilon 2^{p-1}\overline{A}\frac{\|\nabla u\|_p^p}{\|u\|_p^p} + \frac{d}{p}\varepsilon 2^{p-1}\mathrm{Vol}(M)^{p/p^*}\frac{|\overline{u}|^p}{\|u\|_p^p} - \log\varepsilon.\end{aligned}$$

Indeed, we first used Jensen inequality for the probability measure $\frac{|u(x)|^p}{\|u\|_p^p}\,\mathrm{d}x$, then the inequality (2.4) and finally the numerical inequality $\log(t) \le \varepsilon t - \log\varepsilon$, which holds for any $\varepsilon, t > 0$.
The proof is thus complete. □

3. Preliminary Results

We first recall two facts proved in [5] for the case $p > d$ remarking that their proof do not depend upon the range of p.

Lemma 3.1. *The semigroup $\{T_t\}_{t\ge 0}$ associated with the functional $\mathcal{E}_p$ satisfies the properties:*

- $\overline{T_t u} = \overline{u}$ *for any $u \in L^1(M)$ and any $t \ge 0$;*
- $T_t u = T_t(u - \overline{u}) + \overline{u}$ *for all $u \in L^1(M)$.*

In view of the above lemma it is clear that it suffices to prove theorem (1.1) for data with zero mean.

Lemma 3.2. *Let u be a weak solution to the equation (1.2) corresponding to an essentially bounded initial datum $u_0 \in \mathrm{L}^\infty(M)$ with zero mean. Let also $r : [0,t) \to [2,+\infty)$ be a monotonically non-decreasing C^1 function. Then*

$$\frac{\mathrm{d}}{\mathrm{d}s}\log\|u(s)\|_{r(s)} = \frac{\dot r(s)}{r(s)}J\left(r(s),u(s)\right) - \left(\frac{p}{r(s)+p-2}\right)^p \frac{(r(s)-1)}{\|u(s)\|_{r(s)}^{r(s)}}\left\|\nabla\left(|u(s)|^{\frac{r(s)+p-2}{p}}\right)\right\|_p^p \tag{3.1}$$

Lemma 3.3. *Let u be a weak solution to the equation (1.2) corresponding to an essentially bounded initial datum $u_0 \in \mathrm{L}^\infty(M)$ with zero mean.*
Let also $r : [0,t) \to [2,+\infty)$ be a monotonically non-decreasing C^1 function. Then

$$\begin{aligned}\frac{\mathrm{d}}{\mathrm{d}s}\log\|u(s)\|_{r(s)} \le & -\frac{\dot r(s)}{r(s)}\frac{d(p-2)}{pr(s)+d(p-2)}\log\|u(s)\|_{r(s)} + \\ & -\frac{\dot r(s)}{r(s)}\frac{d}{pr(s)+d(p-2)}\log\left(\frac{p^{p+2}}{dA}\frac{r(s)^3(r(s)-1)}{\dot r(s)(r(s)+p-2)^p(pr(s)+d(p-2))}\right) \\ & + K\|u(0)\|_2^{p-2}\end{aligned} \tag{3.2}$$

where $K = \mathrm{Vol}(M)^{(3/2)}p/A$.

Proof. We can rewrite the LSI (2.2) in the following form:

$$\|\nabla f\|_p^p \ge \frac{p\|f\|_p^p}{\varepsilon A d}\left[J(1,f^p) + \frac{d}{p}\log(\varepsilon)\right] - \frac{|\overline{f}|^p}{A}.$$

Then we apply it to the function $f = |u(s,x)|^{(r(s)+p-2)/p}$ and obtain:

$$\begin{aligned}\left\|\nabla|u(s)|^{(r(s)+p-2)/p}\right\|_p^p \ge & \frac{p\|u(s)\|_{r(s)+p-2}^{r(s)+p-2}}{\varepsilon A d}\left[J(1,u(s)^{r(s)+p-2}) + \frac{d}{p}\log(\varepsilon)\right] \\ & - \frac{\left|\overline{|u(s)|^{(r(s)+p-2)/p)}}\right|^p}{A}\end{aligned} \tag{3.3}$$

since $\||u(s)|^{(r(s)+p-2)/p}\|_p^p = \|u(s)\|_{r(s)+p-2}^{r(s)+p-2}$. Then we apply this to the inequality (3.1) of previous lemma and we obtain:

$$\frac{\mathrm{d}}{\mathrm{d}s}\log\|u(s)\|_{r(s)} = \frac{\dot r(s)}{r(s)}J\left(r(s),u(s)\right) - \left(\frac{p}{r(s)+p-2}\right)^p \frac{r(s)-1}{\|u(s)\|_{r(s)}^{r(s)}}\left\|\nabla\left(|u(s)|^{\frac{r(s)+p-2}{p}}\right)\right\|_p^p \tag{3.4}$$

so that

$$\begin{aligned}\frac{\mathrm{d}}{\mathrm{d}s}\log\|u(s)\|_{r(s)} \le & \frac{\dot r(s)}{r(s)^2}J\left(1,u(s)^{r(s)}\right) - \frac{p^{p+1}(r(s)-1)}{\varepsilon A d(r(s)+p-2)^p}\frac{\|u(s)\|_{r(s)+p-2}^{r(s)+p-2}}{\|u(s)\|_{r(s)}^{r(s)}}\times \\ & \times\left[J\left(1,u(s)^{r(s)+p-2}\right) - \frac{d}{p}\log\varepsilon\right] + R(p,r(s),u(s),A,d)\end{aligned}$$

since $J\left(1, u(s)^{r(s)}\right) = r(s)J\left(r(s), u(s)\right)$, where

$$R = \frac{p^p(r(s)-1)}{A(r(s)+p-2)^p} \frac{\left|\overline{|u(s)|^{(r(s)+p-2)/p}}\right|^p}{\|u(s)\|_{r(s)}^{r(s)}}.$$

Now choose

$$\varepsilon = \varepsilon(s) = \frac{r(s)^3}{\dot r(s)} \frac{p^{p+2}(r(s)-1)}{Ad(r(s)+p-2)^p(p(r(s)+d(p-2)))} \frac{\|u(s)\|_{r(s)+p-2}^{r(s)+p-2}}{\|u(s)\|_{r(s)}^{r(s)}}$$
$$= \varepsilon_1 \frac{\|u(s)\|_{r(s)+p-2}^{r(s)+p-2}}{\|u(s)\|_{r(s)}^{r(s)}}$$

and obtain from (3.4):

$$\begin{aligned}
\frac{\mathrm{d}}{\mathrm{d}s}\log\|u(s)\|_{r(s)} \le{}& \frac{\dot r(s)}{r(s)^2}\left[J\left(1,u(s)^{r(s)}\right) - \frac{pr(s)}{pr(s)+d(p-2)}J\left(1,u(s)^{r(s)+p-2}\right)\right.\\
&\left.-\frac{pdr(s)}{p(pr(s)+d(p-2))}\log\frac{\|u(s)\|_{r(s)+p-2}^{r(s)+p-2}}{\|u(s)\|_{r(s)}^{r(s)}}\right]\\
&-\frac{\dot r(s)}{r(s)^2}\frac{pdr(s)}{p(pr(s)+d(p-2))}\log\varepsilon_1 + R\\
\le{}& \frac{\dot r(s)}{r(s)^2}\left[J\left(1,u(s)^{r(s)}\right) - \frac{pr(s)}{pr(s)+d(p-2)}J\left(1,u(s)^{r(s)+p-2}\right)\right.\\
&\left.-\frac{(p-2)pdr(s)}{p(pr(s)+d(p-2))}J\left(1,u(s)^{r(s)}\right) - \frac{(p-2)pdr(s)}{p(pr(s)+d(p-2))}\log\|u(s)\|_{r(s)}^{r(s)}\right]\\
&-\frac{\dot r(s)}{r(s)^2}\frac{pdr(s)}{p(pr(s)+d(p-2))}\log\varepsilon_1 + R\\
={}& \frac{\dot r(s)}{r(s)^2}\frac{pr(s)}{(pr(s)+d(p-2))}\left[J\left(1,u(s)^{r(s)}\right) - J\left(1,u(s)^{r(s)+p-2}\right)\right]\\
&-\frac{\dot r(s)}{r(s)^2}\frac{(p-2)pdr(s)}{p(pr(s)+d(p-2))}\log\|u(s)\|_{r(s)}
\end{aligned} \tag{3.5}$$

$$\begin{aligned}
&-\frac{\dot r(s)}{r(s)^2}\frac{pdr(s)}{p(pr(s)+d(p-2))}\log\varepsilon_1 + R\\
&\le -\frac{\dot r(s)}{r(s)}\frac{(p-2)d}{pr(s)+d(p-2)}\log\|u(s)\|_{r(s)} - \frac{\dot r(s)}{r(s)}\frac{d}{pr(s)+d(p-2)}\log\varepsilon_1 + R
\end{aligned}$$

We used first the fact that

$$\log\frac{\|u(s)\|_{r(s)+p-2}^{r(s)+p-2}}{\|u(s)\|_{r(s)}^{r(s)}} \ge (p-2)\left[J\left(r(s),u(s)\right) + \log\|u(s)\|_{r(s)}\right] \tag{3.6}$$

which follows from two basic facts. First, the function $N(r,u) = \log \|u\|_r^r$ is convex with respect to the variable $r \geq 1$, so its derivative is an increasing function of $r \geq 1$. Moreover $N^{'}(r,u) = J(r,u) + \log \|u\|_r$, so the following inequality:

$$N(r+p-2,u) - N(r,u) \geq N^{'}(r,u)(p-2) = \left[J(r,u) + \log \|u\|_r\right](p-2)$$

holds if $p \geq 2$ and leads to (3.6).
The last estimate is obtained by the following monotonicity property of the Young functional

$$J\left(1,u^r\right) - J\left(1,u^{r+p-2}\right) \leq 0, \quad \text{if } p \geq 2$$

the proof of the fact that $J\left(1,u^r\right)$ is a non-decreasing function of $r \geq 1$ is a consequence of the convexity (w.r.t. the variable r) of the function:

$$\phi(r,u) = \log \|u\|_{1/r}.$$

We refer to [3] for a proof of such fact, but comment that it is equivalent to the well known interpolation inequality:

$$\|u\|_{1/r} \leq \|u\|_{1/p}^{\theta} \|u\|_{1/q}^{1-\theta}$$

valid when $\frac{1}{r} = \frac{\theta}{p} + \frac{1-\theta}{q}$. Now deriving ϕ respect to r gives us:

$$\frac{\mathrm{d}}{\mathrm{d}r}\phi(r,u) = -\frac{1}{r} J\left(\frac{1}{r},u\right)$$

thus, as derivative of a convex functions, $-\frac{1}{r}J\left(\frac{1}{r},u\right)$ is non-decreasing.
Our next goal will be to give an estimate on the term R. To this end we use an Hölder and an interpolation inequality to yield

$$\begin{aligned}\|u\|_{(r+p-2)/p} &\leq \mathrm{Vol}(M)^{1/(r+p-2)} \|u\|_{(r+p-2)/(p-1)} \\ &\leq \mathrm{Vol}(M)^{1/(r+p-2)} \|u\|_1^{(p-2)/(r+p-2)} \|u\|_r^{r/(r+p-2)}.\end{aligned}$$

Therefore

$$\begin{aligned}\frac{\left\|\overline{|u(s)|^{(r(s)+p-2)/p}}\right\|^p}{\|u(s)\|_{r(s)}^{r(s)}} &= \frac{\|u(s)\|_{(r(s)+p-2)/p}^{r(s)+p-2}}{\|u(s)\|_{r(s)}^{r(s)}} \\ \leq \mathrm{Vol}(M)\frac{\|u(s)\|_{(r(s)+p-2)/(p-1)}^{(r(s)+p-2)}}{\|u(s)\|_{r(s)}^{r(s)}} &\leq \mathrm{Vol}(M)\|u(s)\|_1^{p-2}.\end{aligned}$$

The statement finally follows by the bounds

$$\frac{p^p(r(s)-1)}{A(r(s)+p-2)^p} \leq \frac{p^p}{A(q+p-2)^{p-1}} \leq \frac{p}{A}$$

valid because $r(s) \geq q \geq 1$ and $p \geq 2$ by assumption, together with the Hölder inequality and the L^2 contraction property of the evolution at hand:

$$\|u(s)\|_1 \leq \mathrm{Vol}(M)^{(1/2)} \|u(s)\|_2 \leq \mathrm{Vol}(M)^{(1/2)} \|u(0)\|_2 = \mathrm{Vol}(M)^{(1/2)} \|u_0\|_2$$

which is well known to hold for any $s > 0$. □

Lemma 3.4. *Let u be a weak solution to the equation (1.2) corresponding to an essentially bounded initial datum $u_0 \in L^\infty(M)$ with zero mean.*
Let also $r : [0,t) \to [2,+\infty)$ be a monotonically non-decreasing C^1 function. Then the following differential inequality holds true for any $s \geq 0$:

$$\frac{\mathrm{d}}{\mathrm{d}s} y(s) + p(s)y(s) + q(s) \leq 0 \tag{3.7}$$

With

$$\begin{aligned} y(s) &= \log \|u(s)\|_{r(s)} \\ p(s) &= \frac{\dot{r}(s)}{r(s)} \frac{d(p-2)}{pr(s) + d(p-2)} \\ q(s) &= \frac{\dot{r}(s)}{r(s)} \frac{d}{pr(s)+d(p-2)} \log \left(\frac{p^{p+2} r(s)^3 (r(s)-1)}{Ad\dot{r}(s)(pr(s)+d(p-2))(r(s)+p-2)^p} \right) \\ &\quad - K\|u_0\|_2^{p-2} \end{aligned} \tag{3.8}$$

In particular, choosing $r(s) = \frac{qt}{t-s}$, one gets the bound:

$$y(t) = \lim_{s \to t^-} y(s) \leq \lim_{s \to t^-} y_L(s) = y_L(t)$$

with

$$\begin{aligned} y_L(t) &= \frac{pq}{pq + d(p-2)} y_L(0) \\ &\quad - \frac{d}{pq+d(p-2)} \log(t) + c_2(p,q,d,\mathrm{Vol}(M))\|u_0\|_2^{p-2} t + c_1(p,q,d) \end{aligned} \tag{3.9}$$

Proof. The fact that $y(s)$ satisfies the differential inequality (3.7) follows immediately by the inequality (3.2) of lemma (3.3), by our choice of $p(s)$ and $q(s)$. Therefore $y(s) \leq y_L(s)$ for any $s \geq 0$ provided $y(0) \leq y_L(0)$ where $y_L(s)$ is a solution to:

$$\frac{\mathrm{d}}{\mathrm{d}s} y_L(s) + p(s)y_L(s) + q(s) = 0$$

i.e.

$$y_L(s) = \mathrm{e}^{-P(s)} \left[y_L(0) - \int_0^s q(\lambda) \mathrm{e}^{P(\lambda)} \mathrm{d}\lambda \right] = \mathrm{e}^{-P(s)} \left[y_L(0) - Q(s) \right]$$

where

$$P(s) = \int_0^s p(\lambda)\mathrm{d}\lambda, \quad Q(s) = \int_0^s q(\lambda)\mathrm{e}^{P(\lambda)}\mathrm{d}\lambda.$$

Choosing $r(s)$ as in the statement one gets, after straightforward calculations and beside noticing that $r(0) = q$ and $r(s) \to +\infty$ as $s \to t^-$:

$$\mathrm{e}^{-P(t)} = \lim_{s \to t^-} \mathrm{e}^{-P(s)} = \frac{pq}{pq + d(p-2)}$$

and

$$
\begin{aligned}
Q(t) &= \lim_{s\to t^-} \mathrm{Q(s)} \\
&= \frac{d}{pq+d(p-2)}\frac{pq+d(p-2)}{pq}\log\left(\frac{p^{p+2}qt}{Ad}\right) \\
&+ c_0(p,q,d) + c_2(p,q,d,\mathrm{Vol}(M))\|u_0\|_2^{p-2}t
\end{aligned}
$$

for suitable numerical constants $c_0(p,q,d)$ and $c_2(p,q,d,\mathrm{Vol}(M))$. □

End of proof of Theorem 1.1.
The following contractivity property holds true for all $0 \le s \le t$:

$$\|u\,(t)\|_r \le \|u\,(s)\|_r$$

Therefore by the previous results one has, for all such s and t :

$$\|u\,(t)\|_{r(s)} \le \|u\,(s)\|_{r(s)} = \exp\left(\log\|u\,(s)\|_{r(s)}\right) = e^{y(s)} \le e^{y_L(s)}$$

whence, letting $s \to t^-$, and recalling that $r\,(s) \to +\infty$ as $s \to t^-$, we deduce:

$$
\begin{aligned}
\|u\,(t)\|_\infty &= \lim_{s\to t^-}\|u\,(t)\|_{r(s)} \le \lim_{s\to t^-}\|u\,(s)\|_{r(s)} \\
&= \lim_{s\to t^-} e^{y(s)} \le \lim_{s\to t^-} e^{y_L(s)} = e^{y_L(t)}.
\end{aligned}
$$

By the explicit form for $e^{y_L(t)}$ we can now prove the bound (1.3) for small times: it is sufficient to let $y(t) = \log\|u(t)\|_\infty$, $y(0) = y_L(0) = \log\|u(0)\|_q = \log\|u_0\|_q$. So we obtain:

$$\|u(t)\|_\infty \le \mathrm{e}^{c_1(p,q,d)+c_2(p,q,d,\mathrm{Vol}(M))\|u_0\|_2^{p-2}t}\frac{\|u_0\|_q^{\frac{pq}{pq+d(p-2)}}}{t^{\frac{d}{pq+d(p-2)}}}$$

To conclude the proof for small times, we prove an L^2–L^2 time decay estimate for arbitrary time. We compute, for initial data with zero mean

$$
\begin{aligned}
&\frac{\mathrm{d}}{\mathrm{d}t}\|u(t)\|_2^2 = -2\|\nabla u\|_p^p \\
&\le -2\overline{A}^{-p}\|u(t)\|_{p^*}^p \\
&\le -2\overline{A}^{-p}\mathrm{Vol}(M)^{-p\frac{2p+d(p-2)}{p-d}}\|u(t)\|_2^p
\end{aligned}
$$

where we have used the Sobolev inequality in the first step and the constant $\overline{A}$ appearing in (2.3). Thus, setting $f(t) = \|u(t)\|_2^2$ we have proved that

$$\dot f(t) \le -2\overline{A}^{-p}\mathrm{Vol}(M)^{-p\frac{2p+d(p-2)}{p-d}} f(t)^{p/2}.$$

This yields the bound, valid for all positive t:

$$\|u(t)\|_2 \le \frac{1}{\left(Bt + \|u(0)\|_2^{2-p}\right)^{1/(p-2)}}$$

where we have set

$$B = \frac{(p-2)}{\overline{A}^p \mathrm{Vol}(M)^{p\frac{2p+d(p-2)}{p-d}}}.$$

This last estimate also gives the so called absolute bound:

$$\|u(t)\|_2 \le \frac{1}{(Bt)^{1/(p-2)}}$$

The absolute bound, together with the L^q–L^∞ smoothing property above and with the semigroup property yields the bound:

$$\begin{aligned}
\|u(t)\|_\infty &\le e^{c_1(p,q,d)+c_2(p,q,d,\mathrm{Vol}(M))\|u(t/2)\|_2^{p-2}t/2} \frac{\|u(t/2)\|_q^{\frac{2p}{2p+d(p-2)}}}{(t/2)^{\frac{d}{pq+d(p-2)}}} \\
&\le e^{c_1(p,q,d)+c_2(p,q,d,\mathrm{Vol}(M))B} \frac{\|u(t/2)\|_q^{\frac{2p}{2p+d(p-2)}}}{(t/2)^{\frac{d}{pq+d(p-2)}}} \\
&\le C(p,q,d,\overline{A},\mathrm{Vol}(M)) \frac{\|u(0)\|_q^{\frac{2p}{2p+d(p-2)}}}{t^{\frac{d}{pq+d(p-2)}}}
\end{aligned}$$

in the last step we used the L^q contraction property, which is well known to hold for any $q \ge 1$ and $t \ge 0$ and we obtained the desired bound for small times, at least for essentially bounded initial data.

To deal with the case of general L^q–data, it suffices to refer to the discussion given in [8], which does not depend either upon the value of p or on the Euclidean setting. This concludes the proof for small times.

To deal with the case of large times, we use again the above L^2–L^∞ decay, the L^2–L^2 time decay, together with the above absolute bound and the semigroup property to yield, for all positive t:

$$\begin{aligned}
\|u(t)\|_\infty &\le e^{c_1(p,q,d)+c_2(p,q,d,\mathrm{Vol}(M))\|u(t/2)\|_q^{p-2}t/2} \|u(t/2)\|_2^{\frac{2p}{2p+d(p-2)}} \\
&\le \frac{e^{c_1(p,q,d)+c_2(p,q,d,\mathrm{Vol}(M))2/B}}{\left(B(t/2)+\|u(0)\|_2^{2-p}\right)^{\frac{2p}{(2p+d(p-2))(p-2)}}}
\end{aligned}$$

The latter statement is obtained from the numerical inequality

$$a+b \ge a^\varepsilon b^{1-\varepsilon}$$

valid for all positive a, b and all $\varepsilon \in (0,1)$. Putting $a = Bt$ and $b = \|u(0)\|_2^{2-p}$ we thus get, for all $t > 1$

$$\|u(t)\|_\infty \le \frac{C(p,2,d,\overline{A},\mathrm{Vol}(M))\|u(0)\|_2^{\frac{2p(1-\varepsilon)}{(2p+d(p-2))}}}{(Bt)^{\frac{2p\varepsilon}{(2p+d(p-2))(p-2)}}} \qquad \square$$

References

1. N.D. Alikakos, R. Rostamian, *Lower bound estimates and separable solutions for homogeneous equations of evolution in Banach space*, J. Diff. Equations 43 (1982), 323–344.

2. T. Aubin, *Some Nonlinear Problems in Riemannian Geometry*, Springer Monographs in Mathematics, Springer–Verlag (1998).
3. D. Bakry, T. Coulhon, M. Ledoux, L. Saloff-Coste, *Sobolev inequalities in disguise*, Indiana Univ. Math. J. **44** (1995), 1033–1074.
4. P. Benilan, M.G. Crandall, *Completely Accretive Operators*, in: "Semigroup Theory and Evolution Equations", Lecture Notes in Pure and Applied Mathematics **135**, pp. 41–75, M. Dekker, New York, 1991.
5. M. Bonforte, F. Cipriani, G. Grillo, *Ultracontactivity and Convergence to equilibrium for supercritical quasilinear parabolic equation on Riemannian manifolds*, Adv. Diff. Equations **8** (2003), 843–872.
6. M. Bonforte, G. Grillo, *Asymptotics of the porous media equation via Sobolev inequalities* (2002), preprint.
7. P. Bénilan, C. Picard, *Quelques aspects non linaires du principe du maximum, Sminaire de Thorie du Potentiel* n. 4, pp. 1–37, Lecture Notes in Math., **713**, Springer, Berlin, 1979.
8. F. Cipriani, G. Grillo, *Uniform bounds for solutions to quasilinear parabolic equations*, J. Diff. Equations. **177** (2001), 209–234.
9. F. Cipriani, G. Grillo, *Nonlinear markovian semigroups and nonlinear Dirichlet forms*, J. reine angew. Math. **562** (2003), 201–235.
10. F. Cipriani, G. Grillo, L^p–L^∞ *Hölder continuity for quasilinear parabolic equations associated to Sobolev derivations*, J. Math. Anal. Appl. **270** (2002), 267–290.
11. E.B. Davies, *Heat Kernels and Spectral Theory*, Cambridge University Press, 1989.
12. M. del Pino, J. Dolbeault, *Nonlinear diffusions and optimal constants in Sobolev type inequalities: asymptotic behaviour of equations involving the p–Laplacian*, Compt. Rend. Math. Acad. Sci. Paris **34** (2002), 665–670.
13. M. del Pino, J. Dolbeault, *The optimal Euclidean Sobolev logarithmic inequality*, J. Funct. Anal. **180** (2001), 66–84.
14. M. del Pino, J. Dolbeault, *Best constant for Gagliardo–Nirenberg inequalities and applications to nonlinear diffusions*, J. Math. Pures Appl. **81** (2002), 847–875.
15. M. Del Pino, J. Dolbeault, I. Gentil, *Nonlinear diffusion, hypercontractivity and the optimal* L^p*-Euclidean logarithmic Sobolev inequality*, J. Math. Anal. Appl. **293** (2004), 375–388.
16. E. DiBenedetto, *Degenerate Parabolic Equations*, Springer Verlag, New York, 1993.
17. I. Gentil, *The general optimal* L^p*-Euclidean logarithmic Sobolev inequality by Hamilton-Jacobi equation* J. Funct. Anal. **202** (2003), 591–598.
18. L. Gross, *Logarithmic Sobolev inequalities*, Amer. J. Math **97** (1976), 1061–1083.
19. S. Kamin, J.L. Vazquez *Fundamental solutions and asymptotic behaviour for the p–Laplacian equation*, Re. Mat. Iberoamericana **4** (1988), 339–354.
20. P. Lindqvist, *On nonlinear Rayleigh quotients*, Potential Anal. **2** (1993), 199–218.

Matteo Bonforte and Gabriele Grillo
Dipartimento di Matematica, Politecnico di Torino
Corso Duca degli Abruzzi 24, I–10129 Torino, Italy
e-mail: `bonforte@calvino.polito.it, gabriele.grillo@polito.it`

Progress in Nonlinear Differential Equations
and Their Applications, Vol. 61, 27–35

Weighted L^2-spaces and Strong Solutions of the Navier-Stokes Equations in $\mathbb{R}^3$

Lorenzo Brandolese

Abstract. We consider the velocity field $u(x,t)$ of a Navier-Stokes flow in the whole space.

We give a persistence result in a subspace of $L^2(\mathbb{R}^3, (1+|x|^2)^{5/2}dx)$, which allows us to fill the gap between previously known results in the weighted-L^2 setting and those on the pointwise decay of u at infinity.

Mathematics Subject Classification (2000). 36B40, 76D05, 35Q30.

1. Introduction

In this paper we study the spatial localization of the velocity field $u = (u_1, u_2, u_3)$ of a Navier-Stokes flow in $\mathbb{R}^3$. For an incompressible fluid and in the absence of external forces, the Navier-Stokes equations can be written in the following integral form:

$$\nabla \cdot a = 0, \qquad u(t) = e^{t\Delta}a - \int_0^t e^{(t-s)\Delta}\mathbb{P}\nabla \cdot (u \otimes u)(s), \tag{IE}$$

where $\nabla \cdot a = \sum_{j=1}^3 \partial_j a_j = 0$ is the divergence-free condition and $\mathbb{P}$ is the Leray-Hopf projector onto the soleinoidal vectors field, defined by $\mathbb{P}f = f - \nabla\Delta^{-1}(\nabla \cdot f)$, where $f = (f_1, f_2, f_3)$.

If $a \in L^2(\mathbb{R}^3)$, then we know since a very long time that a *weak* solution to (IE) exists such that $u \in L^\infty(]0,\infty[, L^2(\mathbb{R}^3))$ and $\nabla u \in L^2(]0,\infty[, L^2(\mathbb{R}^3))$. If the initial datum is well localized in $\mathbb{R}^3$, then these conditions, of course, do not give us so much information on the spatial localization of $u(t)$ during the evolution. Then the natural problem arises of finding the functional spaces that would provide the good setting for obtaining such information. Several papers have been written on this topic, see, *e.g.*, [6], [7], [9], [1], [14], [12] and the references therein contained (see also [15]). In particular, it was shown in [10] that the condition $a \in L^2(\mathbb{R}^3, (1+|x|^2)^\delta dx)$ $(0 \le \delta \le \frac{3}{2})$ is conserved during the evolution, for a suitable class of weak solutions. Here and below, this weighted-L^2 space is equipped with its natural norm, namely $\left(\int |a(x)|^2 (1+|x|^2)^\delta \, dx\right)^{1/2}$. Moreover, the bound on δ seems to be optimal, as far as we deal with data belonging to general weighted-L^2 spaces.

When dealing with *strong* solutions, it is not difficult to obtain much sharper conclusions on the localization of u. For example, assuming that $a \in L^1 \cap L^2(\mathbb{R}^3)$, He [9] proves, among other things, that $u(t)$ belongs to $L^2(\mathbb{R}^3, (1+|x|^2)^2 dx)$ at least in some time interval $[0,T]$, $T > 0$ (and uniformly in $[0,+\infty[$, under a supplementary smallness assumption). In a slightly different context, we would like to mention here also the work of Miyakawa [13], in which it is shown that $u(x,t) \sim |x|^{-\alpha} t^{-\beta/2}$ as $|x| \to \infty$ or $t \to \infty$, for all $\alpha, \beta \geq 0$ and $1 \leq \alpha + \beta \leq 4$, under suitable assumptions on a.

If we compare these two results, we see that Miyakawa's result seems to give a slightly better conclusion, at least from the localization point of view. Indeed, it implies that the condition $a(x) \sim |x|^{-4}$ at infinity is conserved during the evolution (furthermore, $|x|^{-4}$ is known to be the optimal decay in the generic case), whereas the fact that $u(t) \in L^2(\mathbb{R}^3, (1+|x|^2)^2 dx)$ simply tells us that $u(t) \sim |x|^{-7/2}$ at infinity.

The purpose of this paper is to "fill this gap" and to obtain a persistence result in suitable subspaces of $L^2(\mathbb{R}^3, (1+|x|^2)^\alpha dx)$, for all $0 \leq \alpha < \frac{5}{2}$ which, at least formally, will allow us to recover the optimal decay of the velocity field. More precisely, let us introduce the space Z_δ of functions (or vector fields) f such that

$$\begin{aligned} f(x) &\in L^2(\mathbb{R}^3, (1+|x|^2)^{\delta-2} dx), & (1.1)\\ \nabla f(x) &\in L^2(\mathbb{R}^3, (1+|x|^2)^{\delta-1} dx), & (1.2)\\ \Delta f(x) &\in L^2(\mathbb{R}^3, (1+|x|^2)^{\delta} dx) & (1.3) \end{aligned}$$

and equipped with its natural norm. Then we have the following

Theorem 1.1. *Let $\frac{3}{2} < \delta < \frac{9}{2}$ $(\delta \neq \frac{5}{2}, \frac{7}{2})$ and let $a \in Z_\delta$ be a soleinoidal vector field. Then there exists $T > 0$ such that (IE) possesses a unique strong solution $u \in C([0,T], Z_\delta)$.*

The restriction $\delta < \frac{9}{2}$ is consistent with the instantaneous spreading of the velocity field described, *e.g.*, in [3]: we cannot have $u \in C([0,T], Z_{9/2})$ unless the initial data have some symmetry properties. On the other hand, the condition $\delta > \frac{3}{2}$ agrees with the limit case $\alpha + \beta = 1$ of Miyakawa's profiles.

In order to describe our second motivation for Theorem 1.1, let us first observe that the condition (1.3) on the Laplacian plays an important role also in a previous work [8] of Furioli and Terraneo. Motivated by the problem of the unicity of mild solutions to (IE) in *critical spaces*, they introduced (see also [11]) "a space of molecules" X_δ (defined below) and proved that the Cauchy problem for (IE) is locally well-posed in this space. But their proof is technical and involves the theory of local Muckenoupt weights and other "hard analysis" tools. However, we will show in the following section that the result of [8] is essentially equivalent to Theorem 1.1 and can thus be proved in a simpler way. Moreover, we feel that making evidence of the connection between the localization problem of the velocity field and Furioli and Terraneo's molecules provides also a better understanding of [8].

2. Proof of Theorem 1.1

Throughout this section we shall assume $\frac{3}{2} < \delta < \frac{9}{2}$ and $\delta \neq \frac{5}{2}, \frac{7}{2}$. The proof of Theorem 1.1 relies on the decomposition into dyadic blocks of the elements of Z_δ. It will be convenient to introduce the space L^2_{δ} of all functions $f \in L^2(\mathbb{R}^3, (1+|x|^2)^\delta dx)$ such that $\int x^\alpha f(x)\, dx = 0$ for all $\alpha \in \mathbb{N}^3$, with $0 \le |\alpha| \le [\delta - \frac{3}{2}]$ (where $[\cdot]$ denotes the integer part and $|\alpha| = \alpha_1 + \alpha_2 + \alpha_3$). Note that L^2_{δ} is well defined because of the embedding of $L^2(\mathbb{R}^3, (1+|x|^2)^\delta dx)$ into $L^1(\mathbb{R}^3, (1+|x|)^{[\delta-3/2]} dx)$. We start stating a very simple lemma.

Lemma 2.1. *We have $f \in L^2_{\delta}$ if and only if f can be decomposed as*

$$f = g + \sum_{j=0}^{\infty} f_j,$$

where g and f_j belong to $L^2(\mathbb{R}^3)$, $\operatorname{supp} g \subset \{|x| \le 1\}$, $\operatorname{supp} f_j \subset \{2^{j-1} \le |x| \le 2^{j+1}\}$ *and, moreover,*

$$\begin{aligned} &\|f_j\|_2 \le \epsilon_j 2^{-j\delta}, && \text{with} \quad \epsilon_j \in \ell^2(\mathbb{N}), \\ &\int x^\alpha g(x)\, dx = \int x^\alpha f_j(x)\, dx = 0, && \text{if} \quad 0 \le |\alpha| \le [\delta - \tfrac{3}{2}]. \end{aligned} \tag{2.1}$$

Proof. We start with a bad choice, namely,

$$\tilde{g}(x) = f(x) I_{|x|\le 1} \quad \text{and} \quad \tilde{f}_j(x) = f(x) I_{2^j \le |x| \le 2^{j+1}} \qquad (j = 0, 1, \ldots),$$

where I denotes the indicator function. Letting $\tilde{f}_{-1} = \tilde{g}$, we set

$$J(j,\alpha) = \int x^\alpha \tilde{f}_j(x)\, dx.$$

Since $|\alpha| < \delta - \frac{3}{2}$, the series $J(j,\alpha)$ converges and $\sum_{j=-1}^{\infty} J(j,\alpha) = 0$. We now introduce a family of functions $\psi_\beta \in C_0^\infty(\mathbb{R}^3)$, supported in $\frac{1}{2} \le |x| \le 1$ and such that

$$\int x^\alpha \psi_\beta(x)\, dx = \delta_{\alpha,\beta} \qquad (\alpha, \beta \in \mathbb{N}^3),$$

(with $\delta_{\alpha,\beta} = 0$ or 1 if $\alpha \neq \beta$ or $\alpha = \beta$ respectively) and we define

$$c(j,\alpha) \equiv J(j,\alpha) + J(j+1,\alpha) + \cdots .$$

We finally set, for $j = -1, 0, \ldots$

$$\begin{aligned} f_j(x) = \tilde{f}_j(x) - \sum_\beta \Big(& c(j,\beta) 2^{-(3+|\beta|)j} \psi_\beta(2^{-j}x) \\ & - c(j+1,\beta) 2^{-(3+|\beta|)(j+1)} \psi_\beta(2^{-j-1}x) \Big), \end{aligned}$$

the summation being taken over all $\beta \in \mathbb{N}^3$ such that $0 \le |\beta| \le [\delta - \frac{3}{2}]$.

Since $|J(j,\alpha)| \leq 2^{j|\alpha|}2^{3j/2-j\delta}\tilde{\epsilon}_j$ for some $\tilde{\epsilon}_j \in \ell^2(\mathbb{N})$, we have $|c(j,\alpha)| \leq 2^{j|\alpha|}2^{3j/2-j\delta}\bar{\epsilon}_j$, with $\bar{\epsilon}_j \in \ell^2(\mathbb{N})$. One now easily checks that

$$\sum_{j=-1}^{\infty} f_j = \sum_{j=-1}^{\infty} \tilde{f}_j = f$$

and that g and f_j satisfy (2.1). Lemma 2.1 thus follows. □

Let us show that a similar decomposition applies for Z_δ.

Lemma 2.2. *We have $f \in Z_\delta$ if and only if*

$$f = g + \sum_{j=0}^{\infty} f_j,$$

with

$$\begin{array}{lll} \operatorname{supp} g \subset \{|x| \leq 1\}, & \operatorname{supp} f_j \subset \{2^{j-2} \leq |x| \leq 2^{j+2}\} & \\ g \in L^2(\mathbb{R}^3), & ||f_j||_2 \leq \epsilon_j 2^{2j} 2^{-j\delta}, & \epsilon_j \in \ell^2. \\ ||\nabla f_j||_2 \leq \bar{\epsilon}_j 2^j 2^{-j\delta}, & ||\Delta f_j||_2 \leq \tilde{\epsilon}_j 2^{-j\delta}, & \bar{\epsilon}_j, \tilde{\epsilon}_j \in \ell^2. \end{array} \tag{2.2}$$

Proof. It is obvious that if (2.2) holds true then $f = g + \sum_{j=0}^{\infty} f_j$ belongs to Z_δ. Conversely, let φ and ψ be two compactly supported smooth functions, such that 0 does not belong to the support of ψ and $1 \equiv \varphi(x) + \sum_{j=0}^{\infty} \psi(2^{-j}x)$. If we set $g(x) = f(x)\varphi(x)$ and $f_j(x) = f(x)\psi(2^{-j}x)$, then we have $\nabla f_j(x) = \psi(2^{-j}x)\nabla f(x) + 2^{-j}(\nabla\psi)(2^{-j}x)f(x)$ and

$$\Delta f_j(x) = \psi(2^{-j}x)\Delta f(x) + 2^{-j+1}(\nabla\psi)(2^{-j}x)\cdot\nabla f(x) + 2^{-2j}(\Delta\psi)(2^{-j}x)f(x).$$

Decomposition (2.2) then directly follows from the definition of Z_δ. □

Remark 2.3. If $\frac{7}{2} < \delta < \frac{9}{2}$, then Z_δ is embedded in $L^1(\mathbb{R}^3)$. In this case, using the same arguments as in the proof of Lemma 2.1, one easily sees that an element of Z_δ has a vanishing integral if and only if in (2.2) we may choose g and f_j such that $\int g = \int f_j = 0$ $(j = 0, 1, \ldots)$.

Following Furioli and Terraneo [8], we now denote by X_δ the set of all tempered distributions f vanishing at infinity, such that $\Delta f \in L^2(\mathbb{R}^3, (1 + |x|^2)^\delta)dx)$ and $\int x^\alpha \Delta f(x)\,dx = 0$ for all $\alpha \in \mathbb{N}^3$ such that $|\alpha| \leq [\delta - \frac{3}{2}]$. The norm of X_δ is defined by

$$||f||_{X_\delta}^2 \equiv \int |\Delta f(x)|^2 (1 + |x|^2)^\delta\,dx.$$

This space is useful to model the fact the Laplacian of the velocity field is a non-normalized molecule for the Hardy space $\mathbb{H}^1$ (see [8], see also [12], [4] for other applications of the Hardy spaces to the Navier-Stokes equations).

Our next result shows that X_δ can be characterized by means of a dyadic decompositions very similar to the previous one:

Proposition 2.4. *We have $f \in X_\delta$, with $\frac{3}{2} < \delta < \frac{7}{2}$, $\delta \neq \frac{5}{2}$, if and only if there exist g and f_j ($j = 0, 1, \ldots$) satisfying (2.2), such that $f = g + \sum_{j=0}^{\infty} f_j$. In the case $\frac{7}{2} < \delta < \frac{9}{2}$ we have $f \in X_\delta$ if and only if, in addition, we can choose g and f_j satisfying, $\int g = 0$ and $\int f_j = 0$ for all j.*

Proof. Let us assume that $f \in X_\delta$. Applying Lemma 2.1 to Δf and using the fact that f vanishes at infinity, we see that we may write

$$f = \frac{c}{|x|} * p + \sum_{j=0}^{\infty} \frac{c}{|x|} * q_j, \tag{2.3}$$

c being an absolute constant. Here p and q_j are compactly supported L^2-functions, satisfying

$$\begin{aligned} &\operatorname{supp} q_j \subset \{2^{j-1} \le |x| \le 2^{j+1}\}, \\ &||q_j||_2 \le \epsilon_j 2^{-j\delta} \qquad (\epsilon_j \in \ell^2), \\ &\textstyle\int x^\alpha p(x)\, dx = \int x^\alpha q_j(x)\, dx = 0, \qquad \text{if } |\alpha| \le [\delta - 3/2]. \end{aligned} \tag{2.4}$$

Let us show that, for all $f \in X_\delta$ and $2^j \le |x| \le 2^{j+1}$, we have

$$|f(x)| \le \bar{\epsilon}_j 2^{j/2} 2^{-j\delta}, \qquad \text{with } \bar{\epsilon}_j \in \ell^2. \tag{2.5}$$

To prove (2.5) we set $P = \frac{1}{|\cdot|} * p$, $Q_j = \frac{1}{|\cdot|} * q_j$ ($j = 0, 1, \ldots$) and we pose $d = [\delta - \frac{3}{2}]$. Then we have

$$|Q_j(x)| \le C\epsilon_j 2^{-j(\delta - 1/2)}, \qquad \text{if } |x| \le 4 \cdot 2^j, \tag{2.6}$$

$$|Q_j(x)| \le C|x|^{-(d+2)} \epsilon_j 2^{(d + \frac{5}{2} - \delta)j}, \qquad \text{if } |x| \ge 4 \cdot 2^j. \tag{2.7}$$

The first bound follows from the localization of q_j and Hölder's inequality and the second from the last of (2.4), Taylor's formula and, again, Hölder's inequality. Similar arguments allow us to see that $|P(x)| \le C(1 + |x|)^{-(d+2)}$. Summing upon these inequalities immediately yields (2.5).

Another consequence of (2.3) is the following:

$$\left(\int_{2^j \le |x| \le 2^{j+1}} |\nabla f(x)|^2 \, dx \right)^{1/2} \le C \tilde{\epsilon}_j 2^j 2^{-j\delta}, \qquad \text{with } \tilde{\epsilon}_j \in \ell^2(\mathbb{N}). \tag{2.8}$$

To prove (2.8) we start from $-\nabla f(x) = \frac{cx}{|x|^3} * p + \sum_{j=0}^{\infty} \frac{cx}{|x|^3} * q_j$ and we set $R_j = (x/|x|^3) * q_j$, for all $j = 0, 1, \ldots$. Then,

$$\left(\int_{|x| \le 4 \cdot 2^j} |R_j(x)|^2 \, dx \right)^{1/2} \le C\epsilon_j 2^{-j(\delta - 1)}, \qquad \text{if } |x| \le 4 \cdot 2^j, \tag{2.9}$$

$$|R_j(x)| \le C|x|^{-(d+3)} \epsilon_j 2^{(d + \frac{5}{2} - \delta)j}, \qquad \text{if } |x| \ge 4 \cdot 2^j. \tag{2.10}$$

Indeed, the proof of (2.10) again easily follows using the vanishing of the moments of q_j and the Taylor formula. The proof of (2.9) deserves a more detailed explanation: for $|x| \leq 4 \cdot 2^j$ we write

$$\frac{x}{|x|^3} * q_j(x) = \theta_j * q_j(x), \qquad \text{where} \quad \theta_j(x) = \frac{x}{|x|^3} I_{\{|x| \leq 6 \cdot 2^j\}}.$$

Then (2.9) comes from $||\theta_j * q_j||_2 \leq ||\theta_j||_1 ||q_j||_2 \leq C2^j ||q_j||_2$. Now, (2.8) follows from (2.9) and (2.10) by summation.

We are now ready to show that, if $f \in X_\delta$, then (2.2) holds true. Indeed, using (2.5) and (2.8) we get $X_\delta \subset Z_\delta$ and our claim then follows from Lemma 2.2. Note that, in the case $\frac{7}{2} < \delta < \frac{9}{2}$, the moments of p and q_j vanish up to the order two (see (2.4)). We thus see via the Fourier transform that $\int P(x)\,dx = \int Q_j(x)\,dx = 0$ $(j = 0, 1, \ldots)$. This in turn implies $\int f = 0$ and the first part of Proposition 2.4 immediately follows using Remark 2.3.

Conversely, let $f = g + \sum_{j=0}^{\infty} f_j$, such that (2.2) holds (in the case $\frac{7}{2} < \delta < \frac{9}{2}$ we assume, in addition, $\int g = \int f_j = 0$ for all j). Then the bound $||f||_{X_\delta} < \infty$ is obvious. Moreover, by Hölder's inequality,

$$\sum_{j=0}^{\infty} ||x^\alpha \Delta f_j||_1 < \infty, \qquad 0 \leq |\alpha| \leq [\delta - 3/2].$$

But, integrating by parts shows that $\int x^\alpha \Delta g(x)\,dx = \int x^\alpha \Delta f_j(x)\,dx = 0$ for all $j = 0, 1, \ldots$ and $|\alpha| \leq [\delta - \frac{3}{2}]$ (in the case $|\alpha| = 2$, we need to use $\int g = \int f_j = 0$). Hence $\int x^\alpha \Delta f(x)\,dx = 0$ and we have indeed $f \in X_\delta$. □

Remark 2.5. Comparing the results of Lemma 2.2 and Proposition 2.4, we see that $X_\delta = Z_\delta$, for $\frac{3}{2} < \delta < \frac{7}{2}$ $(\delta \neq \frac{5}{2})$ and $X_\delta = Z_\delta \cap \{f : \int f = 0\}$ for $\frac{7}{2} < \delta < \frac{9}{2}$.

We finish our study of Z_δ with the following lemma

Lemma 2.6. *With the above restrictions on δ, the space Z_δ is an algebra with respect to the pointwise product.*

Proof. For $f \in Z_\delta$ we have $|f(x)| \leq C(1+|x|)^{-\delta+1/2}\epsilon(x)$, where $\epsilon(x)$ is a bounded function such that $\epsilon(x) \to 0$ at infinity. Indeed, let us come back to the decomposition (2.2) and observe that $f_j(x) = \int_{\text{supp} f_j} |x-y|^{-1} \Delta f_j(y)\,dy$. By the Hölder inequality, $|f_j(x)| \leq \epsilon_j 2^{-j\delta} 2^{j/2}$ with $\epsilon_j \in \ell^2$ and our claim follows.

Another useful estimate (which follows interpolating ∇f_j between $||\Delta f_j||_2$ and $||f_j||_\infty$) is

$$\left(\int_{2^j \leq |x| \leq 2^{j+1}} |\nabla f(x)|^4 dx \right)^{1/4} \leq \epsilon_j 2^{j/4 - j\delta}, \qquad \text{with } \epsilon_j \in \ell^2(\mathbb{N}).$$

Using this, we immediately see that, if f and h belong to Z_δ, then $fh \in L^2(\mathbb{R}^3, (1+|x|^2)^{2\delta-5/2}dx)$, $\nabla(fh) \in L^2(\mathbb{R}^3, (1+|x|^2)^{2\delta-3/2}dx)$ and $\Delta(fh) \in L^2(\mathbb{R}^3, (1+|x|^2)^{2\delta-\frac{1}{2}}dx)$. Therefore, $fh \in Z_{2\delta-1/2} \subset Z_\delta$ and Z_δ is indeed a pointwise algebra. □

In our last lemma we show that the operator $e^{t\Delta}\mathbb{P}\nabla$ which appears in (IE) is bounded in Z_δ. The matricial structure of this operator will not have any special role in the sequel, since we shall establish all the relevant estimates componentwise.

Lemma 2.7. *The operator $e^{t\Delta}\mathbb{P}\nabla$ is bounded from Z_δ to X_δ for all $t > 0$, with an operator norm bounded by $C/\sqrt{t}$ as $t \to 0$.*

Proof. If $\frac{7}{2} < \delta < \frac{9}{2}$ then we introduce a function h such that

$$f(x) = cg(x) + h(x), \qquad \text{where} \quad g(x) = (4\pi)^{-3/2}e^{-|x|^2/4}$$

and the constant c is chosen in a such way that $\int h(x)\,dx = 0$. If, instead, $\frac{3}{2} < \delta < \frac{7}{2}$, $\delta \neq \frac{5}{2}$, then we simply set $f(x) = h(x)$. In any case, we deduce from Remark 2.5 that $h \in X_\delta$.

We start showing that $e^{t\Delta}\mathbb{P}\nabla g$ belongs to X_δ for all $0 \le \delta < \frac{9}{2}$. Note that the components of $(e^{t\Delta}\mathbb{P}\nabla g)\widehat{\;}(\xi)$ are given by

$$\mathrm{i}\xi_h\left(1 - \frac{\xi_j\xi_k}{|\xi|^2}\right)\exp(-(t+1)|\xi|^2) \qquad (j,h,k = 1,2,3)$$

and the inverse Fourier transform can be easily computed (see, *e.g.*, [13]): we immediately find that $e^{t\Delta}\mathbb{P}\nabla g$ is a smooth function in $\mathbb{R}^3$, such that

$$|\partial^\alpha e^{t\Delta}\mathbb{P}\nabla g(x)| \le C_\alpha(1+|x|)^{-(4+|\alpha|)} \qquad \text{for all } \alpha \in \mathbb{N}^3.$$

This bound implies that $e^{t\Delta}\mathbb{P}\nabla g \in Z_\delta$, for all $\delta < \frac{9}{2}$. But $\int e^{t\Delta}\mathbb{P}\nabla g = 0$ and thus $e^{t\Delta}\mathbb{P}\nabla g$ belongs, more precisely, to X_δ.

Let us now prove that $e^{t\Delta}\mathbb{P}\nabla h$ does also belong to X_δ. We start recalling that the Sobolev space H^δ is defined by

$$||q||^2_{H^\delta} \equiv \int |\widehat{q}(\xi)|^2(1+|\xi|^2)^\delta\,d\xi$$

and that, for $\delta > \frac{3}{2}$, $H^\delta \subset C^{\delta-3/2}$ (the Hölder-Zygmund space). Thus, stating that h belongs to X_δ is equivalent to state that

$$q(\xi) \equiv |\xi|^2\widehat{h}(\xi) \in H^\delta \qquad \text{and} \qquad \partial^\alpha q(0) = 0, \qquad \text{for all } 0 \le |\alpha| \le [\delta - \tfrac{3}{2}].$$

These two conditions on q can be expressed by saying that q belongs to $L^2(\mathbb{R}^3) \cap \dot{H}^\delta_{rel}$, where $\dot{H}^\delta_{rel}$ is the *realization* of the homogeneous Sobolev space $\dot{H}^\delta$ (see Bourdaud, [2]). Recall that $\dot{H}^\delta_{rel}$ can be injected into $S'(\mathbb{R}^3)$ (this would not be true for $\dot{H}^\delta$, which instead is a space of tempered distributions modulo polynomials) and hence the notion of pointwise multipliers makes sense in the realized space. It follows from the result of [2] that $m(\xi) \equiv \xi_h/|\xi|$ is a multiplier for $\dot{H}^\delta_{rel}$ (this would follow also from the general characterization of multipliers of Besov spaces due to Youssfi [16]).

Since we already observed that $h \in X_\delta$, it follows from the above discussion that the components of $|\xi|^2\widehat{\mathbb{P}h}(\xi)$, which are given by $(1 - \xi_j\xi_h|\xi|^{-2})q(\xi)$, belong to $L^2(\mathbb{R}^3) \cap \dot{H}^\delta_{rel}$. Hence, $\mathbb{P}h \in X_\delta$. Moreover, $\mathrm{i}\xi_k e^{-t|\xi|^2} \in S(\mathbb{R}^3)$ is also a multiplier of $\dot{H}^{\delta}_{rel}$ (with norm $c/\sqrt{t}$). Then we get $||e^{t\Delta}\mathbb{P}\nabla f||_{X_\delta} \le C(t^{-1/2} + t^{\delta/2})||f||_{Z_\delta}$ and Lemma 2.7 is thus proven. □

We are now in the position to prove Theorem 1.1. The proof is based on the application of Kato's standard iteration argument in the space $C([0,T], Z_\delta)$. We write (IE) in the compact form $u(t) = e^{t\Delta}a - B(u,u)$, where $B(u,v) = \int_0^t e^{(t-s)\Delta}\mathbb{P}\nabla \cdot (u \otimes v)(s)\,ds$.

By Lemma 2.6 and Lemma 2.7, the bilinear operator B is bounded in $C([0,T], Z_\delta)$ and $||B(u,v)|| \le C_T||u||\,||v||$, where $||w|| \equiv \sup_{t\in[0,T]} ||w(t)||_{Z_\delta}$ and $C_T = O(T^{1/2})$ as $t \to 0$. Since it is straightforward to check that $e^{t\Delta}a$ belongs to $C([0,T], Z_\delta)$, if $a \in Z_\delta$, we see that the fixed point argument applies in $C([0,T], Z_\delta)$, at least if $T > 0$ is small enough. Theorem 1.1 then follows.

Remark 2.8. Theorem 1.1 covers the result of Furioli and Terraneo. This is obvious in the case $\frac{3}{2} < \delta < \frac{5}{2}$ and $\frac{5}{2} < \delta < \frac{7}{2}$ since we identified the spaces X_δ and Z_δ. In the case $\frac{7}{2} < \delta < \frac{9}{2}$, we use the fact that $Z_\delta \subset L^1(\mathbb{R}^3)$. Since we know that divergence-free vector fields in $L^1(\mathbb{R}^3)$ have vanishing integral, we deduce from Remark 2.5 that $u \in Z_\delta$ and $\nabla \cdot u = 0$ if and only if $u \in X_\delta$. Thus, Theorem 1.1 and the result of [8] turn out to be equivalent; however, the fact that the velocity field itself, and not only its Laplacian, does belong to weighted L^2-spaces does not seem to have been noticed in [8].

As claimed in the introduction, the restriction $\delta < \frac{9}{2}$ cannot be removed. Indeed, if u is a solution to (IE) such that $u \in C([0,T], Z_{9/2})$, for some $T > 0$, then the initial datum must satisfy the conditions of Dobrokhotov and Shafarevich: $\int(a_h a_k) = 0$, if $h \neq k$ and $\int a_1^2 = \int a_2^2 = \int a_3^2$. This is due to the fact that the localization condition $a \in L^2(\mathbb{R}^3, (1+|x|^2)^{5/2}dx)$ is not conserved during the evolution (see [3]).

References

[1] C. Amrouche, V. Girault, M.E. Schonbek, T.P. Schonbek *Pointwise Decay of Solutions and of higher derivatives to Navier-Stokes Equations*, SIAM J. Math. Anal., **31**, N. 4, 740–753, (2000).

[2] G. Bourdaud, *Localisation et multiplicateurs des espaces de Sobolev homogènes*, Manuscripta Mathematica, **60**, 93–130 (1988).

[3] L. Brandolese, Y. Meyer, *On the instantaneous spreading for the Navier-Stokes system in the whole space*, ESAIM Contr. Optim. Calc. Var. **8**, 273–285 (2002)

[4] R. Coifman, P.-L. Lions, Y. Meyer, S. Semmes, *Compensated compactness and Hardy spaces*, J. Math. Pures Appl. **72**, 247–286 (1993).

[5] S.Y. Dobrokhotov, A.I. Shafarevich, *Some Integral Identities and Remarks on the Decay at Infinity of the Solutions to the Navier-Stokes Equations in the Entire Space*, Russ. J. Math. Phys. Vol. 2, N. 1, 133–135 (1994).

[6] R. Farwing, H. Sohr, *Weighted energy inequalities for the Navier-Stokes equations in exterior domains.* Appl. Analysis **58** N.1-2, 157–173 (1995).

[7] R. Farwing, H. Sohr, *Global estimates in weighted spaces of weak solutions of the Navier-Stokes equations in exterior domains*, Arch. Math. **67**, 319–330 (1996).

[8] G. Furioli, E. Terraneo, *Molecules of the Hardy space and the Navier-Stokes equations*, Funkcial. Ekvac. **45**, N. 1 141–160 (2002).

[9] C. He, *The Cauchy problem for the Navier-Stokes equations*, J. Math. Analysis Appl. **209**, 422–444 (1998).

[10] C. He, Z. Xin, *On the decay properties of Solutions to the nonstationary Navier-Stokes Equations in* $\mathbb{R}^3$, Proc. Roy. Edinburgh Soc. Sect. A, **131**, N. 3, 597–619 (2001).

[11] P.G. Lemarié-Rieusset, *Recent developments in the Navier-Stokes problem*, Chapman & Hall, CRC Press Boca Raton (2002).

[12] T. Miyakawa, *Application of Hardy Space Techniques to the Time-Decay Problem for Incompressible Navier-Stokes Flows in* $\mathbb{R}^n$, Funkcial. Ekvac., **41**, 383–434 (1998).

[13] T. Miyakawa, *On space time decay properties of nonstationary incompressible Navier-Stokes flows in* $\mathbb{R}^n$, Funkcial. Ekvac., **32**, N. 2, 541–557 (2000).

[14] M.E. Schonbek, T.P. Schonbek, *On the boundedness and decay of moments of solutions of the Navier-Stokes equations*, Advances in Diff. Eq., 7–9 July–September (2000).

[15] F. Vigneron, *Ph. D Dissertation*, Ecole polytechnique, Palaiseau (France), in preparation.

[16] A. Youssfi, *Localisation des espaces de Besov homogènes*, Indiana Univ. Math. J., **37**, 565–587 (1988).

Lorenzo Brandolese
Institut Girard Desargues
Université Claude Bernard Lyon 1
21, avenue Claude Bernard
F-69622 Villeurbanne Cedex, France
e-mail: `brandolese@igd.univ-lyon1.fr`

Progress in Nonlinear Differential Equations
and Their Applications, Vol. 61, 37–44

A Limit Model for Unidirectional Non-Newtonian Flows with Nonlocal Viscosity

L. Consiglieri, J.F. Rodrigues and T. Shilkin

Dedicated to V.A. Solonnikov in his 70th birthday

Abstract. A p-Laplacian flow ($1 < p < \infty$) with nonlocal diffusivity is obtained as an asymptotic limit case of a high thermal conductivity flow described by a coupled system involving the dissipation energy.

Mathematics Subject Classification (2000). Primary 35K; Secondary 76D.

Keywords. non-Newtonian fluids, asymptotic limits, nonlocal models.

1. Introduction

In a previous work [2] we have obtained a nonlocal model for an incompressible viscous flow introduced by Ladyzhenskaya [7], in the two-dimensional periodic case, as a limit problem for large values of the thermal conductivity ($\sigma \to \infty$) and vanishing latent heat ($\delta \to 0$) of a non-isothermal Newtonian flow in presence of the dissipation energy. Contrary to the stationary problem that has only one limit ($\sigma \to \infty$) and has been completely studied in [1] for general non-Newtonian flows in two and three dimensions, the evolutionary problem presents some difficulties in the second passage to the limit $\delta \to 0$, due to the term of the dissipation of energy. In [2] the technical restriction on the viscosity function μ of being monotone nonincreasing was required.

In this note we extend the result of [2] to a scalar case corresponding to the thermal viscous incompressible non-Newtonian flow in a tube, which is described by the system

$$\partial_t u - \operatorname{div}\,(\mu(\theta)|\nabla u|^{p-2}\nabla u) = f \quad \text{in } Q_T = \Omega \times (0,T), \tag{1.1}$$

$$u|_{t=0} = u_0, \quad u|_{\partial\Omega} = 0, \tag{1.2}$$

Partly supported by FCT research project POCTI/34471/MAT/2000 and by RFBR research project no. 03-01-00638.

$$\delta\partial_t\theta - \sigma\Delta\theta + \theta = \mu(\theta)|\nabla u|^p \quad \text{in } Q_T, \tag{1.3}$$

$$\theta|_{t=0} = \theta_0, \quad \frac{\partial\theta}{\partial n}\Big|_{\partial\Omega} = 0. \tag{1.4}$$

Here $\Omega \subset \mathbb{R}^2$ is an arbitrary open and bounded set denoting the cross section of the tube, $u : Q_T \to \mathbb{R}$ is the scalar velocity, $\theta : Q_T \to \mathbb{R}$ is the temperature, μ the viscosity, f the given force incorporating the pressure variations in time, δ and σ are positive constants representing the latent heat and the thermal conductivity, respectively, and $1 < p < \infty$. We assume there exist positive constants μ_0 and μ_1 such that the Lipschitz continuous function $\mu = \mu(s)$ satisfies

$$0 < \mu_0 \le \mu(s) \le \mu_1. \tag{1.5}$$

Passing to the limit first in σ ($\sigma \to +\infty$) and afterwards in δ ($\delta \to 0^+$), the local system (1.1)–(1.4) becomes the nonlocal problem

$$\partial_t u - \nu([\nabla u]_{p,\Omega})\Delta_p u = f \quad \text{in } Q_T, \tag{1.6}$$

with (1.2), where ν is a new viscosity related with the initial one μ by the functional relation

$$\nu = \mu \circ \alpha, \qquad \text{with } \alpha \text{ being the inverse function of } \beta(s) = s/\mu(s), \tag{1.7}$$

$$\Delta_p u = \operatorname{div}(|\nabla u|^{p-2}\nabla u)$$

denotes the p-Laplacian for $1 < p < \infty$, and

$$[\nabla u(t)]_{p,\Omega} = \frac{1}{|\Omega|}\int_\Omega |\nabla u(x,t)|^p dx \quad \text{for a.e. } t \in (0,T)$$

denotes the $W_0^{1,p}(\Omega)$-norm at power p, divided by $|\Omega| = meas(\Omega)$.

In this work we shall assume that

$$\mu(s) > s\mu'(s) \quad \text{a.e. } s \in \mathbb{R}, \tag{1.8}$$

which together with (1.5) implies that α and β given in (1.7) are strictly increasing functions.

While the first limit $\sigma \to \infty$ is based only in energy estimates for weak solutions of (1.1)–(1.4), in particular, in a sharp L^q-estimate ($q < 4/3$) on the gradient of θ in terms of the L^1-norm of the right-hand side of (1.3), as obtained in [2], the second limit $\delta \to 0^+$ uses the uniform local Hölder continuity of the gradient of the solution to the degenerate parabolic equation (1.6) (see [4]).

Here we have considered only scalar flows of pseudo-plastic type ($1 < p < 2$), Newtonian ($p = 2$) or dilatant type ($p > 2$), but it is clear that similar results also hold for other classes of quasi-Newtonian fluids in tubes as considered in [10].

2. The limit problems and their formulations

For $q \geq 1$, we shall use the usual Lebesgue and Sobolev spaces $L^q(\Omega)$, $W^{1,q}(\Omega)$ as well as the vector valued evolutive spaces $L^q(0,T;B)$ (see [11]).

For any given $\delta > 0$ and $\sigma > 0$, we shall be concerned with weak solutions (u,θ) of the system (1.1)-(1.4) in the class

$$u \in C([0,T];L^2(\Omega)) \cap L^p(0,T;W_0^{1,p}(\Omega)), \quad \partial_t u \in L^{p'}(0,T;W^{-1,p'}(\Omega)),$$
$$\theta \in L^\infty(0,T;L^1(\Omega)) \cap L^q(0,T;W^{1,q}(\Omega)), \quad \partial_t\theta \in L^1(0,T;(W^{1,q}(\Omega))^*),$$
$$1 \leq q < \tfrac{4}{3},$$

satisfying

$$\int_{Q_T} (-u\partial_t\varphi + \mu(\theta)|\nabla u|^{p-2}\nabla u \cdot \nabla\varphi)\, dx\, dt = \int_{Q_T} f\varphi\, dx\, dt + \int_\Omega u_0(x)\varphi(x,0)\, dx,$$
$$\forall\varphi \in C^1(\bar{Q}_T):\ \varphi|_{t=T} = 0,\ \varphi|_{\partial\Omega} = 0, \tag{2.1}$$

$$\int_{Q_T} (-\delta\theta\partial_t\eta + \sigma\nabla\theta\cdot\nabla\eta + \theta\eta)\, dx\, dt = \int_{Q_T} \mu(\theta)|\nabla u|^p\eta\, dx\, dt$$
$$+\delta\int_\Omega \theta_0(x)\eta(x,0)\, dx, \quad \forall\eta \in C^1(\bar{Q}_T):\ \eta|_{t=T} = 0, \tag{2.2}$$

and the corresponding limit problems when $\sigma \to \infty$ and $\delta \to 0$.

For given $f \in L^2(Q_T)$, $u_0 \in L^2(\Omega)$ and $\theta_0 \in L^1(\Omega)$, under the assumption (1.5) the first asymptotic result can be formulated in the following theorem.

Theorem 2.1. *Let $\delta > 0$ be fixed and (u_σ,θ_σ) be a solution to (2.1)–(2.2) corresponding to each $\sigma > 0$. Then there exist*

$$u \in C([0,T];L^2(\Omega)) \cap L^p(0,T;W_0^{1,p}(\Omega)), \quad \partial_t u \in L^{p'}(0,T;W^{-1,p'}(\Omega)),$$
$$\xi \in W^{1,1}(0,T),$$

and a subsequence $\sigma \to \infty$, such that

$$u_\sigma \to u \quad \text{in } L^p(0,T;W_0^{1,p}(\Omega)), \quad \partial_t u_\sigma \rightharpoonup \partial_t u \quad \text{in } L^{p'}(0,T;W^{-1,p'}(\Omega)),$$
$$\theta_\sigma \to \xi \quad \text{in } L^q(0,T;W^{1,q}(\Omega)), \quad \nabla\theta_\sigma \to 0 \text{ in } L^q(Q_T),$$

and (u,ξ) satisfy the equations

$$\partial_t u - \mu(\xi)\Delta_p u = f \quad \text{in } Q_T \tag{2.3}$$

with (1.2) and

$$\delta\frac{d\xi}{dt} + \xi = \mu(\xi)[\nabla u]_{p,\Omega}, \quad \text{in } (0,T) \tag{2.4}$$

with the initial condition

$$\xi|_{t=0} = \frac{1}{|\Omega|}\int_\Omega \theta_0\, dx. \tag{2.5}$$

By the theory of nonlinear parabolic equations [9], we know that if $u_0 \in W_0^{1,p}(\Omega)$ a solution to (2.3)–(1.2) can be written in the form

$$\begin{gathered}(\partial_t u, w) + \mu(\xi)(|\nabla u|^{p-2}\nabla u, \nabla w) = (f, w),\\ \forall w \in W_0^{1,p}(\Omega), \quad \text{for a.e. } t \in (0,T),\\ u|_{t=0} = u_0,\end{gathered} \tag{2.6}$$

where $(\cdot,\cdot)$ means the duality between spaces $W^{-1,p'}(\Omega)$ and $W_0^{1,p}(\Omega)$. Rewriting (2.6) as

$$\frac{1}{\mu(\xi)}(\partial_t u, w) + (|\nabla u|^{p-2}\nabla u, \nabla w) = \frac{1}{\mu(\xi)}(f, w),$$

we can formally choose $w = \partial_t u$ as a test function (for instance, using the Faedo-Galerkin method of approximation), obtaining for a.e. $t \in (0,T)$

$$\frac{1}{\mu_1}\|\partial_t u\|_{2,\Omega}^2 + \frac{1}{p}\frac{d}{dt}\|\nabla u\|_{p,\Omega}^p \leq \frac{1}{\mu_0}\|f\|_{2,\Omega}\|\partial_t u\|_{2,\Omega},$$

and the following estimate holds

$$\|\partial_t u\|_{2,Q_T}^2 + \sup_{t\in[0,T]}\|\nabla u(t)\|_{p,\Omega}^p \leq C\left(\|\nabla u_0\|_{p,\Omega}^p + \|f\|_{2,Q_T}^2\right), \tag{2.7}$$

where the constant C only depends on μ_0, μ_1 and p, but is independent of ξ and δ.

By the local regularity theory [4], if $f \in L^\infty(Q_T)$ by assumption (1.5) we may also conclude that ∇u_δ belongs to a bounded subset of $C^\gamma(Q')$, for some γ, $0 < \gamma < 1$, for any compact subset $Q' = \overline{\Omega'} \times [s,t]$, $\Omega' \Subset \Omega$ and $0 < s < t < T$ uniformly in $\delta > 0$.

Then we can formulate the second passage to the limit in $\delta \to 0^+$.

Theorem 2.2. *Suppose*

$$f \in L^\infty(Q_T), \qquad u_0 \in W_0^{1,p}(\Omega), \quad \textit{and } \mu \textit{ satisfies } (1.8).$$

For each $\delta > 0$ let (u_δ, ξ_δ) be a solution of (2.3)–(2.5) *obtained in Theorem* 2.1. *Then there exist u and a subsequence $\delta \to 0^+$ such that*

$$\begin{aligned} u_\delta \to u \quad & \textit{in } \ L^p(0,T; W_0^{1,p}(\Omega)),\\ \nabla u_\delta \to \nabla u \quad & \textit{uniformly in } \ Q' \Subset Q_T,\end{aligned}$$

and u satisfies the equation (1.6) *with* (1.2), *where ν is given by* (1.7).

3. Existence of weak solutions and their convergence

3.1. Proof of Theorem 2.1

The existence of at least one solution to (2.1)–(2.2) can be shown essentially as in [2], so we only sketch its proof. Define the functional

$$\mathcal{T} : \zeta \in L^q(Q_T) \mapsto u_\zeta \mapsto \theta_\zeta \in L^q(0,T; W^{1,q}(\Omega)),$$

where u_ζ is the unique solution of

$$\int_{Q_T} (-u_\zeta \partial_t \varphi + \mu(\zeta)|\nabla u_\zeta|^{p-2}\nabla u_\zeta \cdot \nabla\varphi)\, dx\, dt = \int_{Q_T} f\varphi\, dx\, dt \\ + \int_\Omega u_0(x)\varphi(x,0)\, dx, \quad \forall \varphi \in C^1(\bar{Q}_T) : \varphi|_{t=T} = 0,\ \varphi|_{\partial\Omega} = 0, \tag{3.1}$$

satisfying the estimate (see [9] and note that the embedding $W^{1,p}(\Omega) \hookrightarrow L^2(\Omega)$ is valid for every $p > 1$ since $n = 2$)

$$\|u_\zeta\|_{L^\infty(0,T;L^2(\Omega))} + \|u_\zeta\|_{L^p(0,T;W_0^{1,p}(\Omega))} + \|\partial_t u_\zeta\|_{L^{p'}(0,T;W^{-1,p'}(\Omega))} \\ \le C\Big(\|f\|_{2,Q_T} + \|u_0\|_{2,\Omega}\Big); \tag{3.2}$$

and θ_ζ is the unique solution of the problem

$$\int_{Q_T} (-\delta\theta_\zeta \partial_t \eta + \sigma\nabla\theta_\zeta \cdot \nabla\eta + \theta_\zeta\eta)\, dx\, dt = \int_{Q_T} \mu(\zeta)|\nabla u_\zeta|^p \eta\, dx\, dt \\ +\delta \int_\Omega \theta_0(x)\eta(x,0)\, dx, \quad \forall \eta \in C^1(\bar{Q}_T) : \eta|_{t=T} = 0, \tag{3.3}$$

which satisfies the following estimates (see Proposition 3.2 of [2])

$$\|\theta_\zeta\|_{L^\infty(0,T;L^1(\Omega))} + \sqrt{\sigma}\|\nabla\theta_\zeta\|_{q,Q_T} + \|\theta_\zeta\|_{\frac{3q}{2},Q_T} \le \mathcal{F}\left(\|\nabla u_\zeta\|^p_{p,Q_T}, \|\theta_0\|_{1,\Omega}\right) \tag{3.4}$$

where the constant C and the majorant $\mathcal{F}$ (which is a nondecreasing continuous function of its arguments) depend only on μ_0, μ_1, and δ. Then the Schauder fixed point theorem yields the existence of a $\theta = T\theta$, which together with $u = u_\theta$ guarantees the existence of a solution to (2.1)–(2.2), satisfying the estimates (3.2) and (3.4), independently of $\sigma > 0$.

Let $\sigma \to \infty$. The estimates (3.2) and (3.4) allow us to extract a subsequence such that

$$u_\sigma \rightharpoonup u \quad \text{in } L^p(0,T;W_0^{1,p}(\Omega)), \quad \partial_t u_\sigma \rightharpoonup \partial_t u \quad \text{in } L^{p'}(0,T;W^{-1,p'}(\Omega)), \\ \theta_\sigma \rightharpoonup \theta \quad \text{in } L^q(0,T;W^{1,q}(\Omega)), \quad \nabla\theta_\sigma \to 0 \quad \text{in } L^q(Q_T).$$

Hence $\nabla\theta = 0$ and there exists $\xi \in L^q(0,T)$ such that $\theta(x,t) = \xi(t)$ a.e. $(x,t) \in Q_T$. Denoting

$$\tilde{\theta}_\sigma(t) \equiv \frac{1}{|\Omega|}\int_\Omega \theta_\sigma(x,t)\, dx,$$

we have

$$\tilde{\theta}_\sigma \rightharpoonup \xi \quad \text{in} \quad L^q(0,T),$$

and arguing as in [2] we obtain the convergence

$$\theta_\sigma \to \xi \quad \text{in} \quad L^q(Q_T).$$

Consequently we may pass to the limit as $\sigma \to \infty$ in the equation (2.1) obtaining (2.3). Since $W_0^{1,p}(\Omega) \hookrightarrow L^2(\Omega)$ is compact in $n = 2$ for $p > 1$, we have $u_\sigma(T) \to$

$u(T)$ in $L^2(\Omega)$ and

$$\int_{Q_T} \mu(\theta_\sigma)|\nabla u_\sigma|^p \, dx \, dt = \int_{Q_T} f u_\sigma \, dx \, dt + \frac{1}{2}\int_\Omega u_0^2 \, dx - \frac{1}{2}\int_\Omega u_\sigma^2(T) \, dx$$

$$\longrightarrow \int_{Q_T} f u \, dx \, dt + \frac{1}{2}\int_\Omega u_0^2 \, dx - \frac{1}{2}\int_\Omega u^2(T) \, dx = \int_{Q_T} \mu(\xi)|\nabla u|^p \, dx \, dt,$$

which, in particular, implies the strong convergence $u_\sigma \to u$ in $L^p(0,T;W_0^{1,p}(\Omega))$.

Taking the limit $\sigma \to \infty$ in (2.2) with test functions $\eta = \eta(t)$ depending only on t we obtain (2.4) and (2.5) in weak form, concluding the proof of Theorem 2.1.

3.2. Proof of Theorem 2.2

Let (u_δ, ξ_δ) be a solution of the system (2.3)–(2.5) and consider the equation

$$\delta \frac{d\xi_\delta}{dt} + \xi_\delta = \mu(\xi_\delta)[\nabla u_\delta]_{p,\Omega}. \tag{3.5}$$

Denote by $g_\delta(t) = [\nabla u_\delta(t)]_{p,\Omega} - \beta(\xi(0))$, α the inverse function of $\beta(\xi) = \xi/\mu(\xi)$ as in (1.7), and $a(\zeta) \equiv \int_0^{\alpha(\zeta+\beta(\xi(0)))} ds/\mu(s)$ obtaining

$$\delta \frac{d}{dt} a(\zeta_\delta) + \zeta_\delta = g_\delta \quad \text{for } \zeta_\delta = \beta(\xi_\delta) - \beta(\xi(0)) \Leftrightarrow \xi_\delta = \alpha(\zeta_\delta + \beta(\xi(0))). \tag{3.6}$$

From nonlinear operator theory in $L^1(0,T)$, we know that the resolvent $J_\delta = (I + \delta \frac{d}{dt} a)^{-1}$ satisfies

$$J_\delta \underset{\delta \to 0}{\longrightarrow} I \quad \text{and} \quad \|J_\delta\|_{\mathcal{L}(L^1(0,T);L^1(0,T))} \leq 1$$

since $A = -\frac{d}{dt}a$ is a maximal dissipative operator (see [3, 5.5.1.], for instance) with dense domain $D(A) = \{\zeta \in L^1(0,T) : \ A\zeta \in L^1(0,T), \ \zeta(0) = 0\}$. Indeed, taking into account the assumption (1.8), we have

$$a'(\zeta) = \frac{\alpha'(s)}{\mu(\alpha(s))} = \frac{\mu^2(s)}{\mu(\alpha(s))(\mu(s) - s\mu'(s))} > 0,$$

for $s = \zeta + \beta(\xi(0))$, and therefore

$$\begin{aligned} &\textstyle\int_0^T [\frac{d}{dt}a(\zeta_1) - \frac{d}{dt}a(\zeta_2)]\operatorname{sign}(\zeta_1 - \zeta_2) \\ &= \textstyle\int_0^T [\frac{d}{dt}a(\zeta_1) - \frac{d}{dt}a(\zeta_2)]\operatorname{sign}(a(\zeta_1) - a(\zeta_2)) = |a(\zeta_1(T)) - a(\zeta_2(T))| \geq 0 \end{aligned}$$

yields the dissipative property of A. Since $\mathcal{D}(0,T) \subset D(A)$, the existence and uniqueness of solution to (3.6) implies that A is a maximal operator in $L^1(0,T)$ with dense domain.

For a subsequence $\{u_\delta\}$ weakly convergent in $L^p(0,T;W_0^{1,p}(\Omega))$, since ∇u_δ is uniformly Hölder continuous in each compact subset Q' of Q_T we may suppose that $\nabla u_\delta \to \nabla u$ uniformly in Q', and also the strong convergence $u_\delta \to u$ in $L^p(0,T;W_0^{1,p}(\Omega))$. Therefore we get

$$g_\delta \to g = [\nabla u]_{p,\Omega} - \beta(\xi(0)) \quad \text{a.e. } t \in (0,T),$$

as g_δ is uniformly bounded in $L^\infty(0,T)$ we have

$$g_\delta \underset{\delta\to 0}{\to} g \quad \text{in } L^q(0,T), \quad 1 \le q < \infty,$$

and we obtain the estimate for the solution ζ_δ of (3.6)

$$\|\zeta_\delta - g\|_{L^1(0,T)} \le \|J_\delta\|_{\mathcal{L}(L^1(0,T);L^1(0,T))}\|g_\delta - g\|_{L^1(0,T)} + \|J_\delta g - g\|_{L^1(0,T)}.$$

Then the sequence $\{\xi_\delta\}$ solving (3.5), or equivalently (3.6), is such that

$$\xi_\delta \to \xi = \alpha\Big([\nabla u]_{p,\Omega}\Big) \quad \text{in } L^1(0,T).$$

Passing now to the limit $\delta \to 0$ in the equation (2.3) the conclusion follows immediately.

Acknowledgment

T. Shilkin acknowledges CMAF/Univ. of Lisbon for hospitality and partial support. The authors wish to thank Isabel Simão for bringing their attention to the reference [3].

References

[1] L. Consiglieri, J.F. Rodrigues, *On stationary flows with energy dependent nonlocal viscosities.* Zapiski Nauchn. Seminarov POMI **295** (2003), 99–117.

[2] L. Consiglieri, J.F. Rodrigues, T. Shilkin, *On the Navier-Stokes equations with the energy-dependent nonlocal viscosities.* Zapiski Nauchnyh Seminarov POMI **306** (2003), 71–91.

[3] G. DaPrato, J. Zabczyk, *Ergodicity for infinite-dimensional systems*, Cambridge Univ. Press, Cambridge 1996.

[4] E. DiBenedetto, *Degenerate parabolic equations*, Springer-Verlag, New York 1993.

[5] D. Hilhorst, J.F. Rodrigues, *On a nonlocal diffusion equation with discontinuous reaction*, Adv. Differential Equations **5** (2000) 4–6, 657–680.

[6] O.A. Ladyzhenskaya, *Mathematical problems in the dynamics of a viscous incompressible fluid.* 2nd rev. aug. ed., "Nauka", Moscow, 1970: English transl. of 1st ed., *The mathematical theory of viscous incompressible flow.* Gordon and Breach. New York 1969.

[7] O.A. Ladyzhenskaya, *On new equations for the description of the motion of viscous incompressible fluids and on the solvability "in large" of some boundary value problems for them.* Proceedings of Steklov Math. Inst., **102** (1967), 85–104.

[8] O.A. Ladyzhenskaya, V.A. Solonnikov, N.N. Uraltseva, *Linear and quasilinear equations of parabolic type.* Translations of Mathematical Monographs, Vol. 23 American Mathematical Society, Providence, R.I. 1967.

[9] J.L. Lions, *Quelques méthodes de résolution des problèmes aux limites non linéaires.* Gauthier-Villars, Paris 1969.

[10] J.F. Rodrigues, *Thermoconvection with dissipation of quasi-Newtonian fluids in tubes*, in Navier-Stokes equations and related nonlinear problems. A. Sequeira (ed). Plenum Press, New York 1995, 279–288.

[11] J. Simon, *Compact sets in the space* $L^p(0,T;B)$, Ann. Mat. Pura Appl. **146** (4) (1987), 65–96.

L. Consiglieri
CMAF and FCUL
Campo Grande
University of Lisboa
1749-016 Lisboa, Portugal
e-mail: glieri@ptmat.fc.ul.pt

J.F. Rodrigues
FCUL and CMAF
Av. Prof. Gama Pinto, 2
University of Lisboa
1649-003 Lisboa
Portugal
and
CMUC
University of Coimbra
Portugal
e-mail: rodrigue@ptmat.fc.ul.pt

T. Shilkin
V.A. Steklov Mathematical Institute
Fontanka 27
191023 St. Petersburg, Russia
e-mail: shilkin@pdmi.ras.ru

Progress in Nonlinear Differential Equations
and Their Applications, Vol. 61, 45–64

On the Problem of Thermocapillary Convection for Two Incompressible Fluids Separated by a Closed Interface

Irina Vlad. Denisova

In honor of the jubilee of Vsevolod A. Solonnikov

Abstract. We consider the unsteady motion of a drop in another incompressible fluid. On the unknown interface between the liquids, the surface tension is taken into account. Moreover, the coefficient of surface tension depends on the temperature. We study this problem of the thermocapillary convection by M.V. Lagunova and V.A. Solonnikov's technique developed for a single liquid.

The local existence theorem for the problem is proved in Hölder classes of functions. The proof is based on the fact that the solvability of the problem with a constant coefficient of surface tension was obtain earlier. For a given velocity vector field of the fluids, we arrive at a diffraction problem for the heat equation which solvability is established by well-known methods. Existence of a solution to the complete problem is proved by successive approximations.

Mathematics Subject Classification (2000). Primary 35Q30,76D05; Secondary 35R35.

Keywords. Thermocapillary convection, Two viscous incompressible fluids, Interface problem, Navier-Stokes system, Heat equation.

1. Statement of the problem and formulation of the main result

This paper deals with unsteady motion of two viscous incompressible fluids separated by a closed unknown interface Γ_t. Both liquids have finite volume, they are bounded by a given surface S where the adhesion condition holds. We assume that boundaries Γ_t and S have no intersection. On the interface Γ_t, we take into account the surface tension depending on the temperature.

Partly supported by the US Civilian Reaserch and Development Foundation (CRDF) through grant number RU-M1-2596-ST-04.

The main result of the study is local (in time) unique solvability of the problem described in Hölder spaces of functions. The proof of this fact uses the technique of M.V. Lagunova and V.A. Solonnikov developed for the investigation of the thermocapillary convection problem for a drop in vacuum [1], and it is based on the existence theorem for the case of constant temperature [2, 3, 4].

Now give a mathematical formulation of the problem of thermocapillary convection for two liquids in a reservoir [5, 6].

Let, at the initial moment $t = 0$, a fluid with the viscosity $\nu^+ > 0$ and the density $\rho^+ > 0$ occupy a bounded domain $\Omega_0^+ \subset \mathbb{R}^3$. We denote $\partial\Omega_0^+$ by Γ_0. And there let be a fluid with the viscosity $\nu^- > 0$ and the density $\rho^- > 0$ in the"exterior" domain Ω_0^-. The boundary $S \equiv \partial(\Omega_0^+ \cup \Gamma_0 \cup \Omega_0^-)$ is a given closed surface, $S \cap \Gamma_0 = \emptyset$.

For every $t > 0$, it is necessary to find the interface Γ_t between the domains Ω_t^+ and Ω_t^-, as well as the velocity vector field $\boldsymbol{v}(x,t) = (v_1, v_2, v_3)$, the pressure function p and the temperature θ of both fluids satisfying the following initial-boundary value problem:

$$\begin{aligned}
&\mathcal{D}_t\boldsymbol{v} + (\boldsymbol{v}\cdot\nabla)\boldsymbol{v} - \nu^\pm\nabla^2\boldsymbol{v} + \frac{1}{\rho^\pm}\nabla p = \boldsymbol{f}, \quad \nabla\cdot\boldsymbol{v} = 0,\\
&\mathcal{D}_t\theta + (\boldsymbol{v}\cdot\nabla)\theta - k^\pm\nabla^2\theta = 0 \quad \text{in} \quad \Omega_t^-\cup\Omega_t^+, \quad t>0,\\
&\boldsymbol{v}|_{t=0} = \boldsymbol{v}_0, \quad \theta|_{t=0} = \theta_0 \quad \text{in} \quad \Omega_0^-\cup\Omega_0^+, \qquad\qquad (1.1)\\
&[\boldsymbol{v}]|_{\Gamma_t} = \lim_{\substack{x\to x_0\in\Gamma_t,\\ x\in\Omega_t^+}} \boldsymbol{v}(x) - \lim_{\substack{x\to x_0\in\Gamma_t,\\ x\in\Omega_t^-}} \boldsymbol{v}(x) = 0, \quad [\theta]|_{\Gamma_t} = 0,\\
&\boldsymbol{v}|_S = 0, \quad \theta|_S = a,\\
&[\mathbb{T}\boldsymbol{n}]|_{\Gamma_t} = \sigma(\theta)H\boldsymbol{n} + \nabla_{\Gamma_t}\sigma(\theta), \qquad \left[k^\pm\frac{\partial\theta}{\partial\boldsymbol{n}}\right]\bigg|_{\Gamma_t} + \kappa\theta\nabla_{\Gamma_t}\cdot\boldsymbol{v} = 0 \ \text{on}\ \Gamma_t.
\end{aligned}$$

Here $\mathcal{D}_t = \partial/\partial t$, $\nabla = (\partial/\partial x_1, \partial/\partial x_2, \partial/\partial x_3)$, $\nu^\pm, \rho^\pm$ are the step functions of viscosity and density, respectively, $\boldsymbol{f}$ is the given vector field of mass forces, $\boldsymbol{v}_0$, θ_0 are the initial data, a is the given temperature on the surface S, $\mathbb{T}$ is the stress tensor with the elements

$$T_{ik} = -\delta_i^k p + \mu^\pm(\partial v_i/\partial x_k + \partial v_k/\partial x_i), \qquad i,k = 1,2,3;$$

$\mu^\pm = \nu^\pm\rho^\pm$, δ_i^k is the Kronecker symbol, $\sigma(\theta) = \sigma_1 - \kappa(\theta-\theta_1) > 0$ is the coefficient of the surface tension, $\sigma_1, \kappa, \theta_1$ are the positive constants, $\boldsymbol{n}$ is the outward normal to Ω_t^+, $H(x,t)$ is twice the mean curvature of Γ_t ($H < 0$ at the points where Γ_t is convex towards Ω_t^-), $k^\pm$ is the step function of thermal conductivity, ∇_{Γ_t} is the gradient on Γ_t. We suppose that a Cartesian coordinate system $\{x\}$ is introduced in $\mathbb{R}^3$. The centered dot denotes the Cartesian scalar product.

We imply the summation from 1 to 3 with respect to repeated indexes. We mark the vectors and the vector spaces by boldface letters.

Moreover, to exclude the mass transportation through Γ_t, we assume that the liquid particles do not leave Γ_t. It means that Γ_t consists of the points $x(\xi,t)$

such that the corresponding vector $\boldsymbol{x}(\xi,t)$ solves the Cauchy problem

$$\mathcal{D}_t\boldsymbol{x} = \boldsymbol{v}(x(t),t), \quad \boldsymbol{x}|_{t=0} = \boldsymbol{\xi}, \quad \xi \in \Gamma_0, \ t > 0. \tag{1.2}$$

Hence, $\Gamma_t = \{x(\xi,t)|\ \xi \in \Gamma_0\}$, $\Omega_t^{\pm} = \{x(\xi,t)|\ \xi \in \Omega_0^{\pm}\}$.

Let us pass from the Eulerian to Lagrangian coordinates by the formula

$$\boldsymbol{x} = \boldsymbol{\xi} + \int_0^t \boldsymbol{u}(\xi,\tau)\mathrm{d}\tau \equiv X_{\boldsymbol{u}}(\xi,t) \tag{1.3}$$

(here $\boldsymbol{u}(\xi,t)$ is the velocity vector field in the Lagrangian coordinates). Next, we apply the well-known relation

$$H\boldsymbol{n} = \Delta(t)\boldsymbol{x},$$

where $\Delta(t)$ denotes the Beltrami-Laplace operator on Γ_t. This transformation leads us to the problem for $\boldsymbol{u}$ and $q = p(X_{\boldsymbol{u}},t)$ with the given interface $\Gamma \equiv \Gamma_0$. If the angle between $\boldsymbol{n}$ and the exterior normal $\boldsymbol{n}_0$ to Γ is acute this problem is equivalent to the following system:

$$\begin{aligned}
&\mathcal{D}_t\boldsymbol{u} - \nu^{\pm}\nabla_{\boldsymbol{u}}^2\boldsymbol{u} + \frac{1}{\rho^{\pm}}\nabla_{\boldsymbol{u}}q = \boldsymbol{f}(X_{\boldsymbol{u}},t),\\
&\nabla_{\boldsymbol{u}}\cdot\boldsymbol{u} = 0, \qquad \mathcal{D}_t\hat{\theta} - k^{\pm}\nabla_{\boldsymbol{u}}^2\hat{\theta} = 0 \quad \text{in } Q_T^{\pm} = \Omega_0^{\pm}\times(0,T),\\
&\boldsymbol{u}|_{t=0} = \boldsymbol{v}_0, \qquad \hat{\theta}|_{t=0} = \theta_0 \quad \text{in } \Omega_0^{-}\cup\Omega_0^{+}, \qquad\qquad (1.4)\\
&[\boldsymbol{u}]|_{G_T} = 0, \quad [\hat{\theta}]|_{G_T} = 0, \quad \boldsymbol{u}|_{S_T} = 0, \quad \hat{\theta}|_{S_T} = \hat{a} \quad (S_T = S\times(0,T)),\\
&[\Pi_0\Pi\mathbb{T}_{\boldsymbol{u}}(\boldsymbol{u})\boldsymbol{n}]|_{G_T} = \Pi_0\Pi\nabla_{\boldsymbol{u}}\sigma(\hat{\theta}),\\
&[\boldsymbol{n}_0\cdot\mathbb{T}_{\boldsymbol{u}}(\boldsymbol{u},q)\boldsymbol{n}]|_{G_T} - \sigma(\hat{\theta})\boldsymbol{n}_0\cdot\Delta(t)\int_0^t \boldsymbol{u}\big|_{\Gamma}\mathrm{d}\tau = \sigma(\hat{\theta})H_0(\xi) +\\
&\qquad\qquad + \sigma(\hat{\theta})\boldsymbol{n}_0\cdot\int_0^t \dot{\Delta}(\tau)\boldsymbol{\xi}\big|_{\Gamma}\mathrm{d}\tau + \boldsymbol{n}_0\cdot\Pi\nabla_{\boldsymbol{u}}\sigma(\hat{\theta}),\\
&[k^{\pm}\boldsymbol{n}\cdot\nabla_{\boldsymbol{u}}\hat{\theta}]|_{G_T} + \kappa\hat{\theta}\Pi\nabla_{\boldsymbol{u}}\cdot\boldsymbol{u} = 0 \quad \text{on } G_T = \Gamma\times(0,T).
\end{aligned}$$

Here we have used the notation: $\nabla_{\boldsymbol{u}} = \mathbb{A}\nabla$, $\mathbb{A}$ is the matrix of cofactors A_{ij} to the elements

$$a_{ij}(\xi,t) = \delta_i^j + \int_0^t \frac{\partial u_i}{\partial \xi_j}\mathrm{d}t'$$

of the Jacobian matrix of the transformation (1.3), the vector $\boldsymbol{n}$ is related to $\boldsymbol{n}_0$ as $\boldsymbol{n} = \mathbb{A}\boldsymbol{n}_0/|\mathbb{A}\boldsymbol{n}_0|$; $\Pi\boldsymbol{\omega} = \boldsymbol{\omega} - \boldsymbol{n}(\boldsymbol{n}\cdot\boldsymbol{\omega})$, $\Pi_0\boldsymbol{\omega} = \boldsymbol{\omega} - \boldsymbol{n}_0(\boldsymbol{n}_0\cdot\boldsymbol{\omega})$ are the projections of a vector $\boldsymbol{\omega}$ onto the tangent plane to Γ_t and to Γ, respectively. The tensor $\mathbb{T}_{\boldsymbol{u}}(\boldsymbol{w},q)$ has the elements

$$(\mathbb{T}_{\boldsymbol{u}}(\boldsymbol{w},q))_{ij} = -\delta_j^i q + \mu^{\pm}(A_{jk}\partial w_i/\partial\xi_k + A_{ik}\partial w_j/\partial\xi_k),$$

$H_0(\xi) = \boldsymbol{n}_0\cdot\Delta(0)\boldsymbol{\xi}$ is twice the mean curvature of Γ.

We remind the definition of Hölder spaces. Let Ω be a domain in $\mathbb{R}^n$, $n \in \mathbb{N}$; for $T > 0$ we put $\Omega_T = \Omega\times(0,T)$; finally, let $\alpha \in (0,1)$. By $C^{\alpha,\alpha/2}(\Omega_T)$ we denote the set of functions f in Ω_T having norm

$$\|f\|^{(\alpha,\alpha/2)}_{\Omega_T} = \|f\|_{\Omega_T} + \langle f\rangle^{(\alpha,\alpha/2)}_{\Omega_T},$$

where

$$\|f\|_{\Omega_T} = \sup_{t\in(0,T)}\sup_{x\in\Omega}|f(x,t)|, \qquad \langle f\rangle^{(\alpha,\alpha/2)}_{\Omega_T} = \langle f\rangle^{(\alpha)}_{x,\Omega_T} + \langle f\rangle^{(\alpha/2)}_{t,\Omega_T},$$

$$\begin{aligned}\langle f\rangle^{(\alpha)}_{x,\Omega_T} &= \sup_{t\in(0,T)}\sup_{x,y\in\Omega}|f(x,t)-f(y,t)||x-y|^{-\alpha},\\ \langle f\rangle^{(\mu)}_{t,\Omega_T} &= \sup_{x\in\Omega}\sup_{t,\tau\in(0,T)}|f(x,t)-f(x,\tau)||t-\tau|^{-\mu}, \quad \mu\in(0,1).\end{aligned}$$

We introduce the following notation:

$$\begin{aligned}\mathcal{D}^{\boldsymbol{r}}_x &= \partial^{|\boldsymbol{r}|}/\partial x_1^{r_1}\dots\partial x_n^{r_n}, \quad \boldsymbol{r}=(r_1,\dots,r_n), \quad r_i\geq 0,\ |\boldsymbol{r}|=r_1+\dots+r_n,\\ \mathcal{D}^s_t &= \partial^s/\partial t^s, \quad s\in\mathbb{N}\cup\{0\}.\end{aligned}$$

Let $k\in\mathbb{N}$. By definition, the space $C^{k+\alpha,(k+\alpha)/2}(\Omega_T)$ consists of functions f with finite norm

$$\|f\|^{(k+\alpha,\frac{k+\alpha}{2})}_{\Omega_T} = \sum_{|\boldsymbol{r}|+2s\leq k}\|\mathcal{D}^{\boldsymbol{r}}_x\mathcal{D}^s_t f\|_{\Omega_T} + \langle f\rangle^{(k+\alpha,\frac{k+\alpha}{2})}_{\Omega_T},$$

where

$$\langle f\rangle^{(k+\alpha,\frac{k+\alpha}{2})}_{\Omega_T} = \sum_{|\boldsymbol{r}|+2s=k}\langle\mathcal{D}^{\boldsymbol{r}}_x\mathcal{D}^s_t f\rangle^{(\alpha,\frac{\alpha}{2})}_{\Omega_T} + \sum_{|\boldsymbol{r}|+2s=k-1}\langle\mathcal{D}^{\boldsymbol{r}}_x\mathcal{D}^s_t f\rangle^{(\frac{1+\alpha}{2})}_{t,\Omega_T}.$$

The symbol $C^{k+\alpha,(k+\alpha)/2}(\Omega_T)$ denotes the subspace of $C^{k+\alpha,(k+\alpha)/2}(\Omega_T)$ whose elements f have the property: $\mathcal{D}^i_t f\big|_{t=0}=0$, $i=0,\dots,\left[\frac{k+\alpha}{2}\right]$.

We define $C^{k+\alpha}(\Omega)$, $k\in\mathbb{N}\cup\{0\}$, as the space of functions $f(x)$, $x\in\Omega$, with the norm

$$\|f\|^{(k+\alpha)}_{\Omega} = \sum_{|\boldsymbol{r}|\leq k}\|\mathcal{D}^{\boldsymbol{r}}_x f\|_{\Omega} + \langle f\rangle^{(k+\alpha)}_{\Omega}.$$

Here

$$\langle f\rangle^{(k+\alpha)}_{\Omega} = \sum_{|\boldsymbol{r}|=k}\langle\mathcal{D}^{\boldsymbol{r}}_x f\rangle^{(\alpha)}_{\Omega} = \sum_{|\boldsymbol{r}|=k}\sup_{x,y\in\Omega}|\mathcal{D}^{\boldsymbol{r}}_x f(x)-\mathcal{D}^{\boldsymbol{r}}_y f(y)||x-y|^{-\alpha}.$$

We also need the following semi-norm with $\alpha,\gamma\in(0,1)$:

$$|f|^{(1+\alpha,\gamma)}_{\Omega_T} = \langle f\rangle^{(1+\alpha,\gamma)}_{\Omega_T} + \langle f\rangle^{(\frac{1+\alpha-\gamma}{2})}_{t,\Omega_T},$$

here

$$\langle f\rangle^{(1+\alpha,\gamma)}_{\Omega_T} = \max_{t,\tau\in(0,T)}\max_{x,y\in\Omega}\frac{|f(x,t)-f(y,t)-f(x,\tau)+f(y,\tau)|}{|x-y|^{\gamma}|t-\tau|^{(1+\alpha-\gamma)/2}}.$$

There exists the estimate

$$\langle f\rangle^{(1+\alpha,\gamma)}_{\Omega_T} \leq c_1\langle f\rangle^{(1+\alpha,\frac{1+\alpha}{2})}_{\Omega_T}.$$

We consider that $f \in C^{(1+\alpha,\gamma)}(\Omega_T)$ if

$$\|f\|_{\Omega_T} + |f|_{\Omega_T}^{(1+\alpha,\gamma)} < \infty.$$

Finally, if a function f has finite norm

$$\|f\|_{\Omega_T}^{(\gamma,\mu)} \equiv \langle f\rangle_{x,\Omega_T}^{(\gamma)} + \|f\|_{t,\Omega_T}^{(\mu)}, \qquad \gamma \in (0,1), \quad \mu \in [0,1),$$

where

$$\|f\|_{t,\Omega_T}^{(\mu)} = \begin{cases} \|f\|_{\Omega_T} + \langle f\rangle_{t,\Omega_T}^{(\mu)} & \text{if } \mu > 0, \\ \|f\|_{\Omega_T} & \text{if } \mu = 0, \end{cases}$$

then it belongs to the Hölder space $C^{\gamma,\mu}(\Omega_T)$.

We suppose that a vector-valued function is an element of a Hölder space if all its components belong to this space, and its norm is defined as the maximal norm of the components.

Let us set $Q_T = Q_T^- \cup Q_T^+$ and

$$\|f\|_{Q_T}^{(k+\alpha)} = \|f\|_{Q_T^-}^{(k+\alpha)} + \|f\|_{Q_T^+}^{(k+\alpha)}, \qquad \|f\|_{\cup\Omega^\pm}^{(k+\alpha)} = \|f\|_{\Omega^-}^{(k+\alpha)} + \|f\|_{\Omega^+}^{k+\alpha)}.$$

Now, we formulate the main result of this paper.

Theorem 1.1. *Suppose that* $\Gamma \in C^{3+\alpha}$, $\boldsymbol{f}, D_x\boldsymbol{f} \in \boldsymbol{C}^{\alpha,(\alpha+\varepsilon)/2}(\mathbb{R}^3 \times (0,T))$, $\boldsymbol{v}_0 \in \boldsymbol{C}^{2+\alpha}(\Omega_0^- \cup \Omega_0^+)$, $\sigma \in C^{3+\alpha}(\mathbb{R}_+)$, $\sigma \geq \sigma_0 > 0$, $a \in C^{2+\alpha,1+\alpha/2}(S)$, $a > 0$, $S \in C^{2+\alpha}$ *with some* $\alpha \in (0,1)$, $\varepsilon \in (0, 1-\alpha)$, $T < \infty$. *Moreover, let it hold the compatibility conditions*

$$\begin{aligned} &\nabla \cdot \boldsymbol{v}_0 = 0, \ \ \boldsymbol{v}_0|_S = 0, \quad [\boldsymbol{v}_0]|_\Gamma = 0, \ \ [\theta_0]|_\Gamma = 0, \ \ \theta_0|_S = a|_{t=0}, \\ &[\Pi_0 \mathbb{T}(\boldsymbol{v}_0)\boldsymbol{n}_0]|_\Gamma = \Pi_0 \nabla\sigma(\theta_0), \qquad [\Pi_0(\nu^\pm \nabla^2 \boldsymbol{v}_0 - \frac{1}{\rho^\pm}\nabla q_0)]|_\Gamma = 0, \\ &(\Pi_S(\nu^- \nabla^2 \boldsymbol{v}_0 - \frac{1}{\rho^-}\nabla q_0))|_S = 0, \qquad k^- \nabla^2 \theta_0|_S = \frac{\partial a}{\partial t}\Big|_{t=0}, \\ &[k^\pm \nabla^2 \theta_0]|_\Gamma = 0, \ \ \left[k^\pm \frac{\partial \theta_0}{\partial \boldsymbol{n}_0}\right]\Big|_\Gamma + \kappa\theta_0(\Pi_0\nabla)\cdot \boldsymbol{v}_0 = 0, \end{aligned} \tag{1.5}$$

where $q_0(\xi) \equiv q(\xi, 0)$ *is a solution of the diffraction problem*

$$\begin{aligned} &\frac{1}{\rho^\pm}\nabla^2 q_0(\xi) = \nabla \cdot (\boldsymbol{f}(\xi,0) - \mathcal{D}_t \mathbb{B}^*\big|_{t=0}\boldsymbol{v}_0(\xi)), \qquad \xi \in \Omega_0^- \cup \Omega_0^+, \\ &[q_0]|_\Gamma = \left[2\mu^\pm \frac{\partial \boldsymbol{v}_0}{\partial \boldsymbol{n}_0}\cdot \boldsymbol{n}_0\right]\Big|_\Gamma \sigma(\theta_0)H_0(\xi), \qquad \xi \in \Gamma, \\ &\left[\frac{1}{\rho^\pm}\frac{\partial q_0}{\partial \boldsymbol{n}_0}\right]\Big|_\Gamma = \left[\nu^\pm \boldsymbol{n}_0 \cdot \nabla^2 \boldsymbol{v}_0\right]\big|_\Gamma \qquad (\frac{\partial}{\partial \boldsymbol{n}_0} = \boldsymbol{n}_0 \cdot \nabla), \\ &\frac{1}{\rho^-}\frac{\partial q_0}{\partial \boldsymbol{n}_S}\Big|_S = \nu^- \boldsymbol{n}_S \cdot \nabla^2 \boldsymbol{v}_0\big|_S \qquad (\frac{\partial}{\partial \boldsymbol{n}_S} = \boldsymbol{n}_S \cdot \nabla). \end{aligned} \tag{1.6}$$

Here $H_0(\xi) = \boldsymbol{n}_0 \cdot \Delta(0)\xi\big|_\Gamma$, $\mathbb{B} = \mathbb{A} - \mathbb{I}$, $\mathbb{I}$ *is the identity matrix*, $\mathbb{B}^*$ *is the transpose to* $\mathbb{B}$, $\boldsymbol{n}_S$ *is the outward normal to* S, $\Pi_S\boldsymbol{\omega} \equiv \boldsymbol{\omega} - \boldsymbol{n}_S(\boldsymbol{n}_S \cdot \boldsymbol{\omega})$.

Then there exists a positive constant $T_0 \leq T$ such that problem (1.4) has a unique solution $(\boldsymbol{u}, q, \hat{\theta})$ with the properties:

$$\boldsymbol{u} \in \boldsymbol{C}^{2+\alpha,1+\alpha/2}(Q_{T_0}),\ q \in C^{(1+\alpha,\gamma)}(Q_{T_0})\ (\gamma = 1-\varepsilon > \alpha),$$
$$\nabla q \in \boldsymbol{C}^{\alpha,\alpha/2}(Q_{T_0}),\ \hat{\theta} \in C^{2+\alpha,1+\alpha/2}(Q_{T_0}).$$

The value of T_0 depends on the data norms and on the curvature of Γ.

The proof of this theorem is based on the solvability of auxiliary linearized problems.

2. Linearized problems

First, in this section, we discuss the following linear problem:

$$\begin{gathered}
\mathcal{D}_t \boldsymbol{w} - \nu^{\pm}\nabla^2_{\boldsymbol{u}}\boldsymbol{w} + \frac{1}{\rho^{\pm}}\nabla_{\boldsymbol{u}} s = \boldsymbol{f}(\xi,t), \qquad \nabla_{\boldsymbol{u}}\cdot\boldsymbol{w} = r \quad \text{in } Q_T,\\
\boldsymbol{w}|_{t=0} = \boldsymbol{w}_0 \qquad \text{in } \Omega^-\cup\Omega^+, \qquad\qquad (2.1)\\
[\boldsymbol{w}]|_{G_T} = 0, \quad \boldsymbol{w}|_S = 0, \qquad [\mu^{\pm}\Pi_0\Pi\mathbb{S}_{\boldsymbol{u}}(\boldsymbol{w})\boldsymbol{n}]|_{G_T} = \Pi_0\boldsymbol{d},\\
[\boldsymbol{n}_0\cdot\mathbb{T}_{\boldsymbol{u}}(\boldsymbol{w},s)\boldsymbol{n}]|_{G_T} - \sigma(\xi)\boldsymbol{n}_0\cdot\Delta(t)\int_0^t \boldsymbol{w}\big|_\Gamma d\tau = b + \int_0^t B d\tau \quad \text{on } G_T,
\end{gathered}$$

where

$$(\mathbb{S}_{\boldsymbol{u}}(\boldsymbol{w}))_{ij} = (A_{jk}\partial w_i/\partial\xi_k + A_{ik}\partial w_j/\partial\xi_k),$$

Problem (2.1) was considered in [2]–[4]. In [4] (Theorem 3.1), it was proved unique solvability for it in any finite time interval when S was absent and Q_T coincided with the whole space $\mathbb{R}^3$. This result was obtained in Hölder spaces with power-like weights at infinity but it is valid also in our case. The proof is only simpler without weight, in addition, the weighted spaces are equivalent to the ordinary Hölder spaces in bounded domains. Now we cite this existence theorem.

Theorem 2.1. *Let $\alpha,\gamma \in (0,1)$, $\gamma > \alpha$, and let $T < \infty$. Assume that $\Gamma, S \in C^{2+\alpha}$, $\sigma \in C^{1+\alpha}(\Gamma)$, $\sigma \geq \sigma_0 > 0$, and that for $\boldsymbol{u} \in \boldsymbol{C}^{2+\alpha,1+\alpha/2}(Q_T)$, $[\boldsymbol{u}]|_{G_T} = 0$, we have*

$$(T + T^{\gamma/2})\|\boldsymbol{u}\|^{(2+\alpha,1+\alpha/2)}_{Q_T} \leq \delta \qquad (2.2)$$

for some sufficiently small $\delta > 0$.

Moreover, we assume that the following four groups of conditions are fulfilled:

1) *there exists a vector $\boldsymbol{g} \in \boldsymbol{C}^{\alpha,\alpha/2}(Q_T)$ and a tensor $\mathbb{G} = \{G_{ik}\}^3_{i,k=1}$ with $G_{ik} \in C^{(1+\alpha,\gamma)}(Q_T) \cap C^{\gamma,0}(Q_T)$ such that*

$$\mathcal{D}_t r - \nabla_{\boldsymbol{u}}\cdot\boldsymbol{f} = \nabla\cdot\boldsymbol{g}, \quad g_i = \partial G_{ik}/\partial\xi_k, \quad i = 1,2,3,$$

(these equalities are understood in a weak sense) and, moreover,

$$[(\boldsymbol{g} + \mathbb{A}^*\boldsymbol{f})\cdot\boldsymbol{n}_0]|_{G_T} = 0;$$

2) $\boldsymbol{f} \in \boldsymbol{C}^{\alpha,\alpha/2}(Q_T)$, $r \in C^{1+\alpha,\frac{1+\alpha}{2}}(Q_T)$, $\boldsymbol{w}_0 \in \boldsymbol{C}^{2+\alpha}(\Omega^- \cup \Omega^+)$, $\boldsymbol{a} \in \boldsymbol{C}^{1+\alpha,\frac{1+\alpha}{2}}(G_T)$, $b \in C^{(1+\alpha,\gamma)}(G_T)$, $B \in C^{\alpha,\alpha/2}(G_T)$;

3)
$$\nabla \cdot \boldsymbol{w}_0(\xi) = r(\xi,0) = 0, \quad [\boldsymbol{w}_0]|_\Gamma = 0,$$

$$[\Pi_0 \mathbb{T}(\boldsymbol{w}_0(\xi))\boldsymbol{n}_0]|_{\xi\in\Gamma} = \Pi_0 \boldsymbol{a}(\xi,0), \quad \xi \in \Gamma,$$

$$\Big[\Pi_0\Big(\boldsymbol{f}(\xi,0) - \frac{1}{\rho^\pm}\nabla s(\xi,0) + \nu^\pm \nabla^2 \boldsymbol{w}_0(\xi)\Big)\Big]\Big|_{\xi\in\Gamma} = 0,$$

$$\Pi_S\Big(\boldsymbol{f}(\xi,0) - \frac{1}{\rho^-}\nabla s(\xi,0) + \nu^- \nabla^2 \boldsymbol{w}_0(\xi)\Big)\Big|_{\xi\in S} = 0;$$

4) $s_0(\xi) = s(\xi,0)$ *is a solution of the problem*

$$\frac{1}{\rho^\pm}\nabla^2 s_0(\xi) = \nabla \cdot (\mathcal{D}_t \mathbb{B}^*|_{t=0}\boldsymbol{w}_0(\xi) - \boldsymbol{g}(\xi,0)) \quad \text{in} \quad \Omega^- \cup \Omega^+,$$

$$[s_0]|_\Gamma = \Big[2\mu^\pm \frac{\partial \boldsymbol{w}_0}{\partial \boldsymbol{n}_0} \cdot \boldsymbol{n}_0\Big]\Big|_\Gamma - b|_{t=0},$$

$$\Big[\frac{1}{\rho^\pm}\frac{\partial s_0}{\partial \boldsymbol{n}_0}\Big]\Big|_\Gamma = [\boldsymbol{n}_0 \cdot (\boldsymbol{f}|_{t=0} + \nu^\pm \nabla^2 \boldsymbol{w}_0)]|_\Gamma,$$

$$\frac{1}{\rho^-}\frac{\partial s_0}{\partial \boldsymbol{n}_S}\Big|_S = \nu^- \boldsymbol{n}_S \cdot \nabla^2 \boldsymbol{w}_0|_S.$$

Under all these assumptions, the problem (2.1) *has a unique solution* $(\boldsymbol{w}, s)$ *with the properties:* $\boldsymbol{w} \in \boldsymbol{C}^{2+\alpha,1+\alpha/2}(Q_T)$, $s \in C^{(1+\alpha,\gamma)}(Q_T)$, $\nabla s \in \boldsymbol{C}^{\alpha,\alpha/2}(Q_T)$; *moreover, this solution satisfies the inequality*

$$\begin{aligned} N_{t'}[\boldsymbol{w}, s] \equiv\ & \|\boldsymbol{w}\|_{Q_{t'}}^{(2+\alpha,1+\alpha/2)} + \|\nabla s\|_{Q_{t'}}^{(\alpha,\alpha/2)} + \|s\|_{t,Q_{t'}}^{(\frac{1+\alpha-\gamma}{2})} + \langle s\rangle_{Q_{t'}}^{(1+\alpha,\gamma)} \\ \leq\ & c_1(t')\Big\{\|\boldsymbol{f}\|_{Q_{t'}}^{(\alpha,\alpha/2)} + \|r\|_{Q_{t'}}^{(1+\alpha,\frac{1+\alpha}{2})} + \|\boldsymbol{w}_0\|_{\cup\Omega^\pm}^{(2+\alpha)} + \|\boldsymbol{g}\|_{Q_{t'}}^{(\alpha,\alpha/2)} \\ & + |\mathbb{G}|_{Q_{t'}}^{(1+\alpha,\gamma)} + \|\mathbb{G}\|_{Q_{t'}}^{(\gamma,0)} + \|\boldsymbol{d}\|_{G_{t'}}^{(1+\alpha,\frac{1+\alpha}{2})} + \|b\|_{G_{t'}} + |b|_{G_{t'}}^{(1+\alpha,\gamma)} \\ & + \|\nabla_\Gamma b\|_{G_{t'}}^{(\alpha,\alpha/2)} + \|B\|_{G_{t'}}^{(\alpha,\alpha/2)} + P_{t'}[\boldsymbol{u}]\|\boldsymbol{w}_0\|_{\cup\Omega^\pm}^{(1)}\Big\} \\ \equiv\ & c_1(t')\{F(t') + P_{t'}[\boldsymbol{u}]\|\boldsymbol{w}_0\|_{\cup\Omega^\pm}^{(1)}\}, \end{aligned}$$

where $c_1(t')$ *is a monotone nondecreasing function of* $t' \leq T$, $\nabla_\Gamma = \Pi_0 \nabla$, *and*

$$P_t[\boldsymbol{u}] = t^{\frac{1-\alpha}{2}}\|\nabla \boldsymbol{u}\|_{Q_t} + \|\nabla \boldsymbol{u}\|_{Q_t}^{(\alpha,\alpha/2)}.$$

Let us consider also the problem with the unknown temperature function ψ:

$$\begin{aligned} & \mathcal{D}_t\psi - k^\pm \nabla_{\boldsymbol{u}}^2 \psi = f \quad \text{in} \quad Q_T, \\ & \psi|_{t=0} = \psi_0 \quad \text{in} \quad \Omega^- \cup \Omega^+, \\ & [\psi]|_{G_T} = 0, \qquad \psi|_{S_T} = \varphi, \\ & [k^\pm \boldsymbol{n} \cdot \nabla_{\boldsymbol{u}}\psi]|_{G_T} + \kappa\psi(\Pi\nabla_{\boldsymbol{u}}) \cdot \boldsymbol{w} = d \quad \text{on} \quad G_T. \end{aligned} \tag{2.3}$$

Remark 2.2. We observe that the differential operator $\Pi\nabla_{\boldsymbol{u}} = A_{ij}\frac{\partial}{\partial\xi_j} - n_i n_k A_{kj}\frac{\partial}{\partial\xi_j}$ has not the derivative $\frac{\partial}{\partial\boldsymbol{n}_0}$. Indeed, let us consider the multiplier of $\frac{\partial}{\partial\boldsymbol{n}_0}$ in this expression. It is $A_{ij}n_{0j} - n_i n_k A_{kj} n_{0j} = |\mathbb{A}\boldsymbol{n}_0|(n_i - n_i n_k n_k) = 0$. Hence, the quantity $(\Pi\nabla_{\boldsymbol{u}})\cdot\boldsymbol{w}$ is well defined on the boundary Γ, the vector field $\boldsymbol{w}$ being continuous across Γ.

Theorem 2.3. *Let surfaces* $\Gamma, S \in C^{2+\alpha}$, *function* $\sigma \in C^{1+\alpha}(\Gamma)$, $\sigma \geq \sigma_0 > 0$, *and vectors* $\boldsymbol{u}, \boldsymbol{w} \in \boldsymbol{C}^{2+\alpha,1+\alpha/2}(Q_T)$, $[\boldsymbol{u}]|_\Gamma = [\boldsymbol{w}]|_\Gamma = 0$, *and satisfy the inequality* (2.2). *Then for arbitrary* $f \in C^{\alpha,\alpha/2}(Q_T), \varphi \in C^{\alpha,\alpha/2}(S_T), \psi_0 \in C^{2+\alpha}(\Omega^- \cup \Omega^+), d \in C^{1+\alpha,(1+\alpha)/2}(G_T)$ *which satisfy the compatibility conditions*

$$\begin{aligned} &[\psi_0]|_\Gamma = 0, \quad \psi_0|_S = \varphi|_{t=0}, \quad [k^\pm\nabla^2\psi_0]|_\Gamma = [f|_{t=0}]|_\Gamma, \\ &[k^\pm\frac{\partial\psi_0}{\partial\boldsymbol{n}_0}]|_\Gamma + \kappa\psi_0\nabla_\Gamma\cdot\boldsymbol{w}(\xi,0) = d(\xi,0), \quad \xi\in\Gamma, \\ &k^-\nabla^2\psi_0|_S + f|_{S,t=0} = \frac{\partial\varphi}{\partial t}|_{t=0}, \end{aligned} \tag{2.4}$$

problem (2.3) *has a unique solution* $\psi \in C^{2+\alpha,1+\alpha/2}(Q_T)$ *and the estimate*

$$\begin{aligned} \|\psi\|^{(2+\alpha,1+\alpha/2)}_{Q_T} \leq c_2(T)\{&\|f\|^{(\alpha,\alpha/2)}_{Q_T} + \|\psi_0\|^{(2+\alpha)}_{\cup\Omega^\pm} + \|\varphi\|^{(2+\alpha,1+\alpha/2)}_{S_T} + \\ &+ \|d\|^{(1+\alpha,\frac{1+\alpha}{2})}_{G_T} + T^{\frac{1-\alpha}{2}}\|\nabla\boldsymbol{u}\|_{Q_T}\|\nabla\boldsymbol{w}\|_{Q_T}\|\psi_0\|^{(1)}_{\cup\Omega^\pm}\} \end{aligned} \tag{2.5}$$

holds. Here c_2 *is a nondecreasing function of* T.

Proof. We rewrite problem (2.3) in the form

$$\begin{aligned} &\mathcal{D}_t\psi - k^\pm\nabla^2\psi = f + h_1(\psi) \quad \text{in } Q_T, \\ &\psi|_{t=0} = \psi_0 \quad \text{in } \Omega^-\cup\Omega^+, \\ &[\psi]|_{G_T} = 0, \quad \psi|_{S_T} = \varphi, \\ &[k^\pm\boldsymbol{n}_0\cdot\nabla\psi]|_{G_T} + \kappa\psi(\Pi_0\nabla)\cdot\boldsymbol{w}|_{G_T} = d + h_2(\psi) + h_3(\psi,\boldsymbol{w}), \end{aligned} \tag{2.6}$$

where

$$\begin{aligned} h_1(\psi) &= k^\pm(\nabla^2_{\boldsymbol{u}}\psi - \nabla^2\psi), \\ h_2(\psi) &= [k^\pm(\boldsymbol{n}_0\cdot\nabla\psi - \boldsymbol{n}\cdot\nabla_{\boldsymbol{u}}\psi)]\big|_\Gamma, \\ h_3(\psi,\boldsymbol{w}) &= \kappa\psi(\Pi_0\nabla - \Pi\nabla_{\boldsymbol{u}})\cdot\boldsymbol{w}\big|_\Gamma. \end{aligned} \tag{2.7}$$

We note that the function $\kappa(\Pi_0\nabla)\cdot\boldsymbol{w}|_{G_T}$, the multiplier of ψ in the third boundary condition, belongs to $C^{1+\alpha,\frac{1+\alpha}{2}}(G_T)$. Therefore, under hypotheses (2.4), problem (2.6) with $h_1(\psi) = h_2(\psi) = h_3(\psi) = 0$ is solvable in $C^{2+\alpha,1+\frac{\alpha}{2}}(Q_T)$ [7].

Let us write problem (2.6) in the operator form

$$\begin{aligned} \psi &= \mathcal{L}[f + h_1(\psi), \psi_0, \varphi, d + h_2(\psi) + h_3(\psi,\boldsymbol{w})] \\ &\equiv \mathcal{L}[f,\psi_0,\varphi,d] + \mathcal{K}_{\boldsymbol{u},\boldsymbol{w}}(\psi). \end{aligned} \tag{2.8}$$

Here $\mathcal{L}$ is a linear continuous operator from the subspace of $C^{\alpha,\frac{\alpha}{2}}(Q_T)\times C^{2+\alpha}(\Omega^-\cup\Omega^+)\times C^{2+\alpha,1+\frac{\alpha}{2}}(S_T)\times C^{1+\alpha,\frac{1+\alpha}{2}}(\Gamma)$, which elements (f,ψ_0,φ,d) satisfy (2.4), into the Hölder space $C^{2+\alpha,1+\frac{\alpha}{2}}(Q_T)$; $\mathcal{K}_{\boldsymbol{u},\boldsymbol{w}}(\psi) = \mathcal{L}[h_1(\psi),0,0,h_2(\psi)+h_3(\psi,\boldsymbol{w})]$. To every quadruple (f, ψ_0, φ, d), the operator $\mathcal{L}$ poses in correspondence a solution of (2.6) with $h_1(\psi) = h_2(\psi) = h_3(\psi) = 0$.

We take into account that

$$h_1(\psi) = k^{\pm}(\mathbb{A}\nabla \cdot \mathbb{B}\nabla)\psi + k^{\pm}\mathbb{B}\nabla \cdot \nabla\psi,$$
$$h_2(\psi) = [k^{\pm}((\boldsymbol{n}_0 - \boldsymbol{n}) \cdot \nabla\psi - \boldsymbol{n} \cdot \mathbb{B}\nabla\psi)]\big|_{\Gamma},$$
$$h_3(\psi, \boldsymbol{w}) = \kappa\psi\{\boldsymbol{n}(\boldsymbol{n} \cdot \nabla) - \boldsymbol{n}_0(\boldsymbol{n}_0 \cdot \nabla) - (\Pi\mathbb{B}\nabla)\} \cdot \boldsymbol{w}\big|_{\Gamma},$$

where $\mathbb{B} = \mathbb{A} - \mathbb{I}$. Then Lemma 3.1 from [4] implies the inequality

$$\begin{aligned}\|h_1(\psi)\|_{Q_T}^{(\alpha,\alpha/2)} + \|h_2(\psi)\|_{G_T}^{(1+\alpha,\frac{1+\alpha}{2})} &\le c\{\delta\|\psi\|_{Q_T}^{(2+\alpha,1+\alpha/2)} \\ &\quad + T^{\frac{1-\alpha}{2}}\|\nabla \boldsymbol{u}\|_{Q_T}\|\psi_0\|_{\cup\Omega^{\pm}}^{(1)}\}, \\ \|h_3(\psi, \boldsymbol{w})\|_{G_T}^{(1+\alpha,\frac{1+\alpha}{2})} &\le c\{\delta\|\psi\|_{Q_T}^{(2+\alpha,1+\alpha/2)}\|\boldsymbol{w}\|_{Q_T}^{(2+\alpha,1+\frac{\alpha}{2})} \\ &\quad + T^{\frac{1-\alpha}{2}}\|\nabla \boldsymbol{u}\|_{Q_T}\|\psi_0\|_{\cup\Omega^{\pm}}^{(1)}\|\nabla \boldsymbol{w}\|_{Q_T}\}.\end{aligned} \tag{2.9}$$

Hence, the operator $\mathcal{K}_{\boldsymbol{u},\boldsymbol{w}}(\psi)$ is a contraction for small δ, i.e., for arbitrary $\psi,\ \psi' \in C^{2+\alpha,1+\frac{\alpha}{2}}(Q_T)$ such that $\psi - \psi'|_{t=0} = 0,\ \ \psi - \psi'|_{S_T} = 0$, the estimate

$$\|\mathcal{K}_{\boldsymbol{u},\boldsymbol{w}}(\psi) - \mathcal{K}_{\boldsymbol{u},\boldsymbol{w}}(\psi')\|_{Q_T}^{(2+\alpha,1+\alpha/2)} \le \varepsilon\|\psi - \psi'\|_{Q_T}^{(2+\alpha,1+\alpha/2)}, \quad \varepsilon < 1.$$

holds. Consequently, equation (2.8), and hence, problem (2.6) are uniquely solvable. Inequality (2.5) follows from (2.9) and boundedness of the operator $\mathcal{L}$. □

Now we linearized boundary conditions in (1.4):

$$[\Pi_0\Pi\mathbb{T}_{\boldsymbol{u}}(\boldsymbol{w})\boldsymbol{n}]|_{G_T} = \Pi_0\Pi\nabla_{\boldsymbol{u}}\sigma(\hat{\theta}) \equiv \boldsymbol{K}_1(\hat{\theta}),$$

$$[\boldsymbol{n}_0\cdot\mathbb{T}_{\boldsymbol{u}}(\boldsymbol{w}, s)\boldsymbol{n}]|_{G_T} - \sigma(\theta_0)\boldsymbol{n}_0\cdot\Delta(t)\int_0^t \boldsymbol{w}\big|_{\Gamma}\mathrm{d}\tau = K_2(\hat{\theta}) + \int_0^t K_3(\boldsymbol{w}, \hat{\theta})\mathrm{d}\tau \quad \text{on} \quad G_T.$$

Here we use the notation:

$$\begin{aligned} K_2(\hat{\theta}) &= \sigma(\hat{\theta})H_0(\xi) + (\boldsymbol{n}_0 - \boldsymbol{n}) \cdot \Pi\nabla_{\boldsymbol{u}}\sigma(\hat{\theta}), \\ K_3(\boldsymbol{w}, \hat{\theta}) &= \sigma(\hat{\theta})\boldsymbol{n}_0 \cdot \dot{\Delta}(t)\boldsymbol{\xi} + (\sigma(\hat{\theta}) - \sigma(\theta_0))\boldsymbol{n}_0 \cdot \Big\{\Delta(t)\boldsymbol{w}\big|_{\Gamma} + \dot{\Delta}(t)\int_0^t \boldsymbol{w}\big|_{\Gamma}\mathrm{d}\tau\Big\} \\ &\quad + \frac{\partial\sigma(\hat{\theta})}{\partial t}\boldsymbol{n}_0 \cdot \Big\{\int_0^t \dot{\Delta}(\tau)\boldsymbol{\xi}\mathrm{d}\tau + \Delta(t)\int_0^t \boldsymbol{w}\big|_{\Gamma}\mathrm{d}\tau\Big\}. \end{aligned}$$

As our K_i coincide with the operators corresponding to the case of a single liquid, we cite two lemmas concerning them from [1] (see also [8]).

Lemma 2.4. *Let continuous $\boldsymbol{u}$ across G_T be subjected to* (2.2). *Then the increments of the operators K_i with respect to the temperature $\theta \in C^{2+\alpha,1+\alpha/2}(Q_T)$ and to the velocity $\boldsymbol{w} \in \boldsymbol{C}^{2+\alpha,1+\alpha/2}(Q_T)$, which are also continuous across G_T:*

$$\begin{aligned} \boldsymbol{K}_1(\theta) - \boldsymbol{K}_1(\theta') &= \Pi_0\Pi\nabla_{\boldsymbol{u}}(\sigma(\theta) - \sigma(\theta')), \\ K_2(\theta) - K_2(\theta') &= (\sigma(\theta) - \sigma(\theta'))H_0(\xi) + (\boldsymbol{n}_0 - \boldsymbol{n}) \cdot \Pi\nabla_{\boldsymbol{u}}(\sigma(\theta) - \sigma(\theta')), \end{aligned}$$

$$
\begin{aligned}
&K_3(\boldsymbol{w},\theta)-K_3(\boldsymbol{w}',\theta')\\
&\quad=(\sigma(\theta)-\sigma(\theta'))\boldsymbol{n}_0\cdot\Big\{\dot{\Delta}(t)\boldsymbol{\xi}+\Delta(t)\boldsymbol{w}\big|_\Gamma+\dot{\Delta}(t)\int_0^t\boldsymbol{w}\big|_\Gamma \mathrm{d}\tau\Big\}\\
&\qquad+(\sigma(\theta')-\sigma(\theta_0))\boldsymbol{n}_0\cdot\Big\{\Delta(t)(\boldsymbol{w}-\boldsymbol{w}')\big|_\Gamma+\dot{\Delta}(t)\int_0^t(\boldsymbol{w}-\boldsymbol{w}')\big|_\Gamma \mathrm{d}\tau\Big\}\\
&\qquad+\Big(\frac{\partial\sigma(\theta)}{\partial t}-\frac{\partial\sigma(\theta')}{\partial t}\Big)\boldsymbol{n}_0\cdot\Big(\int_0^t\dot{\Delta}(\tau)\boldsymbol{\xi}\mathrm{d}\tau+\Delta(t)\int_0^t\boldsymbol{w}\big|_\Gamma\mathrm{d}\tau\Big)\\
&\qquad+\frac{\partial\sigma(\theta')}{\partial t}\boldsymbol{n}_0\cdot\Delta(t)\int_0^t(\boldsymbol{w}-\boldsymbol{w}')\big|_\Gamma\mathrm{d}\tau,
\end{aligned}
$$

where $\theta|_{t=0}=\theta'|_{t=0}=\theta_0$, *satisfy the inequalities*

$$
\begin{aligned}
\|\boldsymbol{K}_1(\theta)\ -\ &\boldsymbol{K}_1(\theta')\|_{G_T}^{(1+\alpha,\frac{1+\alpha}{2})}+\|K_2(\theta)-K_2(\theta')\|_{G_T}^{(1+\alpha,\frac{1+\alpha}{2})}\\
&\le c\Big(1+\|\theta\|_{Q_T}^{(2+\alpha,1+\frac{\alpha}{2})}+\|\theta'\|_{Q_T}^{(2+\alpha,1+\frac{\alpha}{2})}\Big)^{2+\alpha}\\
&\qquad\qquad\times\big(1+\|H_0\|_\Gamma^{(1+\alpha)}\big)\|\theta-\theta'\|_{Q_T}^{(2+\alpha,1+\frac{\alpha}{2})},\\
\|K_3(\boldsymbol{w},\theta)\ -\ &K_3(\boldsymbol{w}',\theta')\|_{G_T}^{(\alpha,\frac{\alpha}{2})}\\
&\le c(T+T^{1/2})\Big\{\Big(1+\|\theta\|_{Q_T}^{(2+\alpha,1+\frac{\alpha}{2})}+\|\theta'\|_{Q_T}^{(2+\alpha,1+\frac{\alpha}{2})}\Big)^2\\
&\qquad\times\Big(\|\nabla\boldsymbol{u}\|_{Q_T}+\|\boldsymbol{w}\|_{Q_T}^{(2+\alpha,1+\frac{\alpha}{2})}\Big)\|\theta-\theta'\|_{Q_T}^{(2+\alpha,1+\frac{\alpha}{2})}\\
&\qquad+\Big\|\frac{\partial\theta'}{\partial t}\Big\|_{Q_T}^{(\alpha,\frac{\alpha}{2})}(1+\langle\theta\rangle_{Q_T}^{(\alpha,\frac{\alpha}{2})})\|\boldsymbol{w}-\boldsymbol{w}'\|_{Q_T}^{(2+\alpha,1+\frac{\alpha}{2})}\Big\}\\
&\quad+c\|\theta'\|_{Q_T}^{(2+\alpha,1+\frac{\alpha}{2})}\Big(1+\|\theta'\|_{Q_T}^{(2+\alpha,1+\frac{\alpha}{2})}\Big)M_T[\boldsymbol{w}-\boldsymbol{w}'].
\end{aligned}
$$

Here

$$
M_T(\boldsymbol{v})=\int_0^T\|\boldsymbol{v}\|_{\Omega^-\cup\Omega^+}^{(2+\alpha)}\mathrm{d}t+\sup_{0<\tau<t<T}\tau^{-\alpha/2}\int_{t-\tau}^t(\|\nabla\boldsymbol{v}\|_{\cup\Omega^\pm}+\|\nabla\nabla\boldsymbol{v}\|_{\cup\Omega^\pm})\mathrm{d}\tau'.
$$

Lemma 2.5. *For the operator differences* K_i-K_i' *corresponding to vectors* $\boldsymbol{u}$ *and* $\boldsymbol{u}'$ *which satisfy* (2.2):

$$
\boldsymbol{K}_1(\theta)-\boldsymbol{K}_1'(\theta)=\Pi_0(\Pi\Pi')\mathbb{A}\nabla\sigma(\theta)-\Pi_0\Pi'(\mathbb{B}-\mathbb{B}')\nabla\sigma(\theta),
$$

$$
\begin{aligned}
K_2(\theta)-K_2'(\theta)&=\boldsymbol{n}_0\cdot(\Pi-\Pi')\mathbb{A}\nabla\sigma+(\boldsymbol{n}_0-\boldsymbol{n}')\cdot\Pi'(\mathbb{A}-\mathbb{A}')\nabla\sigma\\
&\qquad-(\boldsymbol{n}_0-\boldsymbol{n}')\cdot\Pi\mathbb{A}\nabla\sigma-\boldsymbol{n}'\cdot(\Pi-\Pi')\mathbb{A}\nabla\sigma,
\end{aligned}
$$

$$
\begin{aligned}
K_3(\boldsymbol{w},\theta) \quad - \quad & K_3'(\boldsymbol{w},\theta) = \sigma(\theta)\boldsymbol{n}_0 \cdot (\dot{\Delta}(t) - \dot{\Delta}'(t))\boldsymbol{\xi} \\
& +(\sigma(\theta) - \sigma(\theta_0))\boldsymbol{n}_0 \cdot \Big\{(\Delta(t) - \Delta'(t))\boldsymbol{w} + (\dot{\Delta}(t) - \dot{\Delta}'(t)) \int_0^t \boldsymbol{w} d\tau\Big\} \\
& + \frac{\partial \sigma(\theta)}{\partial t}\boldsymbol{n}_0 \cdot \Big\{ \int_0^t (\dot{\Delta}(\tau) - \dot{\Delta}'(\tau))\boldsymbol{\xi} d\tau + (\Delta(t) - \Delta'(t)) \int_0^t \boldsymbol{w} d\tau \Big\},
\end{aligned}
$$

the estimates

$$
\begin{aligned}
& \|\boldsymbol{K}_1(\theta) - \boldsymbol{K}_1'(\theta)\|_{G_T}^{(1+\alpha,\frac{1+\alpha}{2})} + \|K_2(\theta) - K_2'(\theta)\|_{G_T}^{(1+\alpha,\frac{1+\alpha}{2})} \\
& \quad \le c\|\nabla\sigma(\theta)\|_{G_T}^{(1+\alpha,\frac{1+\alpha}{2})} M_T[\boldsymbol{u} - \boldsymbol{u}'] + cT^{\frac{1-\alpha}{2}} \|\nabla\sigma\|_{G_T} \|\nabla(\boldsymbol{u} - \boldsymbol{u}')\|_{Q_T},
\end{aligned}
$$

$$
\begin{aligned}
& \|K_3(\boldsymbol{w},\theta) - K_3'(\boldsymbol{w},\theta)\|_{G_T}^{(\alpha,\frac{\alpha}{2})} \le c\|\sigma(\theta)\|_{G_T}^{(\alpha,\frac{\alpha}{2})} \|\nabla(\boldsymbol{u} - \boldsymbol{u}')\|_{Q_T} \\
& \quad + c(T + T^{\frac{1}{2}}) \left\|\frac{\partial\theta}{\partial t}\right\|_{Q_T}^{(\alpha,\frac{\alpha}{2})} (1 + \langle\theta\rangle_{Q_T}^{(\alpha,\frac{\alpha}{2})}) \|\boldsymbol{w}\|_{Q_T}^{(2+\alpha,1+\frac{\alpha}{2})} \|\boldsymbol{u} - \boldsymbol{u}'\|_{Q_T}^{(2+\alpha,1+\frac{\alpha}{2})} \\
& \quad + c\|\theta\|_{Q_T}^{(2+\alpha,1+\frac{\alpha}{2})} (1 + \|\theta\|_{Q_T}^{(2+\alpha,1+\frac{\alpha}{2})}) \\
& \qquad \times \Big\{ M_T[\boldsymbol{u} - \boldsymbol{u}'] + (T + T^{\frac{1}{2}}) \|\boldsymbol{w}\|_{Q_T}^{(2+\alpha,1+\frac{\alpha}{2})} \|\boldsymbol{u} - \boldsymbol{u}'\|_{Q_T}^{(2+\alpha,1+\frac{\alpha}{2})} \Big\}
\end{aligned}
$$

hold.

In a similar way, we can obtain the following proposition.

Lemma 2.6. *Let $\psi \in C^{2+\alpha,1+\alpha/2}(Q_T)$, $[\psi]|_\Gamma = 0$. For the operator differences*

$$
\begin{aligned}
h_1(\psi) - h_1'(\psi) &= k^\pm(\nabla_{\boldsymbol{u}}^2 \psi - \nabla_{\boldsymbol{u}'}^2 \psi) \\
&= k^\pm \Big\{ \mathbb{A}\nabla \cdot (\mathbb{B} - \mathbb{B}')\nabla\psi + (\mathbb{B} - \mathbb{B}')\nabla \cdot \mathbb{A}'\psi \Big\}, \\
h_2(\psi) - h_2'(\psi) &= \Big[k^\pm (\boldsymbol{n}' \cdot \nabla_{\boldsymbol{u}'}\psi - \boldsymbol{n} \cdot \nabla_{\boldsymbol{u}}\psi) \Big]\Big|_\Gamma \\
&= \Big[k^\pm \{ (\boldsymbol{n}' - \boldsymbol{n}) \cdot \mathbb{A}\nabla\psi + \boldsymbol{n} \cdot (\mathbb{B} - \mathbb{B}')\nabla\psi \} \Big]\Big|_\Gamma, \\
h_3(\psi, \boldsymbol{w}) - h_3'(\psi, \boldsymbol{w}) &= \kappa\psi \Big(\Pi'\nabla_{\boldsymbol{u}'} - \Pi\nabla_{\boldsymbol{u}} \Big) \cdot \boldsymbol{w}|_\Gamma \\
&= \kappa\psi \Big\{ (\Pi' - \Pi)\mathbb{A}'\nabla \cdot \boldsymbol{w}\big|_\Gamma + \Pi(\mathbb{B}' - \mathbb{B})\nabla \cdot \boldsymbol{w}\big|_\Gamma \Big\},
\end{aligned}
$$

we have

$$
\begin{aligned}
& \|h_1(\psi) - h_1'(\psi)\|_{Q_T}^{(\alpha,\frac{\alpha}{2})} \le c_6 \|\psi\|_{Q_T}^{(2+\alpha,1+\frac{\alpha}{2})} M_T[\boldsymbol{u} - \boldsymbol{u}'], \\
& \|h_2(\psi) - h_2'(\psi)\|_{G_T}^{(1+\alpha,\frac{1+\alpha}{2})} \\
& \quad \le c_7 \|\psi\|_{Q_T}^{(2+\alpha,1+\frac{\alpha}{2})} M_T[\boldsymbol{u} - \boldsymbol{u}'] + c_8 T^{\frac{1-\alpha}{2}} \|\nabla\psi\|_{Q_T} \|\nabla(\boldsymbol{u} - \boldsymbol{u}')\|_{Q_T},
\end{aligned}
$$

$$\|h_3(\psi,\boldsymbol{w}) - h_3'(\psi,\boldsymbol{w})\|_{G_T}^{(1+\alpha,\frac{1+\alpha}{2})} \le c_9\|\psi\|_{Q_T}^{(1+\alpha,\frac{1+\alpha}{2})}\Big\{\|\nabla\boldsymbol{w}\|_{Q_T}^{(1+\alpha,\frac{1+\alpha}{2})}M_T[\boldsymbol{u}-\boldsymbol{u}'] + T^{\frac{1-\alpha}{2}}\|\nabla\boldsymbol{w}\|_{Q_T}\|\nabla(\boldsymbol{u}-\boldsymbol{u}')\|_{Q_T}\Big\}.$$

3. Solvability of problem (1.4)

In this section we prove Theorem 1.1.

Proof. Put $(\boldsymbol{u}^{(0)}, q^{(0)}, \theta^{(0)}) = (\boldsymbol{v}_0, 0, \theta_0)$. Next, we define $(\boldsymbol{u}^{(m+1)}, q^{(m+1)})$, $m = 0, 1, \ldots$, as a solution of the following problem:

$$\begin{gathered}
\mathcal{D}_t\boldsymbol{u}^{(m+1)} - \nu^{\pm}\nabla_m^2\boldsymbol{u}^{(m+1)} + \frac{1}{\rho^{\pm}}\nabla_m q^{(m+1)} = \boldsymbol{f}(X_m,t), \quad \nabla_m\cdot\boldsymbol{u}^{(m+1)} = 0 \quad \text{in } Q_T,\\
\boldsymbol{u}^{(m+1)}|_{t=0} = \boldsymbol{v}_0 \quad \text{in } \Omega^-\cup\Omega^+,\\
[\boldsymbol{u}^{(m+1)}]|_{G_T} = 0, \quad \boldsymbol{u}^{(m+1)}|_{S_T} = 0, \qquad (3.1)\\
[\mu^{\pm}\Pi_0\Pi_m\mathbb{S}_m(\boldsymbol{u}^{(m+1)})\boldsymbol{n}_m]|_{G_T} = \boldsymbol{K}_1^{(m)}(\hat\theta^{(m)}),\\
[\boldsymbol{n}_0\cdot\mathbb{T}_m(\boldsymbol{u}^{(m+1)}, q^{(m+1)})\boldsymbol{n}_m]|_{G_T} - \sigma(\theta_0)\boldsymbol{n}_0\cdot\Delta_m(t)\int_0^t \boldsymbol{u}^{(m+1)}|_\Gamma\,\mathrm{d}\tau\\
= K_2^{(m)}(\hat\theta^{(m)}) + \int_0^t K_3^{(m)}(\boldsymbol{u}^{(m)}, \hat\theta^{(m)})\,\mathrm{d}\tau \quad \text{on } G_T.
\end{gathered}$$

Here $\nabla_m = \nabla_{\boldsymbol{u}^{(m)}}$, $\Pi_m\boldsymbol{\omega} = \boldsymbol{\omega} - \boldsymbol{n}_m(\boldsymbol{n}_m\cdot\boldsymbol{\omega})$, $\boldsymbol{n}_m$ is the outward normal to $\Gamma_m = \{\boldsymbol{x} = \boldsymbol{X}_m(\xi,t),\ \xi\in\Gamma\}$, $\boldsymbol{X}_m = \boldsymbol{X}_{\boldsymbol{u}^{(m)}}$; $\mathbb{S}_m = \mathbb{S}_{\boldsymbol{u}^{(m)}}$, $\mathbb{T}_m = \mathbb{T}_{\boldsymbol{u}^{(m)}}$; Δ_m is the Beltrami-Laplace operator on Γ_m, $\boldsymbol{K}_1^{(m)} = \Pi_0\Pi_m\nabla_m\sigma$ etc.

Finally, we determine $\hat\theta^{(m+1)}$, $m = 0, 1, \ldots$, as a solution to the problem

$$\begin{gathered}
\mathcal{D}_t\hat\theta^{(m+1)} - k^{\pm}\nabla_m^2\hat\theta^{(m+1)} = 0 \quad \text{in } Q_T,\\
\hat\theta^{(m+1)}|_{t=0} = \theta_0 \quad \text{in } \Omega^-\cup\Omega^+,\\
[\hat\theta^{(m+1)}]|_{G_T} = 0, \qquad \hat\theta^{(m+1)}|_S = \hat a, \qquad (3.2)\\
[k^{\pm}\boldsymbol{n}_m\cdot\nabla_m\hat\theta^{(m+1)}]|_{G_T} + \kappa\hat\theta^{(m+1)}(\Pi_m\nabla_m)\cdot\boldsymbol{u}^{(m+1)} = 0 \ \text{ on } G_T.
\end{gathered}$$

Let us successively apply Theorem 2.1 to problems (3.1). As to the first hypothesis of it, we can put

$$\boldsymbol{g} \equiv \boldsymbol{g}_m = -\mathbb{A}_m^*\boldsymbol{f}, \qquad \mathbb{G} \equiv \mathbb{G}_m(\xi,t) = \nabla\int_{\cup\Omega^{\pm}} \mathcal{E}(\xi,\eta)\mathbb{A}_m^*\boldsymbol{f}(\eta,t)\,\mathrm{d}\eta, \qquad (3.3)$$

where $\mathbb{A}_m = \mathbb{A}(\boldsymbol{u}^{(m)})$, $\mathcal{E}(\xi,\eta) = \frac{1}{4\pi}\frac{1}{|\xi-\eta|}$ is the fundamental solution of the Laplace equation. The first equality in (3.3) follows from the identity $\frac{\partial A_{ij}}{\partial\xi_j} = 0$ which is true for the co-factor matrix of the Jacobi matrix of an arbitrary transformation and which implies $\mathbb{A}\nabla\cdot\boldsymbol{w} = \nabla\cdot\mathbb{A}^*\boldsymbol{w}$. Moreover, it is obvious that $\big[\boldsymbol{n}_0\cdot\big(\boldsymbol{g}_m + \mathbb{A}_m^*\boldsymbol{f}\big)\big]\,\big|_\Gamma = 0$. The third and forth groups of the assumptions of Theorem 2.1 follow from the hypotheses (1.5), (1.6).

Next, we have the inequalities [4, 8]

$$\|\boldsymbol{g}_m\|_{Q_T}^{(\alpha,\alpha/2)} + |\mathbb{G}_m|_{Q_T}^{(1+\alpha,\gamma)} + \|\mathbb{G}_m\|_{Q_T}^{(\gamma,0)}$$
$$\leq c(T)\Big(1 + (1+T)\|\boldsymbol{u}^{(m)}\|_{Q_T}^{(2+\alpha,1+\alpha/2)}\Big)\|\boldsymbol{f}\|_{t,Q_T}^{(\frac{1+\alpha-\gamma}{2})}$$
$$\leq c_1(T)\|\boldsymbol{f}\|_{Q_T}^{(\alpha,\frac{\alpha+\varepsilon}{2})}, \tag{3.4}$$

$$\|\sigma(\hat\theta^{(m)})H_0(\xi)\|_{G_T}^{(1+\alpha,\frac{1+\alpha}{2})} \leq c\|\sigma(\hat\theta^{(m)})\|_{G_T}^{(1+\alpha,\frac{1+\alpha}{2})}\|H_0\|_\Gamma^{(1+\alpha)}$$
$$\leq c_2\Big(1 + \|\hat\theta^{(m)}\|_{Q_T}^{(1+\alpha,\frac{1+\alpha}{2})}\Big)\|H_0\|^{(1+\alpha)}, \tag{3.5}$$
$$\|\sigma(\hat\theta^{(m)})\boldsymbol{n}_0\cdot\dot\Delta_m(\tau)\boldsymbol{\xi}\|_{G_T}^{(\alpha,\alpha/2)} \leq c\|\sigma(\hat\theta^{(m)})\|_{G_T}^{(\alpha,\alpha/2)}\|\nabla\boldsymbol{u}^{(m)}\|_{G_T}^{(\alpha,\alpha/2)}$$
$$\leq c_3\Big(1 + \|\hat\theta^{(m)}\|_{Q_T}^{(\alpha,\alpha/2)}\Big)\|\boldsymbol{u}^{(m)}\|_{Q_T}^{(2+\alpha,1+\alpha/2)}.$$

Thus, Theorem 2.1, inequalities (3.4), (3.5) and Lemma 2.5 with $\boldsymbol{u}'=0$ imply the estimate

$$\begin{aligned}
N_T[\boldsymbol{u}^{(m+1)},q^{(m+1)}] \leq{}& c_4(T)\Big\{\big(1+\|\boldsymbol{u}^{(m)}\|_{Q_T}^{(2+\alpha,1+\frac{\alpha}{2})}\big)\|\boldsymbol{f}\|_{Q_T}^{(\alpha,\frac{\alpha+\varepsilon}{2})} + \|\boldsymbol{v}_0\|_{\cup\Omega^\pm}^{(2+\alpha)}\Big\}\\
&+c_5\big(1+\|\hat\theta^{(m)}\|_{Q_T}^{(1+\alpha,\frac{1+\alpha}{2})}\big)\Big(\|\boldsymbol{u}^{(m)}\|_{Q_T}^{(2+\alpha,1+\frac{\alpha}{2})} + \|H_0\|_\Gamma^{(1+\alpha)}\Big)\\
&+c_6T^{\frac{1-\alpha}{2}}\|\nabla\sigma(\hat\theta^{(m)})\|_{G_T}\|\nabla\boldsymbol{u}^{(m)}\|_{Q_T}\\
&+c_7\|\hat\theta^{(m)}\|_{Q_T}^{(2+\alpha,1+\frac{\alpha}{2})}\Big(1+\|\hat\theta^{(m)}\|_{Q_T}^{(2+\alpha,1+\frac{\alpha}{2})}\Big)\\
&\qquad\times\Big\{M_T[\boldsymbol{u}^{(m)}] + (T+T^{1/2})\Big(\|\boldsymbol{u}^{(m)}\|_{Q_T}^{(2+\alpha,1+\frac{\alpha}{2})}\Big)^2\Big\}\\
&+c_8\Big(T^{\frac{1-\alpha}{2}}\|\nabla\boldsymbol{u}^{(m)}\|_{Q_T} + \|\nabla\boldsymbol{u}^{(m)}\|_{Q_T}^{(\alpha,\frac{\alpha}{2})}\Big)\|\boldsymbol{v}_0\|_{\cup\Omega^\pm}^{(2+\alpha)}.
\end{aligned} \tag{3.6}$$

Hence, for $m=0$ we can conclude that

$$\begin{aligned}
N_T[\boldsymbol{u}^{(1)},q^{(1)}] \leq{}& c_4(T)\Big\{\big(1+\|\boldsymbol{v}_0\|_{\cup\Omega^\pm}^{(2+\alpha)}\big)\|\boldsymbol{f}\|_{Q_T}^{(\alpha,\frac{\alpha+\varepsilon}{2})} + \|\boldsymbol{v}_0\|_{\cup\Omega^\pm}^{(2+\alpha)}\Big\}\\
&+c_9\Big\{\Big(1+\|\theta_0\|_{\cup\Omega^\pm}^{(2+\alpha)}\Big)\Big(\|\boldsymbol{v}_0\|_{\cup\Omega^\pm}^{(2+\alpha)} + \|H_0\|_\Gamma^{(1+\alpha)}\Big)\\
&+T^{\frac{1-\alpha}{2}}\|\boldsymbol{v}_0\|_{\cup\Omega^\pm}^{(1)}\Big(\|\theta_0\|_{\cup\Omega^\pm}^{(1)} + \|\boldsymbol{v}_0\|_{\cup\Omega^\pm}^{(1+\alpha)}\Big)\\
&+(T+T^{1/2})\|\theta_0\|_{\cup\Omega^\pm}^{(2+\alpha)}\Big(1+\|\theta_0\|_{\cup\Omega^\pm}^{(2+\alpha)}\Big)\\
&\qquad\times\|\boldsymbol{v}_0\|_{\cup\Omega^\pm}^{(2+\alpha)}\Big(1+\|\boldsymbol{v}_0\|_{\cup\Omega^\pm}^{(2+\alpha)}\Big)\Big\}.
\end{aligned} \tag{3.7}$$

The compatibility conditions (2.4) for problem (3.2) follow from relations (1.5). Therefore, Theorem 2.3 gives us for all $m = 0, 1, 2, \dots$

$$\begin{aligned} \|\hat{\theta}^{(m+1)}\|_{Q_T}^{(2+\alpha,1+\frac{\alpha}{2})} \le\ & c_{10}(T)\Big\{ \|\hat{a}\|_{S_T}^{(2+\alpha,1+\frac{\alpha}{2})} + \|\theta_0\|_{\cup\Omega^\pm}^{(2+\alpha)} \\ & + \big(\delta + T^{\frac{1-\alpha}{2}}\|\boldsymbol{v}_0\|_{\cup\Omega^\pm}^{(1)}\big)\|\theta_0\|_{\cup\Omega^\pm}^{(1)}\|\boldsymbol{u}^{(m+1)}\|_{Q_T}\Big\} \end{aligned} \tag{3.8}$$

In this way we deduce by induction that for every $m \ge 0$ the systems (3.1), (3.2) have solutions $(\boldsymbol{u}^{(m+1)}, q^{(m+1)}), \hat{\theta}^{(m+1)}$ on a time interval $(0, T_{m+1}] \subset (0, T]$ such that the preceding approximation $(\boldsymbol{u}^{(m)}, q^{(m)}), \hat{\theta}^{(m)}$ is also defined on this interval, while $\boldsymbol{u}^{(m)}$ satisfies (2.2) with small δ on it.

Now, it is necessary to show that there exists such $T' > 0$ that for $\forall m \in \mathbb{N}\ T_m \ge T'$, the norms $N_T[\boldsymbol{u}^{(j)}, q^{(j)}, \hat{\theta}^{(j)}] \equiv N_{T'}[\boldsymbol{u}^{(m)}, q^{(m)}] + \|\hat{\theta}^{(m)}\|_{Q_{T'}}^{(2+\alpha,1+\frac{\alpha}{2})}$ are uniformly bounded and the sequence $\{\boldsymbol{u}^{(m)}, q^{(m)}, \hat{\theta}^{(m)}\}, m > 0$, converges to a solution of problem (1.4).

To this end, we compose the difference between systems (3.1) corresponding to $j+1$ and to j. Let us consider the functions $\boldsymbol{w}^{(j+1)} = \boldsymbol{u}^{(j+1)} - \boldsymbol{u}^{(j)}$, $s^{(j+1)} = q^{(j+1)} - q^{(j)}$, $j = 1, 2, \dots$, which satisfy the problem

$$\begin{aligned} & \mathcal{D}_t \boldsymbol{w}^{(j+1)} - \nu^\pm \nabla_j^2 \boldsymbol{w}^{(j+1)} + \frac{1}{\rho^\pm}\nabla_j s^{(j+1)} \\ & \quad = \boldsymbol{l}_1^{(j)}(\boldsymbol{u}^{(j)}, q^{(j)}) - \boldsymbol{l}_1^{(j-1)}(\boldsymbol{u}^{(j)}, q^{(j)}) + \boldsymbol{f}(X_j, t) - \boldsymbol{f}(X_{j-1}, t) \equiv \boldsymbol{f}^{(j)}, \\ & \nabla_j \cdot \boldsymbol{w}^{(j+1)} = l_2^{(j)}(\boldsymbol{u}^{(j)}) - l_2^{(j-1)}(\boldsymbol{u}^{(j)}) \equiv r^{(j)} \quad \text{in} \quad Q_{T_{m+1}}, \\ & \qquad \boldsymbol{w}^{(j+1)}|_{t=0} = 0 \qquad \text{in } \Omega^- \cup \Omega^+, \\ & \qquad [\boldsymbol{w}^{(j+1)}]|_{G_{T_{m+1}}} = 0, \quad \boldsymbol{w}^{(j+1)}|_{S_{T_{m+1}}} = 0, \\ & [\mu^\pm \Pi_0 \Pi_j \mathbb{S}_j(\boldsymbol{w}^{(j+1)})\boldsymbol{n}_j]|_{G_{T_{m+1}}} = \boldsymbol{l}_3^{(j)}(\boldsymbol{u}^{(j)}) - \boldsymbol{l}_3^{(j-1)}(\boldsymbol{u}^{(j)}) \\ & \qquad\qquad + \boldsymbol{K}_1^{(j)}(\hat{\theta}^{(j)}) - \boldsymbol{K}_1^{(j-1)}(\hat{\theta}^{(j-1)}), \\ & [\boldsymbol{n}_0 \cdot \mathbb{T}_j(\boldsymbol{w}^{(j+1)}, s^{(j+1)})\boldsymbol{n}_j]|_{G_{T_{m+1}}} - \sigma(\theta_0)\boldsymbol{n}_0 \cdot \Delta_j(t)\int_0^t \boldsymbol{w}^{(j+1)}|_\Gamma \mathrm{d}\tau \\ & \quad = l_4^{(j)}(\boldsymbol{u}^{(j)}, q^{(j)}) - l_4^{(j-1)}(\boldsymbol{u}^{(j)}, q^{(j)}) + K_2^{(j)}(\hat{\theta}^{(j)}) \\ & \quad - K_2^{(j-1)}(\hat{\theta}^{(j-1)}) + \int_0^t (l_5^{(j)}(\boldsymbol{u}^{(j)}) - l_5^{(j-1)}(\boldsymbol{u}^{(j)}))\mathrm{d}\tau \\ & \quad + \int_0^t \Big\{K_3^{(j)}(\boldsymbol{u}^{(j)}, \hat{\theta}^{(j)}) - K_3^{(j-1)}(\boldsymbol{u}^{(j-1)}, \hat{\theta}^{(j-1)})\Big\}\mathrm{d}\tau \quad \text{on } G_{T_{m+1}}. \end{aligned} \tag{3.9}$$

Here

$$\begin{aligned}
\boldsymbol{l}_1^{(j)}(\boldsymbol{w},s) &= \nu^{\pm}(\nabla_j^2-\nabla^2)\boldsymbol{w}+\frac{1}{\rho^{\pm}}(\nabla-\nabla_j)s,\\
l_2^{(j)}(\boldsymbol{w}) &= (\nabla-\nabla_j)\cdot\boldsymbol{w}=-\mathbb{B}_j\nabla\cdot\boldsymbol{w},\\
\boldsymbol{l}_3^{(j)}(\boldsymbol{w}) &= [\Pi_0\mathbb{S}(\boldsymbol{w})\boldsymbol{n}_0-\Pi_0\Pi_j\mathbb{S}_j(\boldsymbol{w})\boldsymbol{n}_j]|_\Gamma,\\
l_4^{(j)}(\boldsymbol{w},s) &= [\boldsymbol{n}_0\cdot(\mathbb{T}(\boldsymbol{w},s)\boldsymbol{n}_0-\mathbb{T}_j(\boldsymbol{w},s)\boldsymbol{n}_j)]|_\Gamma\\
&= [s\boldsymbol{n}_0\cdot(\boldsymbol{n}_j-\boldsymbol{n}_0)+\boldsymbol{n}_0\cdot(\mathbb{S}(\boldsymbol{w})\boldsymbol{n}_0-\mathbb{S}_j(\boldsymbol{w})\boldsymbol{n}_j)]|_\Gamma,\\
l_5^{(j)}(\boldsymbol{w}) &= \sigma(\theta_0)\boldsymbol{n}_0\cdot\mathcal{D}_t\Big\{(\Delta(t)-\Delta(0))\int_0^t\boldsymbol{w}|_\Gamma\mathrm{d}t'\Big\}\\
&= \sigma(\theta_0)\boldsymbol{n}_0\cdot\Big\{(\Delta(t)-\Delta(0))\boldsymbol{w}+\dot{\Delta}(t)\int_0^t\boldsymbol{w}|_\Gamma\,\mathrm{d}t'\Big\},
\end{aligned}$$

Moreover, we observe that the function $\psi^{(j+1)}=\hat{\theta}^{(j+1)}-\hat{\theta}^{(j)}$, $j=1,\dots,$ is a solution of the problem, the result of subtraction of systems (3.2) corresponding to the neighboring indices $j+1$ and j:

$$\begin{aligned}
\mathcal{D}_t\psi^{(j+1)}-k^{\pm}\nabla_j^2\psi^{(j+1)} &= k^{\pm}\nabla_j^2\hat{\theta}^{(j)}-k^{\pm}\nabla_{j-1}^2\hat{\theta}^{(j)}\\
&\equiv h_1^{(j)}(\hat{\theta}^{(j)})-h_1^{(j-1)}(\hat{\theta}^{(j)})\quad\text{in } Q_{T_{m+1}},\\
\psi^{(j+1)}|_{t=0}=0\quad&\text{in }\Omega^-\cup\Omega^+,\\
[\psi^{(j+1)}]|_\Gamma=0,\qquad&\psi^{(j+1)}|_S=0,\\
[k^{\pm}\boldsymbol{n}_j\cdot\nabla_j\psi^{(j+1)}]|_\Gamma+\kappa\psi^{(j+1)}(\Pi_j\nabla_j)\cdot\boldsymbol{u}^{(j+1)} &\\
=-[k^{\pm}\boldsymbol{n}_j\cdot\nabla_j\hat{\theta}^{(j)}]|_\Gamma&-\kappa\hat{\theta}^{(j)}(\Pi_j\nabla_j)\cdot\boldsymbol{u}^{(j+1)}\\
+[k^{\pm}\boldsymbol{n}_{j-1}\cdot\nabla_{j-1}\hat{\theta}^{(j)}]|_\Gamma&+\kappa\hat{\theta}^{(j)}(\Pi_{j-1}\nabla_{j-1})\cdot\boldsymbol{u}^{(j)}\\
\equiv h_2^{(j)}(\hat{\theta}^{(j)})-h_2^{(j-1)}(\hat{\theta}^{(j)})&+h_3^{(j)}(\hat{\theta}^{(j)},\boldsymbol{w}^{(j+1)})\\
+h_3^{(j)}(\hat{\theta}^{(j)},\boldsymbol{u}^{(j)})&-h_3^{(j-1)}(\hat{\theta}^{(j)},\boldsymbol{u}^{(j)})\ \text{ on } G_{T_{m+1}}.
\end{aligned}\tag{3.10}$$

Here $h_i^{(j)}$ are calculated by (2.7), where $\boldsymbol{u}$ replace with $\boldsymbol{u}^{(j)}$.

In order to apply Theorem 2.1 to problem (3.9), we need verify the first hypothesis of this theorem. Following [4], we set

$$\boldsymbol{g}=\boldsymbol{g}^{(j)}=(\mathbb{B}_{j-1}^*-\mathbb{B}_j^*)\mathcal{D}_t\boldsymbol{u}^{(j)}+\mathcal{D}_t(\mathbb{B}_{j-1}^*-\mathbb{B}_j^*)\boldsymbol{u}^{(j)}-\mathbb{A}_j^*\boldsymbol{f}^{(j)},$$

and we have

$$\mathcal{D}_tr^{(j)}-\nabla_j\cdot\boldsymbol{f}^{(j)}=\nabla\cdot\boldsymbol{g}^{(j)},\qquad[(\boldsymbol{g}^{(j)}+\mathbb{A}_j^*\boldsymbol{f}^{(j)})\cdot\boldsymbol{n}_0]|_{G_t}=0.$$

Since

$$\boldsymbol{f}^{(j)}=\partial\big(\boldsymbol{L}_{1k}^{(j)}(\boldsymbol{u}^{(j)},q^{(j)})-\boldsymbol{L}_{1k}^{(j-1)}(\boldsymbol{u}^{(j)},q^{(j)})\big)/\partial\xi_k+\boldsymbol{f}(X_j,t)-\boldsymbol{f}(X_{j-1},t)$$

and

$$\mathcal{D}_t\boldsymbol{u}^{(j)}=\partial\boldsymbol{M}_k^{(j)}/\partial\xi_k+\boldsymbol{f}(X_{j-1},t),$$

where $\boldsymbol{M}_k^{(j)} = \nu^\pm(\mathbb{A}_{j-1}^* \boldsymbol{e}_k \cdot \nabla_{j-1})\boldsymbol{u}^{(j)} - \mathbb{A}_{j-1}^* \boldsymbol{e}_k q^{(j)}/\rho^\pm$, we can write

$$\boldsymbol{g}^{(j)} = \nabla \cdot \mathbb{G}^{(j)} \equiv \partial \boldsymbol{G}_k^{(j)}/\partial \xi_k,$$

where

$$\begin{aligned}
\boldsymbol{G}_k^{(j)} &= (\mathbb{B}_{j-1}^* - \mathbb{B}_j^*)\boldsymbol{M}_k^{(j)} - \mathbb{A}_j^*\big(\boldsymbol{L}_{1k}^{(j)}(\boldsymbol{u}^{(j)}, q^{(j)}) - \boldsymbol{L}_{1k}^{(j-1)}(\boldsymbol{u}^{(j)}, q^{(j)})\big) + \frac{\partial \boldsymbol{W}^{(j)}}{\partial \xi_k},\\
\boldsymbol{W}^{(j)} &= -\int_{\Omega^-\cup\Omega^+} \mathcal{E}(\xi,\eta)\Big\{\frac{\partial}{\partial \eta_i}(\mathbb{B}_{j-1}^* - \mathbb{B}_j^*)\boldsymbol{M}_i^{(j)} - (\mathbb{B}_{j-1}^* - \mathbb{B}_j^*)\boldsymbol{f}(X_{j-1},t)\\
&\quad - \mathcal{D}_t(\mathbb{B}_{j-1}^* - \mathbb{B}_j^*)\boldsymbol{u}^{(j)} - \partial \mathbb{A}_j^*/\partial \eta_i\big(\boldsymbol{L}_{1i}^{(j)}(\boldsymbol{u}^{(j)}, q^{(j)}) - \boldsymbol{L}_{1i}^{(j-1)}(\boldsymbol{u}^{(j)}, q^{(j)})\big)\\
&\quad + \mathbb{A}_j^*\big(\boldsymbol{f}(X_j,t) - \boldsymbol{f}(X_{j-1},t)\big)\Big\}\, d\eta.
\end{aligned}$$

We assume that $(\boldsymbol{u}^{(j)}, q^{(j)})$, $\hat{\theta}^{(j)}$, $j = 1, \dots, m$, satisfy the inequality

$$(T + T^{\gamma/2})N_T[\boldsymbol{u}^{(j)}, q^{(j)}, \hat{\theta}^{(j)}] < \delta, \quad T \le T_{m+1}, \tag{3.11}$$

which is stronger than (2.2).

The norms of the right-hand sides of the system (3.9) were estimated in [4] (Lemmas 4.1–4.4):

$$\begin{aligned}
&\|\boldsymbol{f}^{(j)}\|_{Q_T}^{(\alpha,\alpha/2))} + \|r^{(j)}\|_{Q_T}^{(1+\alpha,\frac{1+\alpha}{2})} + \|\boldsymbol{g}^{(j)}\|_{Q_T}^{(\alpha,\alpha/2))} + |\mathbb{G}^{(j)}|_{Q_T}^{(1+\alpha,\gamma)}\\
&+\|\mathbb{G}^{(j)}\|_{Q_T}^{(\gamma,0)} + \|\boldsymbol{l}_3^{(j)} - \boldsymbol{l}_3^{(j-1)}\|_{G_T}^{(1+\alpha,\frac{1+\alpha}{2})} + \|l_4^{(j)} - l_4^{(j-1)}\|_{G_T}\\
&+\|\nabla_\Gamma(l_4^{(j)} - l_4^{(j-1)})\|_{G_T}^{(\alpha,\alpha/2))} + \|l_4^{(j)} - l_4^{(j-1)}|_{G_T}^{(1+\alpha,\gamma)} + \|l_5^{(j)} - l_5^{(j-1)}\|_{G_T}^{(\alpha,\alpha/2))}\\
&\le c_{11}\Big\{(T + T^{\gamma/2})\|\boldsymbol{w}^{(j)}\|_{Q_T}^{(2+\alpha,1+\alpha/2)} N_T[\boldsymbol{u}^{(j)}, q^{(j)}] + \|\nabla \boldsymbol{w}^{(j)}\|_{Q_T}^{(\alpha,\alpha/2)}\\
&+(T + T^{1/2})\|\boldsymbol{w}^{(j)}\|_{Q_T}^{(\alpha,\alpha/2)} + P_T[\boldsymbol{w}^{(j)}]\|\boldsymbol{u}^{(j)}(\cdot,0)\|_{\Omega^-\cup\Omega^+}^{(1)}\Big\}.
\end{aligned} \tag{3.12}$$

The rest norms are evaluated by Lemmas 2.4, 2.5:

$$\begin{aligned}
&\|\boldsymbol{K}_1^{(j)}(\hat{\theta}^{(j)}) - \boldsymbol{K}_1^{(j-1)}(\hat{\theta}^{(j-1)})\|_{G_T}^{(1+\alpha,\frac{1+\alpha}{2})} + \|K_2^{(j)}(\hat{\theta}^{(j)}) - K_2^{(j-1)}(\hat{\theta}^{(j-1)})\|_{G_T}^{(1+\alpha,\frac{1+\alpha}{2})}\\
&\quad\le c_{12}\Big\{\Big(\|\hat{\theta}^{(j)}\|_{Q_T}^{(2+\alpha,1+\frac{\alpha}{2})} + \|\hat{\theta}^{(j-1)}\|_{Q_T}^{(2+\alpha,1+\frac{\alpha}{2})}\Big)^{2+\alpha}\Big(1 + \|H_0\|_\Gamma^{(1+\alpha)}\Big)\\
&\qquad\times \|\psi^{(j)}\|_{Q_T}^{(2+\alpha,1+\frac{\alpha}{2})} + \|\nabla\sigma(\hat{\theta}^{(j)})\|_{G_T}^{(1+\alpha,\frac{1+\alpha}{2})} M_T[\boldsymbol{w}^{(j)}]\\
&\qquad+ T^{\frac{1-\alpha}{2}}\|\nabla\sigma(\hat{\theta}^{(j)})\|_{G_T}\|\nabla \boldsymbol{w}^{(j)}\|_{Q_T}\Big\},
\end{aligned} \tag{3.13}$$

$$\begin{aligned}
\|K_3^{(j)}(\boldsymbol{u}^{(j)}, \hat{\theta}^{(j)}) &- K_3^{(j-1)}(\boldsymbol{u}^{(j-1)}, \hat{\theta}^{(j-1)})\|_{G_T}^{(\alpha,\frac{\alpha}{2})}\\
&\le \|K_3^{(j)}(\boldsymbol{u}^{(j)}, \hat{\theta}^{(j)}) - K_3^{(j-1)}(\boldsymbol{u}^{(j)}, \hat{\theta}^{(j)})\|_{G_T}^{(\alpha,\frac{\alpha}{2})}\\
&\quad+\|K_3^{(j-1)}(\boldsymbol{u}^{(j)}, \hat{\theta}^{(j)}) - K_3^{(j-1)}(\boldsymbol{u}^{(j-1)}, \hat{\theta}^{(j-1)})\|_{G_T}^{(\alpha,\frac{\alpha}{2})}
\end{aligned}$$

$$\begin{aligned}
\le\ & c_{13}(T+T^{1/2})\Big\{\Big(1+\|\hat\theta^{(j)}\|_{Q_T}^{(2+\alpha,1+\frac{\alpha}{2})}+\|\hat\theta^{(j-1)}\|_{Q_T}^{(2+\alpha,1+\frac{\alpha}{2})}\Big)^2\\
&\qquad\times\Big(\|\nabla \boldsymbol{u}^{(j-1)}\|_{Q_T}+\|\boldsymbol{u}^{(j)}\|_{Q_T}^{(2+\alpha,1+\frac{\alpha}{2})}\Big)\|\psi^{(j)}\|_{Q_T}^{(2+\alpha,1+\frac{\alpha}{2})}\\
&\qquad+\|\hat\theta^{(j-1)}\|_{Q_T}^{(2+\alpha,1+\frac{\alpha}{2})}\Big(1+\|\hat\theta^{(j)}\|_{Q_T}^{(2+\alpha,1+\frac{\alpha}{2})}\Big)\|\boldsymbol{w}^{(j)}\|_{Q_T}^{(2+\alpha,1+\frac{\alpha}{2})}\Big\}\\
+\ & c_{14}\Big\{\|\sigma(\hat\theta^{(j)})\|_{G_T}^{(\alpha,\frac{\alpha}{2})}\|\nabla\boldsymbol{w}^{(j)}\|_{Q_T}\\
&\quad+\|\hat\theta^{(j-1)}\|_{Q_T}^{(2+\alpha,1+\frac{\alpha}{2})}\Big(1+\|\hat\theta^{(j-1)}\|_{Q_T}^{(2+\alpha,1+\frac{\alpha}{2})}\Big)\\
&\qquad\times\Big\{M_T[\boldsymbol{w}^{(j)}]+(T+T^{1/2})\|\boldsymbol{u}^{(j)}\|_{Q_T}^{(2+\alpha,1+\frac{\alpha}{2})}\|\boldsymbol{w}^{(j)}\|_{Q_T}^{(2+\alpha,1+\frac{\alpha}{2})}\Big\}\Big\}
\end{aligned}\tag{3.14}$$

We estimate the solution of problem (3.10) using Theorem 2.3, Lemma 2.6 and inequality (2.9):

$$\begin{aligned}
\|\psi^{(j+1)}\|_{Q_T}^{(2+\alpha,1+\frac{\alpha}{2})}&\le\|h_1^{(j)}(\hat\theta^{(j)})-h_1^{(j-1)}(\hat\theta^{(j)})\|_{Q_T}^{(\alpha,\frac{\alpha}{2})}\\
&+\|h_2^{(j)}(\hat\theta^{(j)})-h_2^{(j-1)}(\hat\theta^{(j)})\|_{G_T}^{(1+\alpha,\frac{1+\alpha}{2})}+\|h_3^{(j)}(\hat\theta^{(j)},\boldsymbol{w}^{(j+1)})\|_{G_T}^{(1+\alpha,\frac{1+\alpha}{2})}\\
&+\|h_3^{(j)}(\hat\theta^{(j)},\boldsymbol{u}^{(j)})-h_3^{(j-1)}(\hat\theta^{(j)},\boldsymbol{u}^{(j)})\|_{G_T}^{(1+\alpha,\frac{1+\alpha}{2})}\\
&\le c_{15}\Big\{\Big(\|\hat\theta^{(j)}\|_{Q_T}^{(2+\alpha,1+\frac{\alpha}{2})}+\|\hat\theta^{(j)}\|_{Q_T}^{(1+\alpha,\frac{1+\alpha}{2})}\|\nabla\boldsymbol{u}^{(j)}\|_{Q_T}^{(1+\alpha,\frac{1+\alpha}{2})}\Big)M_T[\boldsymbol{w}^{(j)}]\\
&\qquad+T^{\frac{1-\alpha}{2}}\|\hat\theta^{(j)}\|_{Q_T}^{(1+\alpha,\frac{1+\alpha}{2})}(1+\|\nabla\boldsymbol{u}^{(j)}\|_{Q_T})\|\nabla\boldsymbol{w}^{(j)}\|_{Q_T}\\
&\qquad+\delta\|\hat\theta^{(j)}\|_{Q_T}^{(2+\alpha,1+\frac{\alpha}{2})}\|\boldsymbol{w}^{(j+1)}\|_{Q_T}^{(2+\alpha,1+\frac{\alpha}{2})}\\
&\qquad+T^{\frac{1-\alpha}{2}}\|\nabla\boldsymbol{u}^{(j)}\|_{Q_T}\|\theta_0\|_{\cup\Omega^+}^{(1)}\|\nabla\boldsymbol{w}^{(j+1)}\|_{Q_T}\Big\},\qquad j=1,2,\ldots
\end{aligned}\tag{3.15}$$

For $j=0$ we have

$$\begin{aligned}
\|\psi^{(1)}\|_{Q_T}^{(2+\alpha,1+\frac{\alpha}{2})}&\le\|\hat\theta^{(1)}\|_{Q_T}^{(2+\alpha,1+\frac{\alpha}{2})}+\|\theta_0\|_{\cup\Omega^\pm}^{(2+\alpha)}\\
&\le c(T)\Big\{\|\hat a\|_{S_T}^{(2+\alpha,1+\frac{\alpha}{2})}+\|\theta_0\|_{\cup\Omega^\pm}^{(2+\alpha)}\\
&\qquad+\big(\delta+T^{\frac{1-\alpha}{2}}\|\boldsymbol{v}_0\|_{\cup\Omega^\pm}^{(1)}\big)\|\theta_0\|_{\cup\Omega^\pm}^{(1)}\|\boldsymbol{u}^{(1)}\|_{Q_T}^{(1)}\Big\}\equiv\Psi(T)
\end{aligned}\tag{3.16}$$

Since $\boldsymbol{w}^{(j)}|_{t=0}=0$, $\nabla\boldsymbol{w}^{(j)}|_{t=0}=0$, we obtain

$$P_T[\boldsymbol{w}^{(j)}]\|\boldsymbol{v}_0\|_{\cup\Omega^\pm}^{(1)}\le c_{13}(T+T^{1/2})\|\boldsymbol{w}^{(j)}\|_{Q_T}^{(2+\alpha,1+\frac{\alpha}{2})}\|\boldsymbol{v}_0\|_{\cup\Omega^\pm}^{(1)}\le c_{13}\delta\|\boldsymbol{w}^{(j)}\|_{Q_T}^{(2+\alpha,1+\frac{\alpha}{2})},$$

$$M_T[\boldsymbol{w}^{(j)}]\le 2(T+T^{1/2})\|\boldsymbol{w}^{(j)}\|_{Q_T}^{(2+\alpha,1+\frac{\alpha}{2})},$$

$$\|\boldsymbol{w}^{(j)}\|_{Q_T}\le\int_0^T\|\mathcal{D}_t\boldsymbol{w}^{(j)}(\cdot,t)\|_{\cup\Omega^\pm}\,dt\le\int_0^T N_t[\boldsymbol{w}^{(j)},s^{(j)}]\,dt.\tag{3.17}$$

Next, we use the interpolation inequality

$$\|\boldsymbol{w}^{(j)}\|_{Q_T}^{(\alpha,\alpha/2)} + \|\nabla \boldsymbol{w}^{(j)}\|_{Q_T}^{(\alpha,\alpha/2)} \le \varepsilon \|\boldsymbol{w}^{(j)}\|_{Q_T}^{(2+\alpha,1+\frac{\alpha}{2})} + c_{16}(\varepsilon)\|\boldsymbol{w}^{(j)}\|_{Q_T}.$$

For $j = 0, 1, \ldots$, the estimates (3.8),(3.11) yield

$$\|\hat{\theta}^{(j)}\|_{Q_T}^{(2+\alpha,1+\frac{\alpha}{2})} \le c_{10}(T)\Big\{\|\hat{a}\|_{S_T}^{(2+\alpha,1+\frac{\alpha}{2})} + \|\theta_0\|_{\cup\Omega^\pm}^{(2+\alpha)}$$
$$+(\delta^2 + \delta\|\boldsymbol{v}_0\|_{\cup\Omega^\pm}^{(1)} + T^{\frac{1-\alpha}{2}}\|\boldsymbol{v}_0\|_{\cup\Omega^\pm}^{(1)}\|\boldsymbol{v}_0\|_{\cup\Omega^\pm})\|\theta_0\|_{\cup\Omega^\pm}^{(1)}\Big\} \equiv \Theta[T].$$

Collecting all of the inequalities beginning with (3.12), we arrive at the estimate:

$$\begin{aligned}
N_T^{(j+1)} \equiv\ & N_T[\boldsymbol{w}^{(j+1)}, s^{(j+1)}, \psi^{(j+1)}] \\
\le\ & \Big\{c_{11}\{(1+c_{13})\delta + (1+T+T^{1/2})\varepsilon\} + c_{14}(c_{17}+\Theta[T])\varepsilon + 2\delta(c_{12}+c_{14}) \\
& + \delta(1+\Theta[T])\{c_{13} + c_{14}\Theta[T] + 3c_{15}\}\Big\}\|\boldsymbol{w}^{(j)}\|_{Q_T}^{(2+\alpha,1+\frac{\alpha}{2})} \\
& + c_{18}(\delta,\Theta,H_0)\|\psi^{(j)}\|_{Q_T}^{(2+\alpha,1+\frac{\alpha}{2})} \\
& + 2c_{15}\delta\Theta[T]\|\boldsymbol{w}^{(j+1)}\|_{Q_T}^{(2+\alpha,1+\frac{\alpha}{2})} + (c_{11}+c_{14})c_{16}(\varepsilon)\|\boldsymbol{w}^{(j)}\|_{Q_T},
\end{aligned} \tag{3.18}$$

where $c_{18}(\delta,\Theta,H_0) = c_{12}(2\Theta[T])^{2+\alpha}(1+\|H_0\|_\Gamma^{(1+\alpha)}) + c_{13}\delta(1+\Theta[T])^2$. We choose δ so small that

$$2c_{15}\delta\Theta[T_{m+1}] < \frac{1}{2}$$

and we take (3.15)–(3.17) for $j-1$ into account. As a result, we have:

$$N_T^{(j+1)} \le \kappa_1 N_T^{(j)} + \kappa_2 N_T^{(j-1)} + c_{19}\int_0^T N_t^{(j)}\,dt, \tag{3.19}$$

where $\kappa_1(T) = 2\Big\{c_{11}\{(1+c_{13})\delta + (1+T+T^{1/2})\varepsilon\} + 2c_{12}\delta + c_{13}\delta(1+\Theta) + c_{14}\{(c_{17}+\Theta)\varepsilon + 2\delta(c_{12}+c_{14}) + \delta(1+\Theta[T])\{c_{13}+c_{14}\Theta[T]+3c_{15}\}\Big\} + 4c_{15}\delta\Theta c_{18}(\delta,\Theta,H_0)$, $\kappa_2(T) = 6c_{15}\delta(1+\Theta)c_{18}(\delta,\Theta,H_0)$ if $j = 2,3,\ldots$.

For $j = 1$, we deduce from (3.16),(3.18) that

$$N_T^{(2)} \le \hat{\kappa}_1 N_T^{(1)} + 2c_{18}(\delta,\Theta,H_0)\Psi[T] + c_{19}\int_0^T N_t^{(1)}\,dt, \qquad \hat{\kappa}_1 < \kappa_1 \tag{3.20}$$

Finally, we take δ,ε so that $\kappa \equiv \kappa_1(T_{m+1}) + \kappa_2(T_{m+1}) < 1$. Summing inequalities (3.19) from $j = 2$ to $j = m$ and adding (3.20), for the expression

$$\Sigma_{m+1}(T) = \sum_{j=1}^{m+1} N_T[\boldsymbol{w}^{(j)}, s^{(j)}]$$

we obtain the estimate

$$\begin{aligned}\Sigma_{m+1}(T) &\leq \kappa_1\Sigma_m(T)+\kappa_2\Sigma_{m-1}(T)+c_{19}\int_0^T \Sigma_m(t)\,dt+F(T)\\ &\leq \kappa\Sigma_{m+1}(T)+c_{19}\int_0^T \Sigma_{m+1}(t)\,dt+F(T),\end{aligned}$$

where $F(T)=N_T^{(1)}+2c_{18}(\delta,\Theta,H_0)\Psi[T]$.

Finally, from the Gronwall lemma applied to the inequality

$$\Sigma_{m+1}(T)\leq c_{20}\int_0^T \Sigma_{m+1}(t)\,dt+\frac{1}{1-\kappa}F(T),$$

it follows that

$$\begin{aligned}N_T[\boldsymbol{u}^{(m+1)},q^{(m+1)},\hat{\theta}^{(m+1)}] &\leq \Sigma_{m+1}(T)+\|\boldsymbol{v}_0\|^{(2+\alpha)}_{\cup\Omega^\pm}+\|\theta_0\|^{(2+\alpha)}_{\cup\Omega^\pm}\\ &\leq \frac{F(T)}{1-\kappa}e^{c_{20}T}+\|\boldsymbol{v}_0\|^{(2+\alpha)}_{\cup\Omega^\pm}+\|\theta_0\|^{(2+\alpha)}_{\cup\Omega^\pm}.\end{aligned}\tag{3.21}$$

As $(\boldsymbol{w}^{(1)},s^{(1)},\psi^{(1)})=(\boldsymbol{u}^{(1)}-\boldsymbol{v}_0,q^{(1)},\hat{\theta}^{(1)}-\theta_0)$, the norm $N_T^{(1)}$ can be evaluated by (3.7), (3.16), hence, $F(T)$ is bounded by the norms of the given functions. If

$$(T+T^{\gamma/2})\Big\{F(T)(1-\kappa)^{-1}e^{c_{20}T}+\|\boldsymbol{v}_0\|^{(2+\alpha)}_{\cup\Omega^\pm}+\|\theta_0\|^{(2+\alpha)}_{\cup\Omega^\pm}\Big\}\leq\delta,\tag{3.22}$$

then condition (3.11) is fulfilled also for $\boldsymbol{u}^{(m+1)},q^{(m+1)},\hat{\theta}^{(m+1)}$. It is clear that there is a number $T=T_0$ satisfying (3.22).

Since the right-hand side of (3.21) is independent of m, for every $m\in\mathbb{N}$ the functions $\boldsymbol{u}^{(m)},q^{(m)},\hat{\theta}^{(m)}$ are defined on the interval $(0,T_0]$ and satisfy the uniform estimate (3.21). It follows that the series $\sum_{j=1}^\infty N_{T_0}^{(j)}$ is convergent, whence we see that the sequence $\boldsymbol{u}^{(m)},q^{(m)},\hat{\theta}^{(m)}$ is also convergent in the norm N_{T_0}. Passing to the limit as $m\to\infty$ in (3.1), (3.2), we make sure that $(\boldsymbol{u},q,\hat{\theta})=\lim_{m\to\infty}(\boldsymbol{u}^{(m)},q^{(m)},\hat{\theta}^{(m)})$ is a solution of (1.4).

Now we prove the uniqueness of the solution obtained. Suppose that $(\boldsymbol{u},q,\hat{\theta})$ and $(\boldsymbol{u}',q',\hat{\theta}')$ are two solutions of (1.4) and consider the difference $\boldsymbol{w}=\boldsymbol{u}-\boldsymbol{u}'$, $s=q-q'$, $\psi=\hat{\theta}-\hat{\theta}'$. The triple $(\boldsymbol{w},s,\psi)$ satisfies a problem of type (3.9), (3.10):

$$\mathcal{D}_t\boldsymbol{w}-\nu^\pm\nabla^2_{\boldsymbol{u}}\boldsymbol{w}+\frac{1}{\rho^\pm}\nabla_{\boldsymbol{u}}s=\boldsymbol{l}_1(\boldsymbol{u}',q')-\boldsymbol{l}_1'(\boldsymbol{u}',q')+\boldsymbol{f}(X_{\boldsymbol{u}},t)-\boldsymbol{f}(X_{\boldsymbol{u}'},t),$$

$$\nabla_{\boldsymbol{u}}\cdot\boldsymbol{w}=l_2(\boldsymbol{u}')-l_2'(\boldsymbol{u}'),\qquad \mathcal{D}_t\psi-k^\pm\nabla^2_{\boldsymbol{u}}\psi=h_1(\hat{\theta}')-h_1'(\hat{\theta}')\quad\text{in } Q_{T_0},$$

$$\boldsymbol{w}|_{t=0}=0,\quad \psi|_{t=0}=0\quad\text{in }\Omega^-\cup\Omega^+,$$

$$[\boldsymbol{w}]|_\Gamma=0,\qquad[\psi]|_\Gamma=0,\qquad \boldsymbol{w}|_S=0,\qquad\psi|_S=0,$$

$$[\Pi_0\Pi\mathbb{S}_{\boldsymbol{u}}(\boldsymbol{w})\boldsymbol{n}]|_\Gamma=\boldsymbol{l}_3(\boldsymbol{u}')-\boldsymbol{l}_3'(\boldsymbol{u}')+\boldsymbol{K}_1(\hat{\theta})-\boldsymbol{K}_1'(\hat{\theta}'),$$

$$[\boldsymbol{n}_0 \cdot \mathbb{T}_{\boldsymbol{u}}(\boldsymbol{w}, s)\boldsymbol{n}]|_\Gamma - \sigma \boldsymbol{n}_0 \cdot \Delta(t) \int_0^t \boldsymbol{w}\, d\tau|_\Gamma = l_4(\boldsymbol{u}', q') - l_4'(\boldsymbol{u}', q')$$

$$+ K_2(\hat{\theta}) - K_2'(\hat{\theta}') + \sigma \int_0^t \left(l_5(\boldsymbol{u}') - l_5'(\boldsymbol{u}') + K_3(\boldsymbol{u}, \hat{\theta}) - K_3'(\boldsymbol{u}', \hat{\theta}')\right) d\tau,$$

$$[k^\pm \boldsymbol{n} \cdot \nabla_{\boldsymbol{u}} \psi]|_\Gamma + \kappa \psi (\Pi \nabla_{\boldsymbol{u}}) \cdot \boldsymbol{u} = h_2(\hat{\theta}') - h_2'(\hat{\theta}') + h_3(\hat{\theta}', \boldsymbol{w})$$

$$+ h_3(\hat{\theta}', \boldsymbol{u}') - h_3'(\hat{\theta}', \boldsymbol{u}') \qquad \text{on } G_{T_0}.$$

Repeating the above arguments, we arrive at an inequality similar to (3.19):

$$N_T[\boldsymbol{w}, s, \psi] \quad \leq \quad \kappa N_T[\boldsymbol{w}, s, \psi] + c_{21} \int_0^t N_t[\boldsymbol{w}, s, \psi] dt, \qquad \kappa < 1,$$

which, by the Gronwall lemma, implies that $\boldsymbol{w} = 0$, $s = 0$, $\psi = 0$.

Theorem 1.3 is completely proved. □

References

[1] M.V. Lagunova and V.A. Solonnikov, Nonstationary problem of thermocapillary convection, Leningrad Branch of Steklov Mathematical Institute (LOMI), preprint E-13-89, Leningrad 1989, p. 28.

[2] I.V. Denisova and V.A. Solonnikov, Solvability in Hölder spaces for a model initial-boundary value problem generated by a problem on the motion of two fluids, *Zap. Nauchn. Sem. Leningrad. Otdel. Mat. Inst. Steklov. (LOMI)*, **188** (1991), 5–44 (Russian) (English transl. in *J. Math. Sci.*, **70** (1994), No. 3, 1717–1746).

[3] I.V. Denisova, Solvability in Hölder spaces of a linear problem concerning the motion of two fluids separated by a closed surface,*Algebra i Analiz*,**5** (1993), No. 4, 122–148 (English transl. in *St. Petersburg Math. J.*, **5** (1994), No. 4).

[4] I.V. Denisova and V.A. Solonnikov, Classical solvability of the problem on the motion of two viscous incompressible fluids, *Algebra i Analiz*, **7** (1995), no. 5, 101–142 (Russian) (English transl. in *St. Petersburg Math. J.*, **7** (1996), no. 5, 755–786).

[5] V.V. Pukhnachov, Motion of viscous fluids with free boundaries, Novosibirsk University, 1989, 97 p.

[6] V.V. Pukhnachov, Thermocapillary convection under low gravity, *Fluid Dynamics Transactions*, Warszawa: PWN, **14** (1989), 145–204.

[7] O.A. Ladyzhenskaya, Boundary value problems of mathematical physics, "Nauka", Moscow, 1973 (English transl., Springer-Verlag, 1985).

[8] I.Sh. Mogilevskiĭ and V.A. Solonnikov, On the solvability of an evolution free boundary problem for the Navier–Stokes equations in Hölder spaces of functions, *Mathematical Problems Relating to the Navier–Stokes Equation*, Ser. Adv. Math. Appl. Sci., World Scientific Publishing Co., Singapore – London, 1992, 105–181.

Irina Vlad. Denisova
Institute of Mechanical Engineering Problems
Russian Academy of Sciences, V.O., Bolshoy pr., 61
St. Petersburg, 199178 Russia
e-mail: `ira@wave.ipme.ru`

Progress in Nonlinear Differential Equations
and Their Applications, Vol. 61, 65–80

Some Mathematical Problems in Visual Transduction

Daniele Andreucci, Paolo Bisegna and Emmanuele DiBenedetto

Abstract. We present a mathematical model for the phototransduction cascade, taking into account the spatial localization of the different reaction processes. The geometric complexity of the problem (set in the rod outer segment) is simplified by a process of homogenization and concentration of capacity.

Mathematics Subject Classification (2000). Primary 35B27, 92C37; Secondary 35K50.

Keywords. homogenization, signal transduction, concentrated capacity, reticular structure.

1. The phototransduction cascade

Signal transduction in living cells occurs by precise, highly regulated localization of key enzimes. The relevant diffusion and reaction processes take place in subcompartments of the cell.

The standard ODE approach, based on Michaelis-Menten well-stirred kinetics averages concentrations within the whole cell volume, and therefore is unable to predict the spatial localization of the signaling processes.

A PDE approach seems more suitable as takes into account the spatial dependence of the signal. However, the geometry of the domain open to diffusion may be so complex as to prevent in practice any affective qualitative or numerical analysis of the problem. This occurs whenever the cell contains structures which impede, or however influence, the diffusion of the signaling molecules, but whose detailed geometry cannot be directly taken into account in the mathematical model.

In [2, 3, 4] we have addressed these issues in the context of the phototransduction cascade in the rod outer segment (ROS) of vertebrates (Figure 1). The ROS is a cylinder of height H and cross section a disc of radius $R + \sigma\varepsilon$, housing

Supported by grant NIH-1-RO1-GM 68953-01.

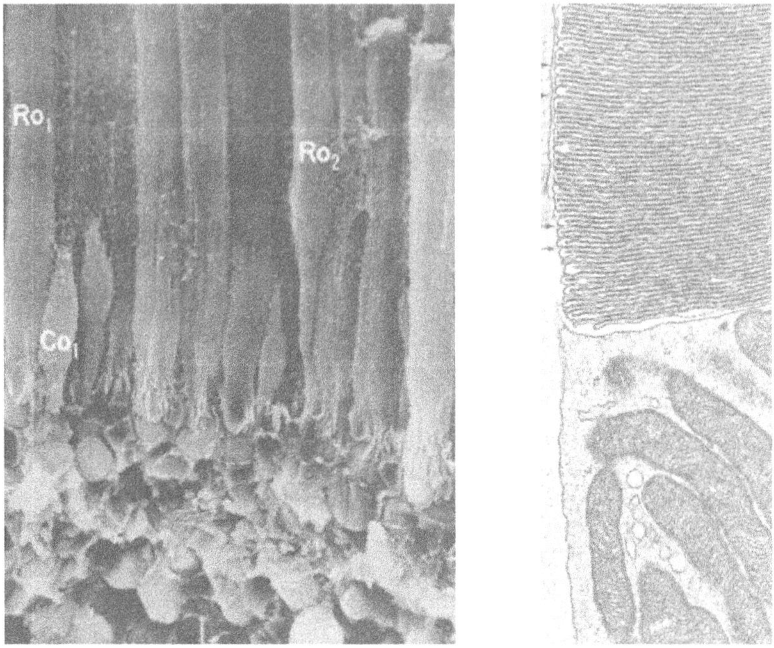

FIGURE 1. On the left: Retina, rods and cones (Frog, Scansion Electron Migrograph). On the right: Retina, rod cells (Transmission Electron Migrograph). From the Internet Atlas of Histology, COM-UIUC

a longitudinal stack of equispaced, parallel, coaxial cylinders C_j, each of radius R and thickness ε, called *discs*. The stack cuts between any two contiguous discs C_j and C_{j+1}, an *inter-discal space* I_j, which is itself a cylinder of cross section a disc of radius R and thickness $\nu\varepsilon$ (Figure 2).[1]

This stack is separated from the plasma membrane by a thin gap S_ε, called *outer shell* of thickness $\sigma\varepsilon$.

Each disc carries a large number of molecules of photoreceptor Rhodopsin, clamped to it. Assume the ROS is in absence of light (dark-adapted). If a photon hits a molecule of Rhodopsin on the disc C_{j_o}, it generates a biochemical cascade whose net result is depletion of cGMP (cyclic-guanosin monophosphate) in the cytosol, and a consequent suppression of ionic current across the outer membrane of the ROS. Such a current is the *response* of the phototransduction cascade and it starts the process of vision by propagating the signal to the brain through the optic nerve. The ROS is sufficiently sophisticated as being able to capture a single photon of light; the corresponding current is called the *single photon response*, from a dark-adapted ROS.

[1] Here ε, σ, ν R and H are given positive parameters. For the Salamander the stack contains about 1000 discs, and $H \approx 22\mu m$, $R \approx 5.5, \mu m$, $\varepsilon \approx 10nm$, $\nu \approx 0.5$ and $\sigma \approx 1$ (Lamb and Pugh [11]).

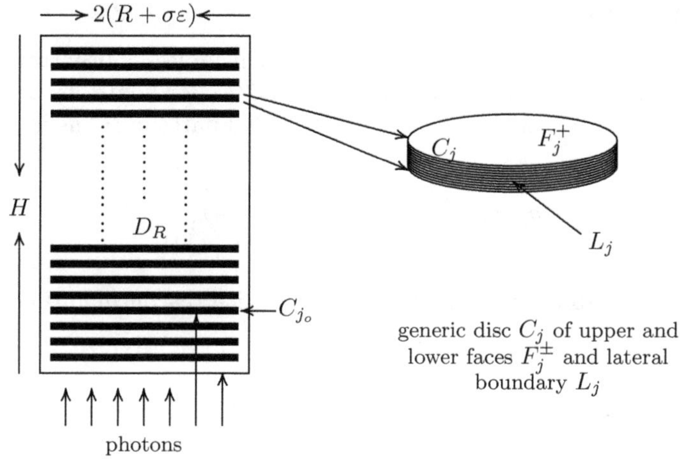

FIGURE 2. Geometry of the ROS and its discs

Diffusion and depletion of cGMP is modulated by the diffusion of Ca^{2+}. Both cGMP and Ca^{2+} are termed *second messengers.*

The diffusion process takes place within the thin layers I_j between the membranous discs (transversal diffusion) and the equally thin outer shell S_ε, (longitudinal diffusion) as indicated in Figure 2.

Diffusion within the interdiscal spaces is important, because these are the only physical spaces through which cGMP can be depleted by the enzyme phosphodiesterase, localized on the faces of the discs. The depletion of cGMP in turn drives the closure of ion channels on the plasma membrane, lowering the influx of Ca^{2+} ions and hyperpolarizing the cell. Therefore diffusion along the outer shell is equally important, because this is the region where the channels reside and Ca^{2+} enters the cytosol.

We propose a mathematical approach that involves homogenized limits and concentrated capacity as $\varepsilon \to 0$, starting from its "initial" physical value ε_o. The discs and the corresponding interdiscal spaces are regarded as becoming thinner and thinner in size and larger and larger in number.

Denote by $\widetilde{\Omega}_\varepsilon$ the domain out of the ROS available for diffusion for a fixed value of the parameter ε. This consists of the ROS from which the closed discs $\overline{C}_j$ have been removed. If $n = n(\varepsilon)$ is their number we let $\varepsilon \to 0$ and $n \to \infty$ in such a way that the volume available for diffusion remains unchanged. Such a condition implies that n and ε are linked by

$$n\varepsilon = H\theta_o \,, \qquad \text{where } \theta_o = 1 + \nu.$$

As $\varepsilon \to 0$, the domain $\widetilde{\Omega}_\varepsilon$ tends roughly speaking to a right circular cylinder (up to a set of measure zero). Thus the original geometrical complexity is resolved into a geometrical simplicity. This in turn lends itself to efficient computational analysis and simulations, and casts light into the local phenomenon of spread of the signal.

2. The physical model

Denote by u_ε and v_ε the ε-approximations of dimensionless concentration of cGMP and Ca^{2+} respectively. Denote also by z the axial coordinate, by $\overline{x} = (x, y)$ the transversal coordinates and by $x = (\overline{x}, z)$ the coordinates in $\mathbb{R}^3$ as in Figure 2. Set

$$a_\varepsilon(x) = \begin{cases} 1 & \text{for } x \in \bigcup\limits_{j \neq j_o} I_j \\ \dfrac{\varepsilon_o}{\varepsilon} & \text{for } x \in I_{j_o} \cup S_\varepsilon \end{cases} \tag{2.1}$$

and consider the system of parabolic equations with discontinuous coefficients,

$$\begin{aligned} &u_\varepsilon,\, v_\varepsilon \in C\big(0,T; L^2(\widetilde{\Omega}_\varepsilon)\big) \cap L^2\big(0,T; W^{1,2}(\widetilde{\Omega}_\varepsilon)\big) \\ &a_\varepsilon(x)\frac{\partial}{\partial t}u_\varepsilon - k_u \operatorname{div} a_\varepsilon(x)\nabla u_\varepsilon = 0 \\ &a_\varepsilon(x)\frac{\partial}{\partial t}v_\varepsilon - k_v \operatorname{div} a_\varepsilon(x)\nabla v_\varepsilon = 0 \end{aligned} \tag{2.2}$$

weakly in $\widetilde{\Omega}_\varepsilon$, where k_u and k_v are given diffusivities. For $\varepsilon = \varepsilon_o$ the coefficients a_ε are identically equal to 1 and the equations represent the classical Fick's diffusion of cGMP and Ca^{2+} in the cytosol. The mathematical model we propose for the diffusion of cGMP and Ca^{2+} is the limiting u and v obtained from (2.2) as $\varepsilon \to 0$, each satisfying some corresponding limiting equation. The rationale is that, by this process, the ROS reduces to a right circular cylinder from which, roughly speaking the discs have been removed, and the outer shell reduces to its outer boundary. As $\varepsilon \to 0$ the coefficients $a_\varepsilon(\cdot)$ become unbounded in the shrinking outer shell S_ε and in the shrinking interdiscal space I_{j_o}. This corresponds to concentrating the capacity of cGMP and Ca^{2+} in these domains to recover their diffusion effect in the limit (see [3]).

The boundary flux for u_ε on the faces of the interdiscal spaces is

$$k_u \nabla u_\varepsilon \cdot \mathbf{n} = -\frac{\nu\varepsilon}{2}\{\gamma_o u_\varepsilon - f(v_\varepsilon)\}\,, \tag{2.3}$$

where $\mathbf{n}$ is the exterior unit normal to $\widetilde{\Omega}_\varepsilon$, γ_o is a given positive constant, and $f(\cdot)$ is a given smooth function defined below in (2.6). Generation of cGMP on the faces of the interdiscal spaces I_j, other than the special space I_{j_o}, is modulated by Ca^{2+} through Guanylyl Ciclase (GC), whereas its depletion is effected by Phosphodiesterase (PDE) (Figure 3). While both PDE and GC are confined to the faces of the discs, in the biological literature their densities [PDE] and [GC] are measured as *volumic* quantities. To account for their surface action they have to be converted into surface densities and distributed on the $2n(\varepsilon)$ faces of the discs. Such a process generates the factor of ε on the right-hand side of (2.3). In turn, such a factor of ε is the correct one to compute the homogenized limit of (2.2). On the face where activation occurs, the flux of cGMP is given by

$$\frac{\varepsilon_o}{\varepsilon}k_u \nabla u \cdot \mathbf{n} = -\frac{\nu\varepsilon_o}{2}\{\gamma_o u_\varepsilon - f(v_\varepsilon)\} - u_\varepsilon f_1(v_\varepsilon, x, t)\,; \tag{2.4}$$

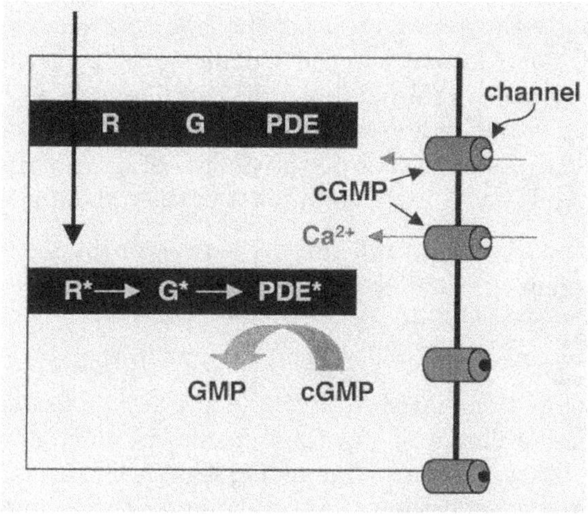

FIGURE 3. A sketch of the phototransduction cascade

where f_1 is a given, smooth, positive, bounded function of its arguments, which takes into account the depletion of cGMP due to the presence of the activated enzyme PDE*. The remaining parts of $\partial\widetilde{\Omega}_\varepsilon$ are impermeable to cGMP; accordingly on these parts, the flux of u_ε vanishes. For further mathematical and physical motivations of the precise meaning of these boundary fluxes we refer to [3] and [4].

Calcium does not penetrate the discs so that v_ε has zero flux on their boundaries. Also Ca^{2+} does not outflow the top and bottom of the ROS. Outflow of Ca^{2+} on the lateral surface of the ROS is proportional to the ionic current it generates. Inflow of Ca^{2+} is, roughly speaking, a function of the number of open ionic channels, which in turn is a function of [cGMP]. These biological facts translate into the flux condition,

$$\frac{\varepsilon_o}{\varepsilon} k_v \nabla v_\varepsilon \cdot \mathbf{n} = -g_1(v_\varepsilon) + g_2(u_\varepsilon)\,, \qquad \text{on } \{|\overline{x}| = R + \sigma\varepsilon\}. \tag{2.5}$$

In visual transduction the form of $f(\cdot)$, $g_1(\cdot)$, and $g_2(\cdot)$ are explicitly given by [12]

$$f(s) = \frac{\gamma_1}{\beta_1^m + s^m}; \quad g_1(s) = \frac{c_1\, s}{d_1 + s}; \qquad g_2(s) = \frac{c_2\, s^\kappa}{d_2^\kappa + s^\kappa}\,, \qquad s > 0\,, \tag{2.6}$$

for given positive constants γ_1, β_1, m, c_1, c_2, d_1, d_2, κ. The initial conditions are those of dark equilibrium, i.e., $u_\varepsilon(\cdot, 0) = \overset{\circ}{u}$ and $v_\varepsilon(\cdot, 0) = \overset{\circ}{v}$ for two given positive constants $\overset{\circ}{u}$, $\overset{\circ}{v}$.

3. The limiting equations

Let z_o be the z-coordinate of the disc C_{j_o} where activation occurs and let $\mathcal{D}_R = \{|\overline{x}| < R\}$. The limiting process is carried out so that z_o remains constant as

$\varepsilon \to 0$ or equivalently the activated face remains at the same level. As $\varepsilon \to 0$ the layered domain $\widetilde{\Omega}_\varepsilon$ tends formally to the cylinder $\Omega_o = \mathcal{D}_R \times (0, H)$, the activated interdiscal spaces I_{j_o}, shrinks formally to the disc $\mathcal{D}_R \times \{z_o\}$, and the outer shell S_ε reduces formally to the lateral boundary $S = \partial\mathcal{D}_R \times (0, H)$ of Ω_o.

The functions u_ε and v_ε generate three pairs of limiting functions, each representing [cGMP] and [Ca^{2+}] in different parts of the rod outer segment. Precisely:

u, v	defined in Ω_o and called the interior limit
$\overset{\circ}{u}, \overset{\circ}{v}$	defined in $\mathcal{D}_R \times \{z_o\}$ and called the limit on the activated level z_o
$\hat{u}, \hat{v}$	defined in S and called the limit in the outer shell.

We next give the equations satisfied by these limiting quantities, each in its own geometric portion, and illustrate how these seemingly different diffusion processes interact with each other. To convey the main ideas we do this in a formal way. The topology of convergence and the precise meaning of these equations is contained in the weak formulations of Section 5. The analogues for v as well as justifications and proofs are in [3].

The limiting equations will contain in various forms the forcing terms generated by (2.3)–(2.6). To simplify the symbolism we will set,

$$F(\overline{x}, z, t) = \gamma_o u - f(v); \quad F_o(\overline{x}, t) = \gamma_o \overset{\circ}{u} - f(\overset{\circ}{v});$$
$$F_*(\overline{x}, t) = \frac{1}{\nu\varepsilon_o} \overset{\circ}{u} f(\overset{\circ}{v}, \overline{x}, t).$$

3.1. Form of the interior limit

The interior limiting u satisfies,

$$u_t - k_u \Delta_{\overline{x}} u = -F \quad \text{in } \Omega_o. \tag{3.1}$$

Here $\Delta_{\overline{x}}$ is the Laplacian acting only on the transversal variables $\overline{x} = (x, y)$. Since u is a function of the transversal variables $\overline{x} = (x, y)$ and the longitudinal variable z, (3.1) can be regarded as a family of diffusion processes parametrized with $z \in (0, H)$, taking place on the disc $\mathcal{D}_R$. Thus the volumic diffusion in (2.2) in the layered structure of the rod, is transformed into a family of 2-dimensional diffusions. Also, the homogenized limit transforms the boundary fluxes in (2.3) into volumic source terms holding in Ω_o.

3.2. The limit at the activated level z_o

The limiting $\overset{\circ}{u}$ satisfies,

$$\overset{\circ}{u}_t - k_u \Delta_{\overline{x}} \overset{\circ}{u} = -F_o - F_* \quad \text{on } \mathcal{D}_R \times \{z_o\}. \tag{3.2}$$

Thus also at the activated level z_o, the volumic diffusion in (2.2) is transformed into a 2-dimensional diffusion on the layer $\mathcal{D}_R \times \{z_o\}$ and the fluxes in (2.4) are transformed into sources defined in the interior of the same disc. Note that the limit equation contains also the term F^* due to activation.

3.3. Limiting equations in the outer shell

The limiting $\hat{u}$ on S is a function of $\theta \in [0, 2\pi)$ and $z \in (0, H)$ and of time. Outside the activated level z_o it must equal the trace on S of the interior limit u. On the activated level z_o it must equal the trace on $\partial\mathcal{D}_R \times \{z_o\}$ of $\overset{\circ}{u}$, i.e.,

$$\begin{aligned} \hat{u}(\theta, z, t) &= u(\overline{x}, z, t) \Big|_{|\overline{x}|=R} \qquad \text{for all } z \neq z_o; \\ \hat{u}(\theta, z_o, t) &= \overset{\circ}{u}(\overline{x}, t) \Big|_{|\overline{x}|=R} . \end{aligned}$$

The interior limit u and the limit $\overset{\circ}{u}$ on the activated level z_o are linked to the limit $\hat{u}$ in the following more essential way. Let Δ_S denote the Laplace-Beltrami operator on S. Then

$$\hat{u}_t - k_u \Delta_S \hat{u} = -\frac{(1-\theta_o)k_u}{\sigma\varepsilon_o} u_\rho \Big|_{|\overline{x}|=R} - \delta_{z_o} \frac{\nu k_u}{\sigma} \overset{\circ}{u}_\rho \Big|_{|\overline{x}|=R} \tag{3.3}$$

in S. Here δ_{z_o} is the Dirac delta on S with mass on z_o and u_ρ and $\overset{\circ}{u}_\rho$ are the normal derivatives of u and $\overset{\circ}{u}$ on $\partial\mathcal{D}_R \times \{z_o\}$. Thus the diffusion of $\hat{u}$ on the limiting outer shell S, is forced by the exterior fluxes on S of the interior limit u and the limit $\overset{\circ}{u}$ on the activated disc $\mathcal{D}_R \times \{z_o\}$.

This is the biophysical law by which the homogenized-concentrated limiting diffusions interact with each other. While it is somewhat intuitive that cGMP coming from the transversal interstices should provide the driving force for the movement of the cGMP on the longitudinal surface S, equations (3.1)–(3.3) provide a precise law by which this occurs. Multiplying (3.3) by $\sigma\varepsilon_o$, the left-hand side represent the pointwise space-time variation of $\sigma\varepsilon_o\hat{u}$. The latter can be regarded as a surface density of cGMP, which has been concentrated on the limiting surface S, starting from the original shell S_{ε_o}. The factor $(1-\theta_o)$ on the right hand side signifies that only a fraction of $(1-\theta_o)$ of S is exposed to the outflow of the homogenized interior limit of cGMP. This is the same fraction of surface exposed to inflow/outflow of cGMP from the interdiscal spaces into the outer shell S_{ε_o} in the original, non homogenized configuration of the ROS.

4. Main ideas in computing the homogenized limit

We continue to limit our discussion to the net $\{u_\varepsilon\}$, the arguments for $\{v_\varepsilon\}$ being similar. By standard energy estimates,[2]

$$0 \le u_\varepsilon(x,t) \le \gamma \qquad \text{for all } (x,t) \in \widetilde{\Omega}_{\varepsilon,T}, \tag{4.1}$$

$$\sup_{0\le t\le T} \left\| \sqrt{a_\varepsilon}\, u_\varepsilon(\cdot,t) \right\|_{2,\widetilde{\Omega}_\varepsilon} + \left\| \sqrt{a_\varepsilon}\nabla u_\varepsilon \right\|_{2,\widetilde{\Omega}_{\varepsilon,T}} \le \gamma, \tag{4.2}$$

$$\int_0^{T-h}\int_{\widetilde{\Omega}_\varepsilon} a_\varepsilon \big[u_\varepsilon(t+h) - u_\varepsilon(t)\big]^2 dxdt \le \gamma h \qquad \text{for all } h \in (0,T), \tag{4.3}$$

for a positive constant γ independent from ε.

[2]For a set $D \subset \mathbb{R}^N$ and a positive number T we set $D_T = D \times (0,T)$.

4.1. The interior limit

As $\varepsilon \to 0$ the discs C_j become thinner while their radius remains R; consequently, as the outer shell S_ε shrinks to S, the resulting domain tends to be disconnected, and the diffusion becomes degenerate, since the discs effectively prohibit transport of mass in the axial direction. For this reason, energy estimates alone are not enough to generate a sufficiently strong notion of limit. It appears that some estimation is needed on the degree of intercommunication of mass between the layers of $\widetilde{\Omega}_\varepsilon$.

Observe that u_ε solves "similar" parabolic problems in any two interdiscal spaces which are at mutual distance h. These two parabolic problems have "similar" boundary data for $\{|\overline{x}| = R\}$, in the following sense. Let Λ_j be the lateral boundary of I_j. Then by the energy estimate (4.2) and the definition of the a_ε,

$$\frac{1}{\varepsilon} \iint_{\Lambda_{j,T}} |u_\varepsilon(\overline{x}, z, t) - u_\varepsilon(\overline{x}, z+h, t)|\, d\eta dt \leq \gamma \sqrt{|h|} \tag{4.4}$$

where $d\eta$ is the surface measure on Λ_j. Here h is an integral multiple of $(1+\nu)\varepsilon$, so that $(\overline{x}, z+h)$ belongs to an interdiscal space whenever $(\overline{x}, z)$ does. Essentially, (4.4) implies that v_ε is continuous in z, in the averaged L^1 sense, on the lateral surfaces Λ_j. As a consequence of the regularizing effect of diffusion, the estimate above implies a stronger pointwise estimate *inside* the interdiscal spaces, i.e.,

$$|u_\varepsilon(\overline{x}, z, t) - u_\varepsilon(\overline{x}, z+h, t)| \leq \gamma |h|^\alpha, \qquad |\overline{x}| < R - \delta, \tag{4.5}$$

where the constant γ depends on δ but not on h, and $\alpha \in (0,1)$ depends only on the data of the problem. This estimate, though not valid when $(\overline{x}, z+h)$ or $(\overline{x}, z)$ belong to the special interdiscal space, is still valid *across* it, i.e., when I_{j_o} falls between z and $z+h$.

Finally, it follows from the regularity theory of parabolic equations that inside each interdiscal space I_j the functions u_ε and v_ε are uniformly smooth, i.e.,

$$|\nabla u_\varepsilon| + |\nabla v_\varepsilon| + |u_{\varepsilon,t}| + |v_{\varepsilon,t}| \leq \gamma, \qquad |\overline{x}| < R - \delta, \quad \gamma = \gamma(\delta). \tag{4.6}$$

Then by the Kirzbraun-Pucci extension theorem [8], we may extend u_ε and v_ε with functions $\overline{u}_\varepsilon$, $\overline{v}_\varepsilon$ which are equi-Hölder continuous in the open cylinder $\Omega_{o,T}$. Therefore, up to a subsequence relabelled with ε,

$$\overline{u}_\varepsilon \to u, \quad \text{and} \quad \overline{v}_\varepsilon \to v \quad \text{uniformly on compact subsets of } \Omega_{o,T}.$$

Take a testing function $\varphi \in C_o^\infty(\Omega_{o,T})$, $\varphi(\cdot, 0) = 0$, $\varphi(\cdot, T) = 0$, and such that $\varphi(\cdot, z, t) \in C_o^\infty(\{|\overline{x}| < (1-\delta)R\})$, for all $z \in (0, H)$, $t \in (0, T]$. We also require $\varphi \equiv 0$ in a neighborhood of I_{j_o}. By the definition, $a_\varepsilon \equiv 1$ on the support of φ. Denote by $\partial^\pm I_j$ respectively the upper and lower base of I_j, and let $\zeta_j^\pm$ be their

z-levels. Using such a φ as a testing function in the weak formulation of (2.2) gives

$$\begin{aligned}
\sum_{j=0}^{n} \int_0^T \int_{I_j} \{ - u_\varepsilon \varphi_t - k_u u_\varepsilon \Delta_{\overline{x}} \varphi \} dx dt + \sum_{j=0}^{n} \int_0^T \int_{I_j} k_u u_{\varepsilon,z} \varphi_z dx dt \\
= -\frac{1}{2} \nu\varepsilon \sum_{j=1}^{n} \int_0^T \int_{\partial I_j^-} (\gamma_o u_\varepsilon - f(v_\varepsilon)) \varphi d\overline{x} dt \\
- \frac{1}{2} \nu\varepsilon \sum_{j=0}^{n-1} \int_0^T \int_{\partial I_j^+} (\gamma_o u_\varepsilon - f(v_\varepsilon)) \varphi d\overline{x} dt \qquad (4.7) \\
= - \sum_{j=0}^{n} \int_0^T \int_{I_j} (\gamma_o u_\varepsilon - f(v_\varepsilon)) \varphi dx dt \\
+ \sum_{j=0}^{n} \int_0^T \int_{I_j} [(\gamma_o u_\varepsilon - f(v_\varepsilon))\varphi]_z (\zeta_j^+ + \zeta_j^- - 2z) dx dt \,.
\end{aligned}$$

Due to the energy estimates, the last sum above is $O(\varepsilon)$. Then (4.7) can be rewritten as

$$\begin{aligned}
\iint_{\Omega_{o,T}} \{ - \overline{u}_\varepsilon \varphi_t - k_u \overline{u}_\varepsilon \Delta_{\overline{x}} \varphi \} \sum_{j=0}^{n} \chi_{I_j} dx dt + \iint_{\Omega_{o,T}} k_u u_{\varepsilon,z} \varphi_z \sum_{j=0}^{n} \chi_{I_j} dx dt \\
= - \iint_{\Omega_{o,T}} (\gamma_o \overline{u}_\varepsilon - f(\overline{v}_\varepsilon)) \varphi \sum_{j=0}^{n} \chi_{I_j} dx dt + O(\varepsilon) \,. \qquad (4.8)
\end{aligned}$$

As $\varepsilon \to 0$ and correspondingly $n \to \infty$,

$$\sum_{j=0}^{n} \chi_{I_j} \longrightarrow (1 - \theta_o) \qquad \text{weakly in } L^2(\Omega_o). \qquad (4.9)$$

The energy estimates ensure that (up to subsequences)

$$k_u u_{\varepsilon,z} \sum_{j=0}^{n} \chi_{I_j} \to \xi \,, \qquad \text{weakly in } L^2(\Omega_{o,T}). \qquad (4.10)$$

A variant of an argument known in the theory of homogenization of stratified structures, yields $\xi \equiv 0$ (see [3], [7]). Therefore, taking the limit $\varepsilon \to 0$ in (4.8) gives

$$(1 - \theta_o) \iint_{\Omega_{o,T}} \{ - u \varphi_t - k_u u \Delta_{\overline{x}} \varphi + (\gamma_o u - f(v)) \varphi \} dx dt = 0 \,. \qquad (4.11)$$

4.2. The limit in the outer shell

Let us introduce the radial average of u_ε in the outer shell

$$\hat{u}_\varepsilon(\theta, z, t) = \frac{1}{\sigma\varepsilon} \int_R^{R+\sigma\varepsilon} u(\rho \cos\theta, \rho \sin\theta, z, t) d\rho \,. \qquad (4.12)$$

Now require that for $|\overline{x}| > R$ the testing function φ be independent of $\overline{x}$, that is

$$\varphi(\theta, z, t) = \varphi \Big|_{|\overline{x}|=R} . \tag{4.13}$$

Require also that φ vanishes in a neighborhood of the special interdiscal space I_{j_o}. By the energy estimates (4.2), the nets $\{\hat{u}_\varepsilon\}$ and $\{\nabla_S \hat{u}_\varepsilon\}$ are equi-bounded in $L^2(S_T)$. Here ∇_S denotes the gradient on the surface S. Moreover taking also into account the uniform time-regularity estimates in (4.3), the net $\{\hat{u}_\varepsilon\}$ is pre-compact in $L^2(S_T)$. Therefore for subnets relabelled with ε,

$$\{\hat{u}_\varepsilon\} \to \hat{u} \text{ in } L^2(S_T) \text{ and } \{\nabla_S \hat{u}_\varepsilon\} \to \nabla_S \hat{u} \text{ weakly in } L^2(S_T). \tag{4.14}$$

Writing the weak formulation for the testing function φ and letting $\varepsilon \to 0$, recalling that $Rd\theta dz$ is the surface measure on S,

$$\begin{aligned} &\lim_{\varepsilon\to 0} \frac{\varepsilon_o}{\varepsilon} \left\{ \iint_{S_{\varepsilon,T}} \{ -u_\varepsilon \varphi_t + k_u \nabla u_\varepsilon \cdot \nabla\varphi \} dx dt - \int_{S_\varepsilon} \overset{\circ}{u} \varphi(x,0) dx \right\}_{\text{outer shell}} \\ &= \sigma\varepsilon_o \left\{ \iint_{S_T} \{ -\hat{u}\varphi_t + k_u \nabla_S \hat{u} \cdot \nabla_S \varphi \} d\eta dt - \int_S \overset{\circ}{u} \varphi(x,0) d\eta \right\}_{\text{outer shell}} . \end{aligned} \tag{4.15}$$

Therefore, in the limit, the concentration of capacity in S_ε leads to a differential equation posed on the surface S. Finally the trace identification

$$\hat{u}(\theta, z, t) = u(\overline{x}, z, t)\Big|_{|\overline{x}|=R} \quad \text{in } L^2(\partial\mathcal{D}_{R,T}) \quad \text{for all } z \in (0, H), \tag{4.16}$$

follows from the energy estimate (4.2) and the uniform Hölder estimates of the Kirzbraun-Pucci extensions $\overline{u}_\varepsilon$ via the triangle inequality.

4.3. The limit on the activated face $\mathcal{D}_R \times \{z_o\}$

Introduce the integral averages

$$\overset{\circ}{u}_\varepsilon (\overline{x}, t) = \frac{1}{\nu\varepsilon} \int_{\zeta_{j_o}^-}^{\zeta_{j_o}^+} u_\varepsilon(\overline{x}, \zeta, t) d\zeta , \qquad \overset{\circ}{v}_\varepsilon (\overline{x}, t) = \frac{1}{\nu\varepsilon} \int_{\zeta_{j_o}^-}^{\zeta_{j_o}^+} v_\varepsilon(\overline{x}, \zeta, t) d\zeta . \tag{4.17}$$

By virtue of the energy estimates, subsequences can be selected and relabelled with ε such that $\{\overset{\circ}{u}_\varepsilon\} \to \overset{\circ}{u}$ strongly in $L^2(\mathcal{D}_{R,T})$ and $\{\nabla_{\overline{x}} \overset{\circ}{u}_\varepsilon\} \to \nabla_{\overline{x}} \overset{\circ}{u}$ weakly in $L^2(\mathcal{D}_{R,T})$. The sequence $\overset{\circ}{v}_\varepsilon$ converges to $\overset{\circ}{v}$ in a similar fashion.

Select a testing function $\varphi(\overline{x}, t)$ not depending on z. The limit of the terms in the equation for u_ε, relevant to the special interdiscal space, is,

$$\nu\varepsilon_o \left\{ \begin{aligned} &\int_{\mathcal{D}_{R,T}} \left\{ - \overset{\circ}{u} \varphi_t + k_u \nabla_{\overline{x}} \overset{\circ}{u} \cdot \nabla_{\overline{x}} \varphi \right\} d\overline{x} dt - \int_{\mathcal{D}_R} \overset{\circ}{u} \varphi(\overline{x}, 0) d\overline{x} \\ &+ \int_{\mathcal{D}_{R,T}} \{ \gamma_o \overset{\circ}{u} - f(\overset{\circ}{v}) \} \varphi d\overline{x} dt + \frac{1}{\nu\varepsilon_o} \int_{\mathcal{D}_{R,T}} \overset{\circ}{u} f_1(\overset{\circ}{v}, \overline{x}, t) \varphi d\overline{x} dt. \end{aligned} \right\}_{\substack{\text{special} \\ \text{level } z_o}}$$

corresponding to a partial differential equation (in weak form) on the disc $\mathcal{D}_R$.

5. Weak formulation of the homogenized problem

The homogenized-concentrated limit can be formulated in the weak integral form (5.3), (5.4) below. Its main feature is that it combines the geometrical properties of the various compartments. This permits one to specialize it under various simplifying assumptions such as transverse or global well-stirred cytosol.

The limit functions $(u, \overset{\circ}{u}, \hat{u})$ and $(v, \overset{\circ}{v}, \hat{v})$ are in the following regularity classes

$$\begin{aligned} &u, v \in C\big(0,T; L^2(\Omega_o)\big);\ |\nabla_{\overline{x}} u|,\ |\nabla_{\overline{x}} v| \in L^2(\Omega_{o,T});\\ &\overset{\circ}{u}, \overset{\circ}{v} \in C\big(0,T; L^2(\mathcal{D}_R)\big);\ |\nabla_{\overline{x}} \overset{\circ}{u}|,\ |\nabla_{\overline{x}} \overset{\circ}{v}| \in L^2(\mathcal{D}_{R,T});\\ &\hat{u}, \hat{v} \in C\big(0,T; L^2(S)\big);\ |(\hat{u}_z, \hat{u}_\theta)|,\ |(\hat{v}_z, \hat{v}_\theta)| \in L^2(S_T), \end{aligned} \tag{5.1}$$

and are mutually related by

$$\begin{aligned} &\hat{u}(\theta, z, t) = u(\overline{x}, z, t)\big|_{|\overline{x}|=R} && \text{in } L^2\big((0,2\pi] \times (0,T]\big) \quad \text{for all } z \neq z_o;\\ &\hat{u}(\theta, z_o, t) = \overset{\circ}{u}(\overline{x}, t)\big|_{|\overline{x}|=R} && \text{in } L^2\big((0,2\pi] \times (0,T]\big);\\ &\hat{v}(\theta, z, t) = v(\overline{x}, z, t)\big|_{|\overline{x}|=R} && \text{in } L^2\big((0,2\pi] \times (0,T]\big) \ \text{for all } z \in (0,H). \end{aligned} \tag{5.2}$$

They satisfy the integral identities

$$\begin{aligned} &(1-\theta_o)\left\{ \iint_{\Omega_{o,T}} \{u_t\varphi + k_u \nabla_{\overline{x}} u \cdot \nabla_{\overline{x}}\varphi\}\, dxdt + \iint_{\Omega_{o,T}} (\gamma_o u - f_1(v))\varphi dxdt \right\}_{\text{interior}}\\ &+\sigma\varepsilon_o \left\{ \iint_{S_T} \{\hat{u}_t\varphi + k_u \nabla_S \hat{u} \cdot \nabla_S \varphi\}\, dSdt \right\}_{\text{outer shell}}\\ &+\nu\varepsilon_o \left\{ \begin{aligned} &\iint_{\mathcal{D}_{R,T}} \left\{ \overset{\circ}{u}_t \varphi + k_u \nabla_{\overline{x}} \overset{\circ}{u} \cdot \nabla_{\overline{x}} \varphi \right\} d\overline{x}dt\\ &\quad + \iint_{\mathcal{D}_{R,T}} \{\gamma_o \overset{\circ}{u} - f_1(\overset{\circ}{v})\}\varphi d\overline{x}dt\\ &\quad + \frac{1}{\nu\varepsilon_o} \iint_{\mathcal{D}_{R,T}} \overset{\circ}{u} f_2(\overset{\circ}{v}, \overline{x}, z_o, t)\varphi d\overline{x}dt \end{aligned} \right\}_{\substack{\text{activated}\\ \text{level } z_o}} = 0 \end{aligned} \tag{5.3}$$

for all testing functions $\varphi \in C^1(\overline{\Omega}_{o,T})$ vanishing for $t = T$;

$$\begin{aligned} &(1-\theta_o)\left\{ \iint_{\Omega_{o,T}} \{v_t\psi + k_v \nabla_{\overline{x}} v \cdot \nabla_{\overline{x}}\psi\}\, dxdt \right\}_{\text{interior}}\\ &+\sigma\varepsilon_o \left\{ \begin{aligned} &\iint_{S_T} \{\hat{v}_t\psi + k_v \nabla_S \hat{v} \cdot \nabla_S \psi\}\, d\eta dt\\ &\quad + \frac{1}{\sigma\varepsilon_o} \iint_{S_T} \{g_1(\hat{v}) - g_2(\hat{u})\}\psi d\eta dt \end{aligned} \right\}_{\text{outer shell}}\\ &+\nu\varepsilon_o \left\{ \iint_{\mathcal{D}_{R,T}} \left\{ \overset{\circ}{v}_t \psi + k_v \nabla_{\overline{x}} \overset{\circ}{v} \cdot \nabla_{\overline{x}} \psi \right\} d\overline{x}dt \right\}_{\substack{\text{activated}\\ \text{level } z_o}} = 0 \end{aligned} \tag{5.4}$$

for all testing functions $\psi \in C^1(\overline{\Omega}_{o,T})$ vanishing for $t = T$.

6. Cytosol well-stirred in the transversal variables (x, y)

Assume the cytosol is well stirred in the transversal variables (x, y). Thus the rod outer segment is ideally lumped on its axis and transversal diffusion effects are immaterial. Such an assumption is suggested by the idea that the system diffuses with infinite speed on each transversal cross section and thereby responds with an instantaneous transversal equilibration. The analysis below will permit one to compare our model to the existing ones based on the assumption of well-stirred.

If u and v are regarded as lumped on the axis of the rod, they depend only on z and t, and are independent of (x, y). Since there is no dependence on the (x, y) variables, by (5.2)

$$u(z,t) = \hat{u}(z,t) \quad \text{and} \quad \overset{\circ}{u}(t) = \hat{u}(z_o, t). \tag{6.1}$$

We insert this information into formula (5.3), and choose a test function φ independent of (x, y), thus obtaining the weak formulation relevant to the case of cytosol well-stirred in the transversal variables (x, y). The corresponding pointwise formal formulation is,

$$\begin{aligned}\{\sigma\varepsilon_o 2\pi R + (1-\theta_o)\pi R^2\}u_t - \sigma\varepsilon_o 2\pi R k_u u_{zz} \\ = -(1-\theta_o)\pi R^2 F - \delta_{z_o}\nu\varepsilon_o\pi R^2\left(\overset{\circ}{u}_t + F_o + F_*\right).\end{aligned} \tag{6.2}$$

Set,

$$f_A = \frac{\sigma\varepsilon_o 2\pi R}{\pi R^2}, \quad f_V = \frac{(1-\theta_o)\pi R^2 H + \sigma\varepsilon_o 2\pi R H}{\pi R^2 H}. \tag{6.3}$$

These two parameters have a geometric and physical significance. Specifically, up to higher-order corrections, f_A is the fraction of the cross-sectional area of the outer segment which is available for longitudinal diffusion, and f_V is the fraction of the total outer segment volume occupied by the cytosol [10, 14]. Then dividing by f_V the previous equation takes the more concise form,

$$u_t - \frac{f_A}{f_V}k_u u_{zz} = -\left(1 - \frac{f_A}{f_V}\right)F - \delta_{z_o}\frac{\nu\varepsilon_o}{f_V}\left(\overset{\circ}{u}_t + F_o + F_*\right). \tag{6.4}$$

This is the law of diffusion of cGMP under the assumption that the cytosol is well stirred in the transversal variables. A key feature is that it distinguishes between diffusion outside the activated level z_o and diffusion at z_o by the action of the Dirac delta function δ_{z_o}. If z is different than the activated level z_o, equation (6.4) implies,

$$u_t - \frac{f_A}{f_V}k_u u_{zz} = -\left(1 - \frac{f_A}{f_V}\right)F, \qquad (z \neq z_o). \tag{6.5}$$

Equation (6.5) is formally similar to a model proposed in [9]. In that work however the term F_* due to activation, is distributed along the longitudinal variable z.

7. Globally well-stirred cytosol

Regard now the rod as a homogeneous bag of cytosol, and [cGMP], [Ca^{2+}], [PDE], [PDE*], [GC] as lumped quantities depending only on time. Thus in particular $u = \mathring{u} = \hat{u}$ and similarly for v. We insert this information into formula (5.3), and choose a test function φ dependent only on t, thus obtaining the weak formulation relevant to the case of globally well-stirred cytosol. The corresponding pointwise formulation, under suitable regularity assumptions, is, to get,

$$\left(1 + \frac{\nu\varepsilon_o}{H\,f_V}\right)\frac{d}{dt}u = -\left(1 - \frac{f_A}{f_V} + \frac{\nu\varepsilon_o}{H\,f_V}\right)F - \frac{\nu\varepsilon_o}{H\,f_V}F_*. \tag{7.1}$$

Now f_V is of the order of one and ε_o is of three orders of magnitude smaller than H and R. Therefore,

$$\left(1 + \frac{\nu\varepsilon_o}{H\,f_V}\right) \approx 1; \qquad \left(1 - \frac{f_A}{f_V} + \frac{\nu\varepsilon_o}{H\,f_V}\right) \approx 1. \tag{7.2}$$

From the expression of f_V and the form (3) of F_*, we obtain the dynamic equation,

$$\frac{d}{dt}u = -F - \frac{1}{Hf_V}uf_1(v,x,t), \tag{7.3}$$

where F is defined in the first of (3). A similar analysis for Calcium gives,

$$\frac{d}{dt}v = \frac{2}{Rf_V}[-g_1(v) + g_2(u)]. \tag{7.4}$$

These formulae coincide with (A3), (A4) of [12], upon identifying the various parameters.

8. Further results and open issues

The homogenized limit lends itself to efficient numerical simulations. We have performed simulations on the full *non-homogenized* system (2.2) with $\varepsilon = \varepsilon_o$ in the original layered geometry of the ROS. Due to computational complexity the simulations where carried with a photon falling exactly at the center of a disc C_j and generating radial solutions. The same simulations were carried also for the corresponding homogenized system and it was found that the two electrical responses differ by no more than 0.1%. We refer to [4, 5] for such an analysis. An advantage of the homogenized model is that simulations can be carried just as easily for solutions that are not necessarily axially symmetric.

A peculiar feature of phototransduction is the *local spread of the response*; that is, if a photon activates a disc at level say z_o then the electrical response is "felt" only in the vicinity of z_o. In particular, two instantaneously activated sites at level z_1 and z_2 sufficiently far apart, and each away from the top and bottom of the ROS, exhibit essentially independent responses. This was theoretically conjectured by McNaughton et al. [10], and some experimental evidence was gathered by Gray-Keller et al. [9]. The numerical simulations on the homogenized model exhibit a

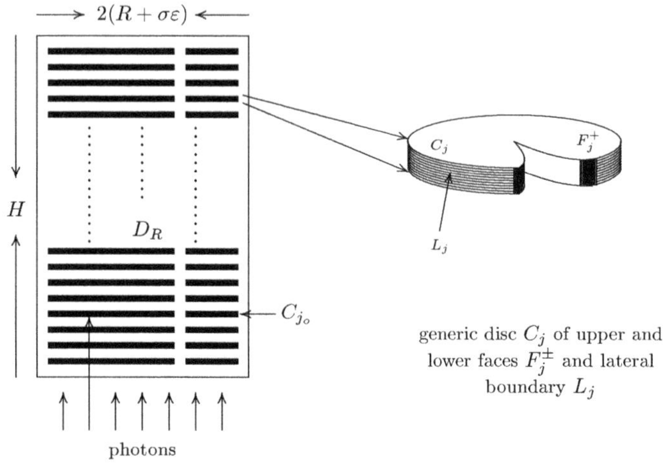

FIGURE 4. Geometry of the incisure of the ROS discs

region of about 4μm, along the z-axis, on both sides of the activating site where the relative current suppression is not less than 1% of the peak response. These numerical experiments appear in [4, 5] to which we refer for a further discussion.

The problem we have presented is all but one aspect of phototransduction. Other central problems include a modelling and mathematical analysis of the *recovery phase* and the *response from a light-adapted steady state.*

The ROS returns to its original state after about 8 sec, i.e., the system *recovers.* While the various biochemical steps of the recovery cascade are known, a suitable mathematical analysis is still lacking.

Light adaptation is the steady-state u_o, v_o after having exposed the ROS to light of a given intensity for a sufficiently long time. The system then *adapts* to that state and a detectable response, from that adapted state, would occur only for a beam of photons of sufficiently high intensity.

Even remaining in the context of a single photon response from a dark-adapted state, the biochemical cascade taking place on the activated disc $\mathcal{D}_R \times \{z_o\}$, and leading to depletion of cGMP in the cytosol, still lacks an organic model and a related mathematical analysis.

The discs C_j bear anatomical "incisures" as in Figure 4. In the Salamander these can be up to 16, whereas in higher vertebrates there is only one incisure. Their presence is believed to enhance the diffusion in the outer shell along the vertical axis of the ROS. Regardless of their number per disc, they are in series as in Figure 4. This suggests to carry the homogenized-concentrated limits for these ROS, where the incisures are permitted to shrink to a "segment". A suitable homogenized limiting system would permit to test the intensity of the response with and without the incisures and thus elucidate their function.

References

[1] D. Andreucci, *Existence and uniqueness of solutions to a concentrated capacity problem with change of phase.* European J. Appl. Math. **1** (1990), 339–351.

[2] D. Andreucci, P. Bisegna, and E. DiBenedetto, *Homogenization and concentrated capacity in reticular almost disconnected structures.* C.R. Acad. Sci. Paris, Série I **335** (2002), 329–332.

[3] D. Andreucci, P. Bisegna, and E. DiBenedetto, *Homogenization and concentrated capacity for the heat equation with non-linear variational data in reticular almost disconnected structures and applications to visual transduction.* Annali di Matematica **182** (2003), 375–407.

[4] D. Andreucci, P. Bisegna, G. Caruso, H.E. Hamm, and E. DiBenedetto, *Mathematical model of the spatio-temporal dynamics of second messengers in visual transduction.* Biophysical Journal **85** (2003), 1358–1376.

[5] G. Caruso, H. Khanal, V. Alexiades, and E. DiBenedetto, *Numerical simulations of single-photon responses,* to appear.

[6] Ph.G. Ciarlet, V. Lods, *Asymptotic analysis of linearly elastic shells. I. Justification of membrane shell equations.* Arch. Rational Mech. Anal. **136** (1996), 119–161.

[7] D. Cioranescu, and J. Saint Jean Paulin, *Homogenization of reticulated structures.* Appl. Math. Sciences **136**, Springer, New York, 1998.

[8] E. DiBenedetto, *Real Analysis.* Birkhäuser, Boston, 2002.

[9] M.P. Gray-Keller, and P.B. Detwiler., *Ca^{2+} dependence of dark- and light-adapted flash responses in rod photoreceptors.* Neuron **17** (1996), 323–331.

[10] T.D. Lamb, P.A. McNaughton, and K.-W. Yau, *Spatial spread of activation and background desensitization in toad rod outer segment.* J. Physiol. **319** (1981), 463–496.

[11] T.D. Lamb, and E.N. Pugh Jr., *A quantitative account of the activation steps involved in phototransduction in amphibian photoreceptors.* J. Physiol. **449** (1992), 719–758.

[12] S. Nikonov, T.D. Lamb, and E.N. Pugh Jr., *The role of steady phosphodiesterase activity in the kinetics and sensitivity of the light-adapted salamander rod photoresponse.* J. Gen. Physiol. **116** (2000), 795–824.

[13] O.A. Oleinik, T.A. Shaposhnikova, *On Homogenization problems for the Laplace operator in partially perforated domains with Neumann's condition on the boundary of cavities.* Rend. Mat. Acc. Lincei (s. 9) **6** (1995), 133–142.

[14] A. Olson, and E.N. Pugh Jr., *Diffusion coefficient of cyclic GMP in salamander rod outer segments estimated with two fluorescent probes.* Biophys. J. **65** (1993), 1335–1352.

Daniele Andreucci
Dipartimento di Metodi e Modelli
Università di Roma La Sapienza
via A.Scarpa 16
I-00161 Roma, Italy
e-mail: andreucci@dmmm.uniroma1.it

Paolo Bisegna
Dipartimento di Ingegneria Civile
Università di Roma Tor Vergata
via del Politecnico 1
I-00133 Roma, Italy
e-mail: bisegna@uniroma2.it

Emmanuele DiBenedetto
Department of Mathematics
Vanderbilt University
Nashville, TN 37240, USA
e-mail: em.diben@Vanderbilt.edu

Progress in Nonlinear Differential Equations
and Their Applications, Vol. 61, 81–89

Global Regularity in Sobolev Spaces for Elliptic Problems with p-structure on Bounded Domains

Carsten Ebmeyer

1. Introduction

Let $\Omega \subset \mathbb{R}^n$, $n \geq 2$, be a bounded domain with a smooth boundary. We consider the Neumann problem $(\mathcal{P}_N)$

$$\begin{aligned} -\textstyle\sum_{i=1}^d \partial_i a_i(x, \nabla u) &= f \qquad \text{in } \Omega, \\ \textstyle\sum_{i=1}^d a_i(x, \nabla u)\, \eta_i &= g_N \quad \text{on } \partial\Omega, \end{aligned}$$

where $\partial_i = \frac{\partial}{\partial x_i}$, $\eta = (\eta_1, \ldots, \eta_n)^T$ is the outward unit normal of $\partial\Omega$, and the functions $a_i : \Omega \times \mathbb{R}^n \to \mathbb{R}$ have p-structure (e.g., $a_i = |\nabla u|^{p-2}\partial_i u$ and $2 < p < \infty$).

Problems with p-structure arise in several physical, engineering and mathematical contexts, such as in elasticity and plasticity theory, or in non-Newtonian fluids; see, e.g., [1, 4, 7]. In this paper we are concerned with the case that $2 < p < \infty$. Our aim is to prove regularity in fractional order Nikolskij spaces for the operator $-\partial_i a_i(x, \nabla u)$. We show that each weak solution u of $(\mathcal{P}_N)$ satisfies

$$u \in \mathcal{N}^{1+\frac{2}{p},p}(\Omega)\,. \tag{1}$$

Due to the embedding theorem of Nikolskij spaces into Sobolev spaces it follows that $u \in W^{s,p}(\Omega)$ for all $s < 1 + \frac{2}{p}$. This result may be seen as a generalization of the $W^{2,2}$-theory of the Laplace operator to problems with p-structure. Recently, it was proved for the p-Laplacian under homogeneous Dirichlet boundary values on convex polygonal domains; see [3].

Moreover, we prove a weighted estimate for the second derivatives of the form

$$\int_\Omega |\nabla u|^{p-2} |\nabla^2 u|^2 < \infty\,. \tag{2}$$

This result is known for the p-Laplace equation $-\Delta_p u := -\mathrm{div}(|\nabla u|^{p-2}\nabla u) = f$, primarily on interior domains $\Omega_0 \subset\subset \Omega$ under some strong assumptions on the data; see [2, 5, 8, 9, 10].

Let us emphasize that we do not make any assumptions on the structure of the coefficients a_i, we only assume growth conditions and an ellipticity condition. In particular, we are able to treat variational as well as non-variational problems. Our method of proof is a difference quotient technique. It requires only minimal regularity on the data and can be applied to a wide range of problems and data, e.g., systems with p-structure for all $1 < p < \infty$, polyhedral domains, Lipschitzian domains with a piecewise smooth boundary, and Dirichlet boundary value conditions.

We conclude this introduction by comparing the results (1) and (2). The function $u = |x|^\alpha$ solves in the weak sense $-\Delta_p u = f$ with $f = c|x|^{\alpha(p-1)-p}$. Iff $\alpha \geq \frac{p+2-n}{p}$ it holds that $u \in \mathcal{N}^{1+\frac{2}{p},p}(\Omega)$. Further, iff $\alpha > \frac{p+2-n}{p}$ we have $u \in W^{1+\frac{2}{p},p}(\Omega)$ and $|\nabla u|^{\frac{p-2}{2}}|\nabla^2 u| \in L^2(\Omega)$.

Moreover, let us note that $1 + \frac{2}{p} < 2$ for $2 < p < \infty$. In general we cannot expect $u \in W^{2,p}(\Omega)$ in the case of $p > 2$.

2. The main results

Let $a_i(x, s)$ be Caratheodory-functions, where $x = (x_1, \ldots, x_n) \in \overline{\Omega}$, $n \geq 2$, and $s \in \mathbb{R}^n$. We set $a_{i,x_k}(x, s) = \frac{\partial}{\partial x_k} a_i(x, s)$ and $a_{i,k}(x, s) = \frac{\partial}{\partial s_k} a_i(x, s)$. The following assumptions on the data are made.

(H1) $\Omega \subset \mathbb{R}^n$, $n \geq 2$, is a bounded domain with a $C^{1,1}$-boundary.

(H2) $f(x) \in W^{1-\frac{2}{p},p'}(\Omega)$ and $g_N \in \mathcal{N}^{\frac{1}{p'},p'}(\partial\Omega)$.

(H3) (*Compatibility condition*) $\int_\Omega f + \int_{\partial\Omega} g_N = 0$.

There are constants $c_1, c_2, c_3 > 0$ and $\kappa \geq 0$ independent of x and s such that

(H4) $|a_i(x, s)| + |a_{i,x_k}(x, s)| \leq g(x) + c_1|s|^{p-1}$ for some $g \in L^{p'}(\Omega)$,

(H5) $|a_{i,k}(x, s)| \leq c_2(\kappa^2 + |s|^2)^{\frac{p-2}{2}}$,

(H6) (*Ellipticity condition*) $c_3(\kappa^2 + |s|^2)^{\frac{p-2}{2}} \leq \sum_{i,k=1}^2 a_{i,k}(x, s)\,\xi_i\,\xi_k$ for all $\xi \in \mathbb{R}^n$.

For simplicity we shall restrict ourselves to the case that $\kappa = 0$.

We use Einstein's summation convention saying that one has to sum over an index that occurs twice. We call a function $u \in W^{1,p}(\Omega)$ a weak solution of $(\mathcal{P}_N)$ if

$$\int_\Omega a_i(x, \nabla u)\,\partial_i\varphi = \int_\Omega f\,\varphi + \int_{\partial\Omega} g_N\,\varphi \tag{3}$$

for all $\varphi \in W^{1,p}(\Omega)$. It is well known that there exists a weak solution that is unique up to a constant. We now state our main results.

Theorem 1. Let $2 < p < \infty$ and u be the weak solution of $(\mathcal{P}_\mathcal{N})$. Then

$$u \in \mathcal{N}^{1+\frac{2}{p},p}(\Omega)\,. \tag{4}$$

Further, it holds that

$$|\nabla u|^{\frac{p-2}{2}} |\nabla^2 u| \in L^2(\Omega) \qquad \text{and} \qquad |\nabla u|^{\frac{p-2}{2}} \nabla u \in W^{1,2}(\Omega;\mathbb{R}^n)\,. \tag{5}$$

We conclude this section by giving the definition of the Nikolskij spaces $\mathcal{N}^{s,p}(\Omega)$ $(1 \le p < \infty,\ s > 0$ no integer); cf. [6]. The space $\mathcal{N}^{s,p}(\Omega)$ consists of all functions for which the norm $\|f\|_{\mathcal{N}^{s,p}(\Omega)} = (\|f\|^p_{L^p(\Omega)} + |f|^p_{\mathcal{N}^{s,p}(\Omega)})^{\frac{1}{p}}$ is finite, where

$$|f|^p_{\mathcal{N}^{s,p}(\Omega)} = \sum_{|\alpha|=m} \sup_{\substack{\delta>0\\ 0<|z|<\delta}} \int_{\Omega^{kh}} \frac{|\partial^\alpha \Delta^k_z f(x)|^p}{|z|^{\sigma p}}\, dx,$$

$s = m + \sigma$, $m \in \mathbb{N}_0$, $k \ge 1$ is an integer, $\sigma > 0$ is no integer, $k > \sigma$, $z \in \mathbb{R}^n$, $\Delta_z f(x) = f(x+z) - f(x)$, $\Delta^k_z = \Delta^{k-1}_z \Delta_z$, and $\Omega^h = \{x \in \Omega : \operatorname{dist}(x,\partial\Omega) \ge h\}$. In particular, putting $m = 1$, $\sigma = \frac{2}{p}$, and $k = 1$ yields a norm of $\mathcal{N}^{1+\frac{2}{p},p}(\Omega)$, that will be used in the sequel. Further, let us note that for all $\varepsilon > 0$ there holds the embedding [6]

$$\mathcal{N}^{s+\varepsilon,p}(\Omega) \hookrightarrow W^{s,p}(\Omega) \hookrightarrow \mathcal{N}^{s,p}(\Omega)\,.$$

3. Proof of the theorem

Let $R > 0$, $B_R = \{x \in \mathbb{R}^n : |x| < R\}$, $\Omega_R = \Omega \cap B_R$, and $\Gamma_R = \partial\Omega \cap B_R$. To begin with we suppose that $\Omega_{4R} = \{x \in B_{4R} : x_n > 0\}$.

Let $0 < h < \frac{R}{2}$, $e_j \in \mathbb{R}^n$ be the jth unit vector, $T^{\pm h}_j v(x) = v(x \pm he_j)$,

$$D^h_j v(x) = \frac{T^h_j v(x) - v(x)}{h}, \quad \text{and} \quad D^{-h}_j v(x) = \frac{v(x) - T^{-h}_j v(x)}{h}\,.$$

Further, let the function $\tau \in W^{2,\infty}(\mathbb{R}^n)$ be a cut-off function satisfying $\tau \equiv 1$ in B_R, $\operatorname{supp}\tau = \overline{B_{2R}}$, and $\partial_n \tau = 0$ in $\{x \in \mathbb{R}^n : |x_n| < \frac{R}{2}\}$.

Notice that e_n is the inner unit normal of $\partial\Omega \cap B_{3R}$. Let $z \in \partial\Omega \cap B_{3R}$, $\lambda > 0$, and $z + \lambda e_n \in \Omega_{3R}$. Thus, $z - \lambda e_n \in B_{3R} \setminus \overline{\Omega}$. We define even extensions of the functions $\partial_i u$ $(1 \le i \le n)$, $a_i(\cdot, s)$, and g onto $B_{3R} \setminus \overline{\Omega}$ by setting

$$\partial_i u(z - \lambda e_n) := \partial_i u(z + \lambda e_n), \tag{6}$$

$a_i(z - \lambda e_n, s) := a_i(z + \lambda e_n, s)$, etc.

We now prove weighted estimates of difference quotients of ∇u in the case that f and g_N are sufficiently smooth. First, we consider directions e_j parallel to the boundary.

Lemma 2. *There is a constant* c *such that for* $1 \le j \le n-1$

$$\sup_{0<h<\frac{R}{2}} \int_{\Omega_{3R}} \tau^2 \left(|T_j^{\pm h}\nabla u|^2 + |\nabla u|^2\right)^{\frac{p-2}{2}} \left|D_j^{\pm h}\nabla u\right|^2 \le c\sigma_R\,,$$

where $\sigma_R = \|\nabla u\|^p_{L^p(\Omega_{3R})} + \|g\|^{p'}_{L^{p'}(\Omega_{3R})} + \|f\|^{p'}_{W^{1,p'}(\Omega_{3R})} + \|g_N\|^{p'}_{\mathcal{N}^{1+\frac{1}{p},p'}(\Gamma_{3R})}$.

Proof. Let $j \in \{1,\dots,n-1\}$ be fixed. We take the test function $\varphi = -\tau^2 D_j^h D_j^{-h} u$ in (3). This yields

$$\begin{aligned} J_0 &:= -\int_{\Omega_{3R}} \tau^2\, a_i(x,\nabla u)\, D_j^h D_j^{-h} \partial_i u \\ &= +\int_{\Omega_{3R}} a_i(x,\nabla u)\, \partial_i \tau^2\, D_j^h D_j^{-h} u - \int_{\Omega_{3R}} \tau^2 f\, D_j^h D_j^{-h} u \\ &\quad - \int_{\Gamma_{3R}} \tau^2 g_N\, D_j^h D_j^{-h} u =: J_1 + J_2 + J_3. \end{aligned}$$

Let us consider J_0. There holds the identity

$$\begin{aligned} 2J_0 &= \int_{\Omega_{3R}} D_j^h(\tau^2 a_i(x,\nabla u))\, D_j^h \partial_i u + \int_{\Omega_{3R}} D_j^{-h}(\tau^2 a_i(x,\nabla u))\, D_j^{-h}\partial_i u \\ &\quad - \int_{\Omega_{3R}} D_j^h(\tau^2 a_i(x,\nabla u)\, D_j^{-h}\partial_i u) - \int_{\Omega_{3R}} D_j^{-h}(\tau^2 a_i(x,\nabla u)\, D_j^h \partial_i u) \\ &=: J_{01} + \cdots + J_{04}. \end{aligned}$$

In view of the fact that $\operatorname{supp}\tau = \overline{B_{2R}}$ we have $J_{03} = J_{04} = 0$. Further, the Leibniz rule $D_j^h(fg) = f\, D_j^h g + D_j^h f\, T_j^h g$ yields

$$J_{01} = \int_{\Omega_{3R}} \tau^2\, D_j^h a_i(x,\nabla u)\, D_j^h \partial_i u + \int_{\Omega_{3R}} D_j^h \tau^2\, T_j^h a_i(x,\nabla u)\, D_j^h \partial_i u.$$

Due to the Taylor expansion of $a_i(x,\cdot)$ and the ellipticity condition we deduce

$$\begin{aligned} &[a_i(x,T_j^h\nabla u) - a_i(x,\nabla u)]\,[T_j^h\partial_i u - \partial_i u] \\ &= [T_j^h\partial_i u - \partial_i u]\,[T_j^h \partial_k u - \partial_k u] \int_0^1 a_{i,k}(x, tT_j^h\nabla u + (1-t)\nabla u)\, dt \\ &\ge c|T_j^h\nabla u - \nabla u|^2 \int_0^1 |tT_j^h\nabla u + (1-t)\nabla u|^{p-2} dt \\ &\ge c\,(|T_j^h\nabla u|^2 + |\nabla u|^2)^{\frac{p-2}{2}}\, |T_j^h \nabla u - \nabla u|^2. \end{aligned}$$

Furthermore, recalling that $\operatorname{supp}\tau = \overline{B_{2R}}$ we get

$$\begin{aligned}
&\int_{\Omega_{3R}} D_j^h\tau^2\, T_j^h a_i(x,\nabla u)\, D_j^h\partial_i u \\
&\quad = \int_{\Omega_{3R}} D_j^h(D_j^{-h}\tau^2\, a_i(x,\nabla u)\,\partial_i u) - \int_{\Omega_{3R}} D_j^h(D_j^{-h}\tau^2\, a_i(x,\nabla u))\,\partial_i u \\
&\quad = -\int_{\Omega_{3R}} D_j^h D_j^{-h}\tau^2\, T_j^h a_i(x,\nabla u)\,\partial_i u - \int_{\Omega_{3R}} D_j^{-h}\tau^2\, D_j^h a_i(x,\nabla u)\,\partial_i u.
\end{aligned}$$

We estimate these integrals by utilizing the hypotheses (H4), (H5), and the fact that $D_j^{-h}\tau^2 = 2T_j^{-h}\tau\, D_j^{-h}\tau + h(D_j^{-h}\tau)^2$. Collecting the results we arrive at

$$\begin{aligned}
J_0 \;\geq\; c\Bigg(&\int_{\Omega_{3R}} \tau^2\, \left(|T_j^h\nabla u|^2 + |\nabla u|^2\right)^{\frac{p-2}{2}}\, \left|D_j^h\nabla u\right|^2 \\
&+\int_{\Omega_{3R}} \tau^2\, \left(|T_j^{-h}\nabla u|^2 + |\nabla u|^2\right)^{\frac{p-2}{2}}\, \left|D_j^{-h}\nabla u\right|^2 \\
&-\|\nabla u\|^p_{L^p(\Omega_{3R})} - \|g\|^{p'}_{L^{p'}(\Omega_{3R})}\Bigg).
\end{aligned}$$

Next, let us note that

$$J_1 = \int_{\Omega_{3R}} D_j^h(a_i(x,\nabla u)\,\partial_i\tau^2)\, D_j^h u + \int_{\Omega_{3R}} D_j^h(a_i(x,\nabla u)\,\partial_i\tau^2\, D_j^h u).$$

The second integral on the right-hand side vanishes. For $\delta > 0$ we deduce

$$\begin{aligned}
|J_1| \;\leq\; \delta &\int_{\Omega_{3R}} \tau^2\, \left(|T_j^h\nabla u|^2 + |\nabla u|^2\right)^{\frac{p-2}{2}}\, \left|D_j^h\nabla u\right|^2 \\
&+c_\delta\left(\|g\|^{p'}_{L^{p'}(\Omega_{3R})} + \|\nabla u\|^p_{L^p(\Omega_{3R})}\right).
\end{aligned}$$

Further, in order to treat $|J_2|$ and $|J_3|$ we integrate by parts (in the discrete sense). We see that we have to estimate the integrals

$$\int_{\Omega_{3R}} \tau^2\, D_j^h f\, D_j^h u \quad\text{and}\quad \int_{\Gamma_{3R}} \tau^2\, D_j^h g_N\, D_j^h u.$$

We find

$$\left\|\tau\, D_j^h f\,\tau\, D_j^h u\right\|_{L^1(\Gamma_{3R})} \leq c\left(\left\|\tau\, D_j^h f\right\|^{p'}_{L^{p'}(\Omega_{3R})} + \left\|\tau\, D_j^h u\right\|^p_{L^p(\Omega_{3R})}\right)$$

and

$$\begin{aligned}
\left\|\tau D_j^h g_N\, \tau D_j^h u\right\|_{L^1(\Gamma_{3R})} \;&\leq\; \left\|\tau\, D_j^h g_N\right\|_{W^{\frac{1}{p},p'}(\Gamma_{3R})} \left\|\tau D_j^h u\right\|_{W^{-\frac{1}{p},p}(\Gamma_{3R})} \\
&\leq\; c\left(\|g_N\|^{p'}_{\mathcal{N}^{1+\frac{1}{p},p'}(\Gamma_{3R})} + \left\|\tau\, D_j^h u\right\|^p_{L^p(\Omega_{3R})}\right).
\end{aligned}$$

Thus, the assertion follows. □

Lemma 3. *There is a constant c such that*

$$\sup_{0<h<\frac{R}{2}} \int_{\Omega_{3R}} \tau^2 \left(|T_n^{\pm h}\nabla u|^2 + |\nabla u|^2\right)^{\frac{p-2}{2}} \left|D_n^{\pm h}\nabla u\right|^2 \le c\sigma_R' ,$$

where $\sigma_R' = \|\nabla u\|^p_{L^p(\Omega_{3R})} + \|g\|^{p'}_{L^{p'}(\Omega_{3R})} + \|f\|^{p'}_{W^{1,p'}(\Omega_{3R})}$.

Proof. We take the test function $\varphi = -\tau^2\psi$ in equation (3), where $\psi(x) = h^{-1}D_n^{-h}\int_0^1 \partial_t u(x+the_n)\,dt$. Let us note that this test function corresponds to that in the previous proof, for $\varphi(x) = -\tau^2(x)\,D_n^{-h}D_n^h u(x)$ if $x_n > h$. More precisely, we have $\psi(x) = h^{-1}D_n^{-h}\int_0^1 \partial_t v(x+the_n)\,dt = D_n^{-h}D_n^h v(x)$, where v is the extension of u defined by reflection at the hyperplane $\{x \in \mathbb{R}^n : x_n = 0\}$. We obtain

$$\begin{aligned} J_0 &:= -\int_{\Omega_{3R}} \tau^2(x)\,a_i(x,\nabla u)\,h^{-1}D_n^{-h}\int_0^1 \partial_i\partial_t u(x+the_n)\,dt\,dx \\ &= \int_{\Omega_{3R}} a_i(x,\nabla u)\,\partial_i\tau^2\,\psi - \int_{\Omega_{3R}} \tau^2\,f\,\psi - \int_{\Gamma_{3R}} \tau^2\,g_N\,\psi \\ &=: J_1 + \cdots + J_3. \end{aligned}$$

Due to the extensions (6) we have $J_0 = -\int_{\Omega_{3R}} \tau^2\,a_i(x,\nabla u)\,D_n^h D_n^{-h}\partial_i u$. Thus, proceeding as above and noting that $\psi = 0$ on Γ_{3R}, the assertion follows. Here we have used the facts that $J_{03} + J_{04} = 0$, for

$$\begin{aligned} J_{03} &= h^{-2}\int_{\Omega_{3R}\setminus\Omega_{3R}+he_n} \tau^2\,a_i(x,\nabla u)\,(\partial_i u - T_n^{-h}\partial_i u) \\ &= -h^{-2}\int_{\Omega_{3R}-he_n\setminus\Omega_{3R}} \tau^2\,a_i(x,\nabla u)\,(T_n^h\partial_i u - \partial_i u) = -J_{04}\,, \end{aligned}$$

$\int_{\Omega_{3R}} D_n^h(D_n^{-h}\tau^2\,a_i(x,\nabla u)\,\partial_i u) + \int_{\Omega_{3R}} D_n^{-h}(D_n^h\tau^2\,a_i(x,\nabla u)\,\partial_i u) = 0$, for

$$-\int_{\Omega_{3R}\setminus\Omega_{3R}+he_n} D_n^{-h}\tau^2\,a_i(x,\nabla u)\,\partial_i u = \int_{\Omega_{3R}-he_n\setminus\Omega_{3R}} D_n^h\tau^2\,a_i(x,\nabla u)\,\partial_i u\,,$$

$\int_{\Omega_{3R}} D_n^h(a_i(x,\nabla u)\,\partial_i\tau^2\,D_n^h v) + \int_{\Omega_{3R}} D_n^{-h}(a_i(x,\nabla u)\,\partial_i\tau^2\,D_n^{-h}v) = 0$, for $\partial_n\tau = 0$ in $\Omega_{3R} - he_n \setminus \Omega_{3R} + he_n$, and $\int_{\Omega_{3R}} D_n^h(\tau^2\,f\,D_n^{-h}v) + \int_{\Omega_{3R}} D_n^{-h}(\tau^2\,f\,D_n^h v) = 0$. □

Lemma 4. *There is a constant c such that*

$$\int_{\Omega_R} \left|\nabla(|\nabla u|^{\frac{p-2}{2}}\nabla u)\right|^2 + \int_{\Omega_R} |\nabla u|^{p-2}|\nabla^2 u|^2 \le c\sigma_R\,,$$

where $\sigma_R = \|\nabla u\|^p_{L^p(\Omega_{3R})} + \|g\|^{p'}_{L^{p'}(\Omega_{3R})} + \|f\|^{p'}_{W^{1,p'}(\Omega_{3R})} + \|g_N\|^{p'}_{\mathcal{N}^{1+\frac{1}{p},p'}(\Gamma_{3R})}$.

Proof. Due to the previous lemmas there is a constant c such that

$$\int_{\Omega_R} \left(|T_j^h\nabla u|^2 + |\nabla u|^2\right)^{\frac{p-2}{2}} \left|D_j^h\nabla u\right|^2 \le c\sigma_R$$

for $1 \le j \le n$ and $0 < h < \frac{R}{2}$. Let $F(r) = |r|^{\frac{p-2}{2}} r$. There are constants $c_1, c_2 > 0$ such that

$$c_1 |F(r) - F(s)|^2 \le (|r|^2 + |s|^2)^{\frac{p-2}{2}} |r-s|^2 \le c_2 |F(r) - F(s)|^2 ;$$

see [3]. Putting $r = T_j^h \nabla u$ and $s = \nabla u$ we get $\left\| D_j^h F(\nabla u) \right\|^2_{L^2(\Omega_R)} \le c\sigma_R$ for $1 \le j \le n$ and $0 < h < \frac{R}{2}$. Utilizing a standard argument (cf. [5]) this implies that $\|\nabla F(\nabla u)\|^2_{L^2(\Omega_R)} \le c\sigma_R$. Noting that $|\nabla F(\nabla u)|^2 \ge \frac{p^2}{4} |\nabla u|^{p-2} |\nabla^2 u|^2$ the assertion follows. □

Theorem 5. Let f and g_N be sufficiently smooth, i.e., $f \in W^{1,p'}(\Omega)$ and $g_N \in \mathcal{N}^{1+\frac{1}{p},p'}(\partial\Omega)$. Then there is a constant c depending only on the data such that

$$\int_\Omega \left| \nabla (|\nabla u|^{\frac{p-2}{2}} \nabla u) \right|^2 + \int_\Omega |\nabla u|^{p-2} |\nabla^2 u|^2 + \|u\|^p_{\mathcal{N}^{1+\frac{2}{p},p}(\Omega)} \le c.$$

Proof. We cover Ω by a finite number of balls $B^i \subset\subset \Omega$ and a finite number of sets Ω^j, where each Ω^j can be mapped in a smooth way onto $\{x \in B_{R_j} : x_n > 0\}$ for some $R_j > 0$. Due to the smoothness of $\partial\Omega$ such a mapping is a $C^{1,1}$- resp. $W^{2,\infty}$-mapping. Thus, the structure of the coefficients does not change. Utilizing Lemma 4 we get

$$\int_\Omega \left| \nabla (|\nabla u|^{\frac{p-2}{2}} \nabla u) \right|^2 + \int_\Omega |\nabla u|^{p-2} |\nabla^2 u|^2 \le c.$$

Moreover, due to the estimate [2]

$$\exists c > 0 : |r-s|^p \le c(|r|^2 + |s|^2)^{\frac{p-2}{2}} |r-s|^2$$

it follows for all $1 \le j \le n$ and $h > 0$ that

$$\int_{\Omega^h} h^{-2} |T_j^h \nabla u - \nabla u|^p \le c \int_{\Omega^h} \left(|T_j^h \nabla u|^2 + |\nabla u|^2 \right)^{\frac{p-2}{2}} \left| D_j^h \nabla u \right|^2 , \qquad (7)$$

and hence $|u|^p_{\mathcal{N}^{1+\frac{2}{p},p}(\Omega)} \le c$. Altogether, we obtain the assertion. □

Proof of Theorem 1. For sufficiently smooth data f and g_N Theorem 5 yields the assertion. Otherwise, we approximate f and g_N by sequences $(f^k)_k \in W^{1,p'}(\Omega)$ and $(g_N^k)_k \in \mathcal{N}^{1+\frac{1}{p},p'}(\partial\Omega)$. For each $k \in \mathbb{N}$ there is a function u^k solving in the weak sense

$$-\partial_i a_i(x, \nabla u^k) = f^k \quad \text{in } \Omega, \qquad a_i(x, \nabla u^k)\, \eta_i = g_N^k \quad \text{on } \partial\Omega.$$

Below we will show there is a constant c independent of k such that

$$\int_\Omega \left| \nabla (|\nabla u^k|^{\frac{p-2}{2}} \nabla u^k) \right|^2 + \int_\Omega |\nabla u^k|^{p-2} |\nabla^2 u^k|^2 \le c. \qquad (8)$$

In view of (7) this implies that the sequence $(u^k)_k$ is uniformly bounded in $\mathcal{N}^{1+\frac{2}{p},p}(\Omega)$. Thus, there is a subsequence, again denoted by $(u^k)_k$, that converges to a function $z \in W^{1,p}(\Omega)$. From the estimates

$$\begin{aligned} & \left\|a_i(x,\nabla u^k) - a_i(x,\nabla z)\right\|_{L^{p'}(\Omega)} \\ \leq\ & c \left\|(|\nabla u^k|^2 + |\nabla z|^2)^{\frac{p-2}{2}} |\nabla u^k - \nabla z|\right\|_{L^{p'}(\Omega)} \\ \leq\ & c\left(\|\nabla u^k\|^{p-2}_{L^p(\Omega)} + \|\nabla z\|^{p-2}_{L^p(\Omega)}\right) \|\nabla u^k - \nabla z\|_{L^p(\Omega)} \end{aligned}$$

it follows that z satisfies (in the weak sense) $-\partial_i a_i(x,\nabla z) = f$ in Ω and $a_i(x,\nabla z)\,\eta_i = g_N$ on $\partial\Omega$. Thus, $z = u$.

Moreover, noting that $\Delta_j^h \Delta_j^{-h} \nabla u^k$ converges to $\Delta_j^h \Delta_j^{-h} \nabla u$ in $L^p(\Omega)$, for any $1 \leq j \leq n$ and $h > 0$ there is a number $k \in \mathbb{N}$ such that

$$\left\|h^{-\frac{2}{p}} \Delta_j^h \Delta_j^{-h} \nabla u\right\|_{L^p(\Omega^h)} \leq 2 \left\|h^{-\frac{2}{p}} \Delta_j^h \Delta_j^{-h} \nabla u^k\right\|_{L^p(\Omega^h)},$$

where $\Omega^h = \{x \in \Omega : \mathrm{dist}(x,\partial\Omega) \geq h\}$. In view of (7) and (8) the sequence $(u^k)_k$ is uniformly bounded in $\mathcal{N}^{1+\frac{2}{p},p}(\Omega)$. Thus, we conclude that $u \in \mathcal{N}^{1+\frac{2}{p},p}(\Omega)$.

Similarly, for any $1 \leq j \leq n$ and $h > 0$ there is a number $k \in \mathbb{N}$ such that

$$\left\|D_j^{\pm h}(|\nabla u|^{\frac{p-2}{2}} \nabla u)\right\|_{L^2(\Omega^h)} \leq 2 \left\|D_j^{\pm h}(|\nabla u^k|^{\frac{p-2}{2}} \nabla u^k)\right\|_{L^2(\Omega^h)}.$$

Due to (8) the right-hand side is uniformly bounded. This implies that

$$\nabla(|\nabla u|^{\frac{p-2}{2}} \nabla u) \in L^2(\Omega).$$

Noting that $\int_\Omega |\nabla u|^{p-2} |\nabla^2 u|^2 \leq c\|\nabla(|\nabla u|^{\frac{p-2}{2}} \nabla u)\|^2_{L^2(\Omega)}$ the assertion (5) follows. Furthermore, in view of (7) we obtain the assertion (4).

It remains to show (8). Therefore, we modify the proofs of the Lemmas 3 and 4: For $\delta > 0$ and $1 \leq j < n$ we estimate

$$\begin{aligned} \left\|\tau D_j^h g_N\, \tau D_j^h u\right\|_{L^1(\Gamma_{3R})} &\leq \left\|\tau\, D_j^h g_N\right\|_{W^{-\frac{1}{p},p'}(\Gamma_{3R})} \left\|\tau D_j^h u\right\|_{W^{\frac{1}{p},p}(\Gamma_{3R})} \\ &\leq c_\delta \left\|g_N\right\|^{p'}_{\mathcal{N}^{\frac{1}{p'},p'}(\Gamma_{3R})} + \delta \left\|u\right\|^p_{\mathcal{N}^{1+\frac{2}{p},p}(\Omega_{3R})}, \end{aligned}$$

for $\|\tau D_j^h u\|_{W^{\frac{1}{p},p}(\Gamma_{3R})} \leq c\|\tau\, D_j^h u\|_{W^{\frac{2}{p},p}(\Omega_{3R})}$,

$$\begin{aligned} \left\|\tau\, f\, \tau\, D_j^h D_j^{-h} u\right\|_{L^1(\Omega_{3R})} &= \frac{1}{2}\left\|\tau\, f\, \tau\, D_j^h D_j^{-h} u\right\|_{L^1(B_{3R})} \\ &\leq \frac{1}{2}\left\|\tau\, f\right\|_{W^{1-\frac{2}{p},p'}(B_{3R})} \left\|\tau\, D_j^h D_j^{-h} u\right\|_{W^{\frac{2}{p}-1,p}(B_{3R})} \\ &\leq c_\delta \left\|f\right\|^{p'}_{W^{1-\frac{2}{p},p'}(\Omega_{3R})} + \delta \left\|u\right\|^p_{\mathcal{N}^{1+\frac{2}{p},p}(\Omega_{3R})}, \end{aligned}$$

and

$$\left\|\tau\, f\, \tau\, \psi\right\|_{L^1(\Omega_{3R})} \leq c_\delta \left\|f\right\|^{p'}_{W^{1-\frac{2}{p},p'}(\Omega_{3R})} + \delta \left\|u\right\|^p_{\mathcal{N}^{1+\frac{2}{p},p}(\Omega_{3R})}.$$

Here we have used the fact that $\psi(x) = h^{-1}D_n^{-h}\int_0^1 \partial_t v(x+the_n)\,dt = D_n^{-h}D_n^h v(x)$, where v is the extension of u defined by reflection at the hyperplane $\{x \in \mathbb{R}^n : x_n = 0\}$, and the estimate $\|v\|^p_{\mathcal{N}^{1+\frac{2}{p},p}(B_{3R})} \le c\,\|u\|^p_{\mathcal{N}^{1+\frac{2}{p},p}(\Omega_{3R})}$. Altogether, we deduce

$$\int_{\Omega_{3R}} \tau^2 \left(|T_j^{\pm h}\nabla u|^2 + |\nabla u|^2\right)^{\frac{p-2}{2}} \left|D_j^{\pm h}\nabla u\right|^2 \le c_\delta \sigma_R'' + \delta\,\|u\|^p_{\mathcal{N}^{1+\frac{2}{p},p}(\Omega_{3R})}$$

for all $1 \le j \le n$ and $0 < h < \frac{R}{2}$, where $\sigma_R'' = \left\|\nabla u^k\right\|^p_{L^p(\Omega_{3R})} + \|g\|^{p'}_{L^{p'}(\Omega_{3R})} + \left\|f^k\right\|^{p'}_{W^{1-\frac{2}{p},p'}(\Omega_{3R})} + \left\|g_N^k\right\|^{p'}_{\mathcal{N}^{\frac{1}{p'},p'}(\Gamma_{3R})}$. Utilizing (7) and arguing as above (8) follows.

□

References

[1] A. Bensoussan – J. Frehse, *Regularity results for nonlinear elliptic systems and applications*, Applied Mathematical Sciences vol. 151, Springer-Verlag, Berlin, 2002.

[2] E. DiBenedetto, *Degenerate parabolic equations*, Springer, New York 1993.

[3] C. Ebmeyer, W.B. Liu and M. Steinhauer, *Global regularity in fractional order Sobolev spaces for the p-Laplace equation on polyhedral domains*, Zeit. Anal. Anwend., to appear.

[4] M. Fuchs – G. Seregin, *Variational methods for problems from plasticity theory and for generalized Newtonian fluids*, Lecture Notes in Mathematics vol. 1749, Springer-Verlag, Berlin, 2000.

[5] E. Giusti, *Metodi diretti nel calcolo delle variazioni*, Unione Matematica Italiana, Bologna 1994. (Italian)

[6] A. Kufner, O. John and S. Fučik, *Function Spaces*, Academia, Prague 1977.

[7] J. Málek – J. Nečas – M. Rokyta – M. Růžička, *Existence and regularity of solutions to evolutional partial differential equations*, Applied Mathematics and Mathematical Computation, vol. 13, Chapman and Hall, London, 1996.

[8] M. Müller and J. Nečas, *Über die Regularität der schwachen Lösungen von Randwertaufgaben für quasilineare elliptische Differentialgleichungen höherer Ordnung*, Czech. Math. J. 25 (100) (1975), pp. 227–239. (German)

[9] J. Nečas, *Introduction to the theory of nonlinear elliptic equations*, Teubner-Texte zur Mathematik, Bd. 52, Leipzig 1983.

[10] P. Tolksdorff, *Everywhere regularity for some quasi-linear systems with lack of ellipticity*, Ann. Mat. Pura Appl. 134 (1983), pp. 241–266.

Carsten Ebmeyer
Mathematisches Seminar
Universität Bonn
Nussallee 15
D-53115 Bonn, Germany
e-mail: `ebmeyerms1@uni-bonn.de`

Progress in Nonlinear Differential Equations
and Their Applications, Vol. 61, 91–108

Temperature Driven Mass Transport in Concentrated Saturated Solutions

Antonio Fasano and Mario Primicerio

Abstract. We study the phenomenon of thermally induced mass transport in partially saturated solutions under a thermal gradient, accompanied by deposition of the solid segregated phase on the "cold" boundary. We formulate a one-dimensional model including the displacement of all species (solvent, solute and segregated phase) and we analyze a typical case establishing existence and uniqueness.

1. Introduction

It is well known that saturation of a solution of a solute S in a solvent Σ is achieved at some concentration c_S depending on temperature T. Typically c_S is a smooth function of T such that $c_S'(T) > 0$. Therefore, if one excludes supersaturation, it is possible to produce the following facts by acting on the thermal field:

(i) cooling a solution of concentration c^* to a temperature T such that $c^* > c_S(T)$, segregation of the substance S is produced as a solid phase, typically in the form of suspended crystals,

(ii) maintaining a thermal gradient in a saturated solution creates a concentration gradient of the solute inducing diffusion.

These phenomena are believed to be the most important origin of the formation of a deposit of solid wax on the pipe wall during the transportation of mineral oils with a high content of heavy hydrocarbons (waxy crude oils) in the presence of significant heat loss to the surroundings (see the survey paper [1]).

In the paper [2] we have illustrated some general features of the behavior of non-isothermal saturated solutions in bounded domains, including the appearance of an unsaturated region and the deposition of solid matter at the boundary.

Work performed in the framework of the cooperation between Enitecnologie (Milano) and I2T3 (Firenze).

The analysis of [2] was based on the following simplifying assumptions:

(a) the three components of the system, namely the solute, the solvent and the segregated phase, have the same density (supposed constant in the range of temperature considered),

(b) the concentrations of the solute and of the segregated phase are small in comparison with the concentration of the solvent.

The consequences of (a) are that gravity has no effect and that the segregation/dissolution process does not change volume.

The consequences of (b) are that solvent can be considered immobile and that the presence of a growing solid deposit has a negligible effect on the mass transport process.

For the specific application to waxy crude oil assumption (a) is reasonable on the basis of experimental evidence, but assumption (b) may not be realistic. Of course eliminating (b) leads to a much more complex situation.

For this reason we want to formulate a new model in which, differently from [2], the displacement of all the components is taken into account, as well as the influence of the growing deposit on the whole process.

In order to be able to perform some mathematical analysis of the problem and to obtain some qualitative results we confine our attention to the one-dimensional case, considering a system confined in the slab $0 < x < L$. Of course the results can be adapted with minor changes to a region bounded by coaxial cylinders (the geometry of some laboratory device devoted to the measure of thickness of deposit layers formed under controlled temperature gradients).

The general features of the model are presented in Section 2. In Section 3 we consider a specific experimental condition in which we pass through three stages: at time $t = 0$ the system is totally saturated with the segregated phase present everywhere, next a desaturation front appears and eventually the saturated zone becomes extinct. The rest of the paper is devoted to the study of the three stages, showing existence and uniqueness and obtaining some qualitative properties.

2. Description of physical system and the governing differential equations

During the evolution of the process we are going to study we can find a saturated and an unsaturated region. Supposing that at any point and at any time the segregated phase is in equilibrium with the solution, there will be no solid component in the unsaturated region. We recall that all the components have the same density ρ, whose dependence on temperature is neglected.

The saturated region is a two-phase system:

- The solid phase is the segregated material. It is made of suspended particles (crystals) having some mobility. We denote its concentration by $\hat{G}(x,t)$
- The liquid phase is a saturated solution. Its concentrantion for the whole system is $\hat{\Gamma}(x,t)$.

In turn, the solution is a two-component system containing

- the solute with concentration $\hat{c}$ (mass of solute per unit volume of the system)
- the solvent with concentration $\hat{\gamma}$ (mass of solvent per unit volume of the system).

In the sequel we will use the non-dimensional quantities

$$G = \hat{G}/\rho, \quad \Gamma = \hat{\Gamma}/\rho, \quad c = \hat{c}/\rho, \quad \gamma = \hat{\gamma}/\rho.$$

Clearly

$$\Gamma = \gamma + c \tag{2.1}$$

$$G + \Gamma = 1. \tag{2.2}$$

We can also introduce the relative non-dimensional concentrations (mass of solute and of solvent per unit mass of the solution)

$$c_{rel} = c/\Gamma \quad \gamma_{rel} = \gamma/\Gamma. \tag{2.3}$$

As we pointed out, saturated region is characterized by the fact that $c_{rel} = c_S(T)$ where the latter quantity is the saturation concentration and depends on the local temperature T only.

On $c_S(T)$ we make the following assumption:

(H1) $$c_S \in C^3, \quad c_S' > 0$$

in a temperature interval $[T_1, T_2]$.

Displacement of the various components is generated by spatial dishomogeneity.

Let J_G, J_Γ be the fluxes of segregated solid and of solution, respectively, in a saturated region.

Let Q be the mass passing, per unit time and per unit volume (rescaled by ρ), from segregated to dissolved phase. Then we have the balance equations

$$\frac{\partial G}{\partial t} + \frac{\partial J_G}{\partial x} = -Q, \tag{2.4}$$

$$\frac{\partial \Gamma}{\partial t} + \frac{\partial J_\Gamma}{\partial x} = Q. \tag{2.5}$$

From (2.2) it follows that

$$\frac{\partial}{\partial x}(J_G + J_\Gamma) = 0, \tag{2.6}$$

expressing bulk volume conservation and implying

$$J_G + J_\Gamma = 0 \tag{2.7}$$

if there is no global mass exchange with the exterior, as we suppose.

At this point we do not take the general view point of mixture theory, but we make the assumption that G is transported by diffusion. Thus

$$J_G = -D_G \frac{\partial G}{\partial x}, \tag{2.8}$$

where D_G is the diffusivity coefficient for the segregated phase.

We notice that (2.8) is consistent with the fact that all the components of the system have the same density, so that we may say that suspended particles do not feel internal rearrangements of the solution components.

Next we have to describe the flow of the components in the solution, and denote by J_γ and J_c the flux of solvent and of solute, respectively. Of course

$$J_\Gamma = J_\gamma + J_c. \tag{2.9}$$

Here too we take a simplification supposing that the solute flow J'_c relative to the solution is of Fickian type, i.e.,

$$J'_c = -D\frac{\partial c_{rel}}{\partial x}, \tag{2.10}$$

where $D > D_G$ is the solute diffusivity so that in the saturated region J'_c is a given function of the thermal gradient.

The flux J_c is the sum of $\Gamma J'_c$ and of the convective flux due to the motion of the solution. Introducing the velocity of the solution

$$V_\Gamma = J_\Gamma/\Gamma \tag{2.11}$$

we have

$$J_c = cV_\Gamma + \Gamma J'_c = c_{rel}J_\Gamma + \Gamma J'_c. \tag{2.12}$$

Consequently we have the following expression for J_c for the saturated and unsaturated case (still retaining the basic assumption of absence of bulk mass transfer, (2.7))

$$J_c = c_S D_G\frac{\partial G}{\partial x} - (1-G)D\frac{\partial c_S}{\partial x}, \quad \text{for the saturated case} \tag{2.13}$$

$$J_c = -D\frac{\partial c_{rel}}{\partial x} = -D\frac{\partial c}{\partial x}, \quad \text{for the unsaturated case } (G=0). \tag{2.14}$$

At this point we can write the balance equation for the solute

$$\frac{\partial c}{\partial t} + \frac{\partial J_c}{\partial x} = Q. \tag{2.15}$$

While for the unsaturated case, (2.15) is nothing but

$$\frac{\partial c}{\partial t} - D\frac{\partial^2 c}{\partial x^2} = 0 \tag{2.16}$$

in the saturated case we have $c = c_S(T)(1-G)$ and hence

$$-c_S\frac{\partial G}{\partial t} + \frac{\partial}{\partial x}\{c_S D_G\frac{\partial G}{\partial x} - (1-G)D\frac{\partial c_S}{\partial x}\} = Q, \tag{2.17}$$

which provides the expression of Q. Thus (2.4) takes the form

$$\frac{\partial G}{\partial t} - D_G\frac{\partial^2 G}{\partial x^2} + \frac{1}{1-c_S}\{(D_G + D)\frac{\partial c_S}{\partial x}\frac{\partial G}{\partial x} - (1-G)D\frac{\partial^2 c_S}{\partial x^2}\} = 0. \tag{2.18}$$

It is also convenient to observe that from $\gamma + c + G = 1$ and $c = c_S(1-G)$ we obtain

$$\gamma = (1-c_S)(1-G), \tag{2.19}$$

from which using (2.7), (2.9), (2.13) we deduce the expression

$$J_\gamma = -D_G \frac{\partial \gamma}{\partial x} + (D - D_G) \frac{\gamma}{1 - c_S} \frac{\partial c_S}{\partial x}. \tag{2.20}$$

We are interested in the in which T is a linear function of x independent of time:

$$T = T_1 + (T_2 - T_1) \frac{x}{L} \tag{2.21}$$

where the boundary temperatures T_1, T_2 (with $T_1 < T_2$) are given in such a way that a saturation phase is present, at least for some time.

Remark 1. *The assumption that temperature has the equilibrium profile* (2.21) *is acceptable if heat diffusivity is much larger than D (which is certainly true), so that thermal equilibrium is achieved before any significant mass transport takes place, and if we may neglect the amount of heat that is released or absorbed during the segregation/dissolution process. In the specific case of waxy crude oils it can be seen that the latter assumption is fulfilled (the influence of latent heat associated to deposition is likewise negligible).*

3. Modelling a specific mass transport process with deposition

We restrict our analysis to the following process, easily reproducible in a laboratory device.

We start with a solution at uniform concentration $\hat{c}^* (< \rho)$ and uniform temperature T^* with $\hat{c}^*$ below saturation. Then we cool the system rapidly to the temperature profile (2.21) in such a way that $c^* = \hat{c}^*/\rho > c_S(T_2)$, so that the whole system becomes saturated with a (non-dimensional) concentration

$$G_0(x) = c^* - c_S(T(x))(1 - G_0(x)) \tag{3.1}$$

of segregated phase, with

$$c_0(x) = c_S(T(x))(1 - G_0(x)) \tag{3.2}$$

being the corresponding concentration of the solute.

These will be our initial conditions. Starting from $t = 0$ the system will evolve through the following stages.

Stage 1. $G > 0$ throughout the system

The mass flow towards the cold wall $x = 0$ produced by the gradient of $c_S(T(x))$ generates various phenomena:

- the solute mass leaving the warm wall $x = L$ has to be replaced by the segregated phase,
- mass exchange occurs between the solid and the liquid phase, as described in the previous section,

- the solute mass liberated at the cold wall has to be segregated: a fraction $\chi \in (0,1]$ of it is used to build up a deposit layer, while the complementary fraction $1-\chi$ is released in the form of suspension.

As we shall see, this stage terminates at a finite time t_1.

Stage 2. The measures of the sets $\{G>0\}$, $\{G=0\}$ are both positive
At time t_1 an unsaturated region appears. This stage is characterized by the simultaneous presence of saturated and unsaturated regions (not necessarily connected for general initial data), separated by one or more free boundaries. Also Stage 2 has to terminate at a finite time t_2, when G becomes zero everywhere.

Stage 3. The whole system is unsaturated
Deposition goes on as long as $c = c_S$ and $\dfrac{\partial c}{\partial x} > 0$ on the deposition front.

Remark 2. *The asymptotic equilibrium is characterized by the absence of segregated phase and uniform solute concentration $c_S(\sigma_\infty)$, where σ_∞ denotes the non-dimensional asymptotic thickness of the deposit. Therefore we can write the trivial mass balance*

$$\sigma_\infty + (1-\sigma_\infty)c_S(\sigma_\infty) = c^*.$$

Since the l.h.s. is a function of σ_∞ increasing from $c_S(0) < c^*$ for $\sigma_\infty = 0$ to $1 > c^*$ for $\sigma_\infty = 1$, there exists one and only one solution $\sigma_\infty \in (0,1)$.

We have to write down the boundary conditions for the three stages.
However, before doing that, we introduce the non-dimensional variables:

$$\xi = x/L,\quad \tau = t/t_D,\quad \delta = D_G/D < 1,\quad \theta = \frac{T-T_1}{T_2-T_1}$$

with $t_D = L^2/D$.

For simplicity we keep the symbols $G(\xi,\tau), \Gamma(\xi,\tau), c(\xi,\tau)$ and $c_S(\theta(\xi))$. Note that $\dfrac{dc_S}{d\xi} = c_S'(\theta)$, $\dfrac{d^2c_S}{d\xi^2} = c_S''(\theta)$, since $\theta(\xi)=\xi$.

With the new variables equations (2.18), (2.16) take the form

$$\frac{\partial G}{\partial \tau} - \delta\frac{\partial^2 G}{\partial \xi^2} + \frac{1}{1-c_S}\{(1+\delta)c_S'(\theta)\frac{\partial G}{\partial \xi} - (1-G)c_S''(\theta)\} = 0, \tag{3.3}$$

$$\frac{\partial c}{\partial \tau} - \frac{\partial^2 c}{\partial \xi^2} = 0. \tag{3.4}$$

Boundary conditions for stage 1
During Stage 1 equation (3.3) must be solved in the domain $D_1 = \{(\xi,\tau)|\sigma(\tau) < \xi < 1, 0<\tau<\tau_1\}$, where $\xi = \sigma(\tau)$ is the deposition front, with initial conditions

$$\sigma(0) = 0,\quad G_0(\xi) = \frac{c^* - c_S(\xi)}{1-c_S(\xi)}, \xi \in (0,1). \tag{3.5}$$

At the boundary $\xi = 1$ we just have $J_\gamma = 0$, meaning

$$\delta\frac{\partial G}{\partial \xi}|_{\xi=1} = -c_S'(1)\frac{1-G}{1-c_S(1)}, 0<\tau<\tau_1. \tag{3.6}$$

Conditions on the deposition front depend on the way the deposit layer is built. As we said, the primary source of the deposit is a fraction $\chi \in (0,1]$ of the incoming solute flux. In addition we expect that the advancing front can capture by adhesion a fraction $\eta \in [0,1]$ of the suspension it finds on its way (in [2] we just considered $\eta = 1$).

Therefore, with the adopted rescaling, the speed $\frac{d\sigma}{d\tau}$ is the sum of two terms

$$\frac{d\sigma}{d\tau} = \chi(1-G)c_S'(\sigma) + \eta \frac{d\sigma}{d\tau} G,$$

yielding

$$\frac{d\sigma}{d\tau} = \chi \frac{1-G}{1-\eta G} c_S'(\sigma), 0 < \tau < \tau_1. \tag{3.7}$$

The most obvious way of deriving the second condition on the deposition front is to impose that the solvent is displaced precisely with the speed $\frac{d\sigma}{d\tau}$.

In the original physical variables the solvent velocity is

$$v_\gamma = \frac{J_\Gamma - J_c}{\gamma} = \frac{D_G \frac{\partial G}{\partial x}(1-c_S) + (1-G)D\frac{\partial c_S}{\partial x}}{(1-G)(1-c_S)} \tag{3.8}$$

from which we deduce the desired condition

$$\frac{d\sigma}{d\tau} = \frac{1}{(1-G)(1-c_S(\sigma))}[\delta \frac{\partial G}{\partial \xi}(1-c_S) + c_S'(\sigma)(1-G)], 0 < \tau < \tau_1. \tag{3.9}$$

Eliminating $\frac{d\sigma}{d\tau}$ between (3.7) and (3.9) we obtain the equivalent equation

$$\frac{\delta}{1-G}\frac{\partial G}{\partial \xi}|_{\xi=\sigma(\tau)} = c_S'[\chi \frac{1-G}{1-\eta G} - \frac{1}{1-c_S}]|_{\xi=\sigma(\tau)} \tag{3.10}$$

For instance, when $\chi = \eta = 1$ (3.10) simply reduces to

$$\frac{\delta}{1-G}\frac{\partial G}{\partial \xi}|_{\xi=\sigma(\tau)} = -\frac{c_S(\sigma)}{1-c_S(\sigma)} c_S'(\sigma), \tag{3.11}$$

while for $\chi = 0$ (no deposition, i.e. $\frac{d\sigma}{d\tau} = 0$ from (3.7)) we find

$$\delta \frac{\partial G}{\partial \xi}|_{\xi=\sigma(\tau)} = -(1-G)\frac{c_S'(\sigma)}{1-c_S'(\sigma)}, 0 < \tau < \tau_1. \tag{3.12}$$

(irrespectively of η which cancels out, having no role).

Thus we have now the complete model for Stage 1, which can be summarized as follows:

Problem 1 ($\eta \neq 1$). *Find the pair* (σ, G) *satisfying the differential equation* (3.3) *in* D_1*, with initial conditions* (3.5)*, boundary condition* (3.6)*, and free boundary conditions* (3.7)*,* (3.10)*, all in the classical sense.*

For $\eta = 1$ condition (3.7) reduces to the o.d.e. $\frac{d\sigma}{d\tau} = \chi c_S'(\sigma)$ and consequently the motion of the deposition front becomes known. The problem is standard in that case.

Boundary conditions for stage 2

Stage 2 differs from Stage 1 because of the simultaneous presence of a saturated and an unsaturated region, separated by a free boundary $\xi = s(\tau)$, which, in the specific case we refer to, is a curve starting from the point $(1, \tau_1)$, where $G(\xi, \tau)$ vanishes for the first time. We shall find sufficient conditions on c_S ensuring that $G(\xi, \tau_1) > 0$ for $\xi \in [\sigma(\tau_1), 1)$ and that the unsaturated region remains connected. We denote by τ_2 the transition time to Stage 3.

In the region $\{s(\tau) < \xi < 1, \tau_1 < \tau < \tau_2\}$, corresponding to concentration below saturation ($G = 0$, and hence $c_{rel} = c < c_S$), the governing equation is (3.4).

The wall $\xi = 1$ is a no-flux boundary, i.e.,

$$\frac{\partial c}{\partial \xi}|_{\xi=1} = 0, \tau_1 < \tau < \tau_2, \tag{3.13}$$

implying of course that also $\frac{\partial \gamma}{\partial \xi}|_{\xi=1} = 0$.

On the desaturation front we have

$$G(s(\tau)^-, \tau) = 0 \tag{3.14}$$

which implies that the absolute solute concentration equals $c_S(s(\tau))$ on both sides of the front:

$$c(s(\tau)+, \tau) = c_S(s(\tau)). \tag{3.15}$$

Continuity of (all) concentrations across the front implies in turn that the total solvent flux has to be continuous, or equivalently that

$$(\delta(1 - c_S)\frac{\partial G}{\partial \xi} + \frac{\partial c_S}{\partial \xi})|_{\xi=s(\tau)-} = \frac{\partial c}{\partial \xi}|_{\xi=s(\tau)+}, \tau_1 < \tau < \tau_2. \tag{3.16}$$

The model for Stage 2 is completed by the conditions

$$G(\xi, \tau_{1+}) = G_1(\xi), \sigma(\tau_1) < \xi < 1, \;\; s(\tau_1) = 1 \tag{3.17}$$

where $G_1(\xi) = G(\xi, \tau_1-)$.

Thus we can state

Problem 2. *Find the functions* (σ, s, G, c) *such that* σ, G *satisfy* (3.3), (3.7), (3.10), (3.17), *and* s, G, c *satisfy* (3.4), (3.13)–(3.16), *all in the classical sense.*

Boundary conditions for stage 3

At time τ_2 the saturated region disappears, i.e., $\sigma(\tau_2) = s(\tau_2)$ (we are still referring to the particular case in which the unsaturated region during Stage 2 is connected). From that time on deposition continues as long as $c(\sigma, \tau) = c_S(\sigma)$, $\frac{\partial c}{\partial \xi}|_{\xi=\sigma(\tau)} > 0$ and necessarily all the incoming mass enters the deposit, irrespectively of the value of χ during Stage 2. Therefore the new conditions on the deposition front are

$$c(\sigma(\tau),\tau) = c_S(\sigma(\tau)), \quad \tau > \tau_2 \tag{3.18}$$

$$\frac{d\sigma}{d\tau} = \frac{\partial c}{\partial \xi}|_{\xi=\sigma(\tau)}, \quad \tau > \tau_2. \tag{3.19}$$

Of course $c(\xi,\tau)$ satisfies (3.13) with initial condition

$$c(\xi,\tau_{2+}) = c_2(\xi), \quad \sigma(\tau_2) < \xi < 1, \tag{3.20}$$

with $c_2(\xi) = c(\xi,\tau_2-)$.

Thus during this stage we have to solve

Problem 3. *Find* (σ, c) *satisfying* (3.4), (3.13), (3.18)–(3.20) *in the classical sense.*

4. Analysis of Stage 1

The overall mass balance during Stage 1 can be expressed by imposing that the solvent mass is conserved, starting from the equation

$$\frac{\partial \gamma}{\partial t} + \frac{\partial J_\gamma}{\partial x} = 0 \tag{4.1}$$

and remembering that $J_\gamma = J_\Gamma - J_c = D_G(1-c_S)\dfrac{\partial G}{\partial x} + D(1-G)\dfrac{\partial c_S}{\partial x}$, so that in non-dimensional variables we have

$$\frac{\partial \gamma}{\partial \tau} + \frac{\partial}{\partial \xi}\{\delta(1-c_S)\frac{\partial G}{\partial \xi} + (1-G)c_S'(\theta)\} = 0. \tag{4.2}$$

Since $G = 1 - \dfrac{\gamma}{1-c_s}$, it is easily seen that the equation above is equivalent to

$$\frac{\partial \gamma}{\partial \tau} - \delta\frac{\partial^2 \gamma}{\partial \xi^2} + (1-\delta)\frac{\partial}{\partial \xi}(\frac{\gamma c_S'}{1-c_S}) = 0. \tag{4.3}$$

Integrating (4.2) over any domain $D_\tau = \{(\xi,\tau')|\sigma(\tau') < \xi < 1, 0 < \tau' < \tau\}$, with $\tau \le \tau_1$, we get

$$\oint_{\partial D_\tau}\{\gamma d\xi - [\delta(1-c_S)\frac{\partial G}{\partial \xi} + (1-G)c_S'(\theta)]d\tau'\} = 0, \tag{4.4}$$

simply expressing $\oint_{\partial D_\tau}\{\gamma d\xi - J_\gamma d\tau\} = 0$.

Since $J_\gamma = 0$ on $\xi = 1$, $J_\gamma = \gamma\dot{\sigma}$ on $\xi = \sigma(\tau)$, we obtain

$$\int_{\sigma(\tau)}^1 \gamma(\xi,\tau)d\xi = \int_0^1 (1-c_s)(1-G_0)d\xi,$$

as expected, which can also be written as

$$\int_{\sigma(\tau)}^1 [G(\xi,\tau) + c(\xi,\tau)]d\xi = c^* - \sigma(\tau) \tag{4.5}$$

having an evident physical meaning.

Remark 3. *For $\eta = 1$ (complete inclusion of the suspended phase)* (3.7) *simplifies to*

$$\frac{d\sigma}{dc} = \chi c_S'(\theta(\sigma)) \tag{4.6}$$

which can be integrated. In this case the deposition front becomes a known function $\sigma^{(1)}(t)$. If we can establish an a priori upper bound $G_{max} < 1$ for G, the factor $\dfrac{1-G}{1-\eta G}$ takes values in $[\dfrac{1-G_{max}}{1-\eta G_{max}}, 1]$. Denoting by $\sigma^{(\eta)}$ the integral of $\dfrac{d\sigma}{d\tau} = \chi \dfrac{1-G_{max}}{1-\eta G_{max}} c_S'(\theta(\sigma))$ with zero initial value, we have the a priori bounds

$$\sigma^{(\eta)}(\tau) \le \sigma(\tau) \le \sigma^{(1)}(\tau) \text{ for } \tau \in (0, \tau_1). \tag{4.7}$$

Proposition 1. *The extinction time τ_1 of Stage 1 is finite. An upper estimate is given by the solution τ^* of*

$$\sigma^{(\eta)}(\tau^*) = \sigma_\infty. \tag{4.8}$$

Proof. Simply use the inequality (4.7) and Remark 2. □

Let us show that G never reaches 1, thus preventing the formation of a solid layer inside the system. To assumption (H1) on c_S we add

(H2) $$c_S'' \le 0.$$

Proposition 2. *Under assumptions* (H1), (H2) *during Stage* 1 *we have $G < 1$ in D_1.*

Proof. We know that $0 < G_0 < 1$, so it will be $G < 1$ at least for some time. Moreover $G > 0$ by definition. Moving the term $(1-G)c_S''(\theta)/(1-c_S)$ to the r.h.s. of (3.3) we see that it is nonpositive, thanks to (H2). Thus G has to take its maximum on the parabolic boundary of D_1. From (3.6) we see that, still for $G < 1$, we have $\dfrac{\partial G}{\partial \xi} < 0$ on $\xi = 1$.

On the boundary $\xi = \sigma(\tau)$ (as long as $G < 1$) we see that for $\eta = 1$

$$\frac{\delta}{1-G}\frac{\partial G}{\partial \xi} = -c_S'(\frac{1}{1-c_S} - \chi) < 0, \quad \forall \chi \in [0,1].$$

Thus $\dfrac{\partial G}{\partial \xi} < 0$ for $\eta \in (0,1)$ too because the r.h.s. of (3.10) is monotone in η.

We conclude that the maximum of G can be taken on $\xi = \sigma(\tau)$. However, if G tends to 1 there, $\dfrac{\partial G}{\partial \xi}$ tends to zero contradicting the boundary point principle for equation (3.3). □

Since in our case we start with $G_0' < 0$, we can have G monotone in ξ if we add the assumption

(H3) $$\left(\frac{c_s''}{1-c_S}\right)' \le 0.$$

Proposition 3. *Under assumptions* (H1)–(H3) *we have* $\frac{\partial G}{\partial \xi} < 0$ *during Stage* 1.

Proof. Set $\omega = \frac{\partial G}{\partial \xi}$. In the previous proposition we have seen that $\frac{\partial G}{\partial \xi} < 0$ on the lateral boundaries. Moreover

$$G_0' = -c_S' \frac{1 - c^*}{(1 - c_S)^2} < 0. \tag{4.9}$$

Differentiating (3.3) w.r.t. ξ we obtain

$$\frac{\partial \omega}{\partial \tau} - \delta \frac{\partial^2 \omega}{\partial \xi^2} + \frac{1+\delta}{1-c_S} c_S' \frac{\partial \omega}{\partial \xi} + \omega \{ \frac{c_S''}{1-c_S}(2+\delta) + \frac{1+\delta}{(1-c_S)^2} c_S'^2 \} = (1-G)(\frac{c_S''}{1-c_S})', \tag{4.10}$$

from which the thesis follows easily using the maximum principle and assumption (H3). □

Remark 4. *An important consequence of the proposition above is that*

$$G(\xi, \tau_1) > 0 \text{ for } \xi \in [\sigma(\tau_1), 1), \tag{4.11}$$

in other words the desaturation front starts from the point $(1, \tau_1)$.

We conclude this section by proving existence and uniqueness of the solution to Problem 1.

Theorem 1. *Problem* 1 *has one unique solution under the assumptions* (H1), (H2).

Proof. We start by noting that from (3.7) we have the obvious a priori estimate

$$0 \leq \frac{d\sigma}{d\tau} \leq \chi \| c_s' \| =: A, \tag{4.12}$$

$\| c_s' \|$ denoting the sup-norm (of course we recall that $0 < G < 1$). □

Now, if we introduce the set

$$\Sigma = \{ \sigma \in C^1([0, \tilde{\tau}]) | \sigma(0) = 0, 0 \leq \dot{\sigma} \leq A, \frac{|\dot{\sigma}(\tau) - \dot{\upsilon}(\tau'')|}{|\tau' - \tau''|^\alpha} \leq B \} \tag{4.13}$$

for some $B > 0$ and $\alpha \in (0, \frac{1}{2})$, and we take any $\sigma \in \Sigma$, we may formulate the problem consisting of equation (3.3), initial condition (3.5) and boundary conditions (3.6), (3.10). For the corresponding solution G of such a problem, whose existence and uniqueness can be proved by means of standard methods, it is not difficult to find $\tilde{\tau}$ such that $G > 0$ for $\tau \in (0, \tilde{\tau})$ irrespectively of the choice of σ in Σ. The inequality $G < 1$ can be established like in Proposition 2. Finally, working on the problem satisfied by $\omega = \frac{\partial G}{\partial \xi}$ we can easily find the bound

$$|\frac{\partial G}{\partial \xi}| \leq B \tag{4.14}$$

with B independent of σ in Σ.

At this point existence can be proved using the following fixed point argument.

Taken $\sigma \in \Sigma$ and computing G we can define $\tilde{\sigma}$ via

$$\frac{d\tilde{\sigma}}{d\tau} = \chi \frac{1-G}{1-\eta G} c_s'(\sigma), \quad \tilde{\sigma}(0) = 0, \tag{4.15}$$

which automatically satisfies $0 \leq \dfrac{d\tilde{\sigma}}{d\tau} \leq A$.

Noting that $|\dfrac{d}{dG}\dfrac{1-G}{1-\eta G}| \leq \dfrac{1}{1-\eta}$ for $\eta < 1$ (while it just vanishes for $\eta = 1$, which however is not the interesting case) for a pair (σ_1, σ_2) of functions in Σ, we have the easy estimate

$$|\frac{d\tilde{\sigma}_1}{d\tau} - \frac{d\tilde{\sigma}_2}{d\tau}| \leq \frac{A}{1-\eta}|G_1(\sigma_1(\tau),\tau) - G_2(\sigma_2(\tau),\tau)| + \chi||c_s''||\ |\sigma_1 - \sigma_2| \tag{4.16}$$

with obvious meaning of the symbols.

Therefore at this point we only need to show that $G(\sigma(t), t)$ depends in a Lipschitz continuous way on σ in the topology of Σ . More precisely, we want to show that

$$||G_1 - G_2||_\tau \leq K_1||\sigma_1 - \sigma_2||_\tau + K_2 \int_0^\tau ||\dot{\sigma}_1 - \dot{\sigma}_2||_{\tau'}(\tau - \tau')^{-1/2} d\tau' \tag{4.17}$$

for some positive constants K_1, K_2, with $||\cdot||_\tau$ denoting the sup-norm restricted to the time interval $(0, \tau)$.

Now, $G(\xi, \tau)$ corresponding to a given $\sigma \in \Sigma$ has the representation

$$\begin{aligned} G(\xi,\tau) = &\int_0^\tau \phi(\tau')\Gamma(\xi,\tau;\sigma(\tau'),\tau')d\tau' \\ &+ \int_0^1 G_0(\xi')\Gamma(\xi,\tau;\xi',0)d\xi' + \int_0^\tau \psi(\tau')\Gamma(\xi,\tau;1,\tau')d\tau' \\ &+ \int_0^\tau \int_{\sigma(\tau')}^1 \Gamma(\xi,\tau;\xi',\tau')\frac{c_s''(\xi')}{1-c_s(\xi')}d\xi' d\tau', \end{aligned} \tag{4.18}$$

with $\Gamma(\xi, \tau; \xi', \tau')$ fundamental solution of the parabolic operator

$$L = \frac{\partial}{\partial\tau} - \delta\frac{\partial^2}{\partial\xi^2} + \frac{1+\delta}{1-c_s}c_s'\frac{\partial}{\partial\xi} + \frac{c_s''}{1-c_s'},$$

with $\tau' < \tau$ and $(\xi, \tau), (\xi', \tau')$ varying in $[0, 1] \times [0, \tilde{\tau}]$. The densities $\phi(\tau), \psi(\tau)$, together with a third unknown $G_\sigma(\tau)$, representing the value of G over $\xi = \sigma(\tau)$, satisfy the system

$$\begin{aligned} \frac{1}{2}\phi(\tau) = &\int_0^\tau \phi(\tau')\Gamma_\xi(\sigma(\tau),\tau;\sigma(\tau'),\tau')d\tau' \\ &+ \int_0^\tau \psi(\tau')\Gamma(\sigma(\tau),\tau);\ 1,\tau')d\tau' + \int_0^1 G_0(\xi')\Gamma_\xi(\sigma(\tau),\tau,\xi',0)d\xi' \end{aligned} \tag{4.19}$$

$$+\int_0^\tau \int_{\sigma(\tau')}^1 \Gamma_\xi(\sigma(\tau),\tau;\xi',\tau') \frac{c_s''(\xi')}{1-c_s(\xi')} d\xi' d\tau'$$
$$- \frac{1-G_\sigma(\tau)}{\delta} c_s'(\sigma)[\chi \frac{1-G_\sigma}{1-\eta G_\sigma} - \frac{1}{1-c_s(\sigma)}]$$

$$\frac{1}{2}\psi(\tau) = -\int_0^\tau \phi(\tau')\Gamma_\xi(1,\tau;\sigma(\tau'),\tau')d\tau' - \int_0^\tau \psi(\tau')\Gamma_\xi(1,\tau;1,\tau')d\tau' \quad (4.20)$$
$$-\int_0^1 G_0(\xi')\Gamma_\xi(1,\tau;\xi',0)d\xi'$$
$$-\int_0^\tau \int_{\sigma(\tau')}^1 \Gamma_\xi(1,\tau;\xi',\tau') \frac{c_s''(\xi')}{1-c_s(\xi')} d\xi' d\tau' - \frac{1}{\delta} \frac{c_s'(1)}{1-c_s(1)} G(1,\tau)$$

$$G_\sigma(\tau) = \int_0^\tau \phi(\tau')\Gamma(\sigma(\tau),\tau;\sigma(\tau'),\tau')d\tau' \quad (4.21)$$
$$+\int_0^1 G_0(\xi')\Gamma(\sigma(\tau),\tau;\xi',0)d\xi' + \int_0^\tau \psi(\tau')\Gamma(\sigma(\tau),\tau;1,\tau')d\tau'$$
$$+\int_0^\tau \int_{\sigma(\tau')}^1 \Gamma(\sigma(\tau),\tau;\xi',\tau') \frac{c_s''(\xi')}{1-c_s(\xi')} d\xi' d\tau',$$

where $G(1,\tau)$ in (4.20) must be replaced with expression obtained from (4.18).

Note that G_σ appears nonlinearly in (4.19) if $\eta < 1$, as we are supposing.

Eliminating G_σ leads to a nonlinear system of Volterra equations with weakly singular kernels. Existence and uniqueness can anyway be proved by standard methods, thanks to the fact that the dependence on G_σ in (4.19) is Lipschitz. Functions ϕ, ψ, G_σ are bounded uniformly for $\sigma \in \Sigma$ and as a consequence of (4.19)–(4.21) they are at least Hölder continuous of order $\frac{1}{2}$ with a Hölder norm uniformly bounded in Σ.

In turn this implies that $\frac{\partial G}{\partial \xi}|_{\xi=\sigma(\tau)}$ has the same type of regularity.

Our task now is to estimate $|G_{\sigma_1} - G_{\sigma_2}|$. Introducing the functions ϕ_i, ψ_i corresponding to $\sigma_i, i = 1, 2$, from (4.21) we see that

$$|G_{\sigma_1}(\tau) - G_{\sigma_2}(\tau)| \le \int_0^\tau |\phi_1(\tau') - \phi_2(\tau')||\Gamma(\sigma_1(\tau),\tau;\sigma_1(\tau'),\tau')|d\tau' \quad (4.22)$$
$$+\int_0^\tau |\psi_1(\tau') - \psi_2(\tau')||\Gamma(\sigma_1(\tau),\tau;\sigma_1(\tau');\tau')|d\tau'$$
$$+M\int_0^\tau |\Gamma(\sigma_1(\tau),\tau;\sigma_1(\tau'),\tau') - \Gamma(\sigma_2(\tau),\tau;\sigma_2(\tau'),\tau')|d\tau'$$

$$+N\Big\{\int_0^\tau |\int_{\sigma_1(\tau)}^{\sigma_2(\tau')} \Gamma(\sigma_1(\tau),\tau;\xi',\tau')d\xi'|d\tau'$$
$$+\int_0^\tau \int_{\sigma_2(\tau')}^1 |\Gamma(\sigma_1(\tau),\tau;\xi',\tau') - \Gamma(\sigma_2(\tau),\tau;\xi',\tau')|d\xi' d\tau'\Big\}.$$

which has to be coupled with similar inequalities for $|\phi_1(\tau)-\phi_2(\tau)|$ and $|\psi_1(\tau)-\psi_2(\tau)|$.

According to the parametrix method [3] the function $\Gamma(\xi,\tau,\xi',\tau')$ is constructed as follows

$$\Gamma(\xi,\tau;\xi',\tau') = Z(\xi,\tau;\xi',\tau') + \int_{\tau'}^{\tau}\int_0^1 Z(\xi,\tau,\eta,\theta)\Phi(\eta,\theta;\xi',\tau')d\eta d\theta \tag{4.23}$$

where

$$Z(\xi,\tau;\xi',\tau') = \frac{1}{2\sqrt{\pi\delta(\tau-\tau')}}\exp[-\frac{(\xi-\xi')^2}{4\delta(\tau-\tau')}] \tag{4.24}$$

is the fundamental solution of the heat operator $\frac{\partial}{\partial\tau} - \delta\frac{\partial^2}{\partial\xi^2}$, and $\Phi(\xi,\tau;\xi',\tau')$ is the solution of the integral equation

$$\Phi(\xi,\tau;\xi',\tau') = LZ(\xi,\tau;\xi',\tau') + \int_{\tau'}^{\tau} LZ(\xi,\tau;\eta,\theta)\Phi(\eta,\theta;\xi',\tau')d\eta d\theta. \tag{4.25}$$

Since the operator L is particularly simple, we can calculate LZ explicitly:

$$LZ(\xi,\tau;\xi',\tau') = \{-\frac{1+\delta}{1-c_s}c_s'\frac{\xi-\xi'}{2\delta(\tau-\tau')} + \frac{c_s''}{1-c_s}\}Z(\xi,\tau;\xi',\tau'), \tag{4.26}$$

so that the development of the parametrix method is largely simplified.

In particular

$$|LZ(\xi,\tau;\xi',\tau')| \le \frac{const.}{(\tau-\tau')^\mu}\frac{1}{|\xi-\xi'|^{1-\mu}}, \quad \mu\in(\frac{1}{2},1) \tag{4.27}$$

so that the kernel in (4.26) is integrable. Moreover Φ is continuous, Hölder continuous w.r.t. the first argument, for $\tau'<\tau$.

When calculating the differences

$$\phi_1-\phi_2 \qquad \text{or} \qquad \Gamma(\sigma_1(\tau),\tau;\sigma_1(\tau'),\tau') - \Gamma(\sigma_2(\tau),\tau;\sigma_2(\tau'),\tau')$$

the main term which comes into play is

$$\Gamma_\xi(\sigma_1(\tau),\tau;\sigma_1(\tau'),\tau') - \Gamma_\xi(\sigma_2(\tau),\tau;\sigma_2(\tau'),\tau')$$

which in turn requires the computation of

$$\Omega(\tau,\tau') = Z_\xi(\sigma_1(\tau),\tau;\sigma_1(\tau'),\tau') - Z_\xi(\sigma_2(\tau),\tau;\sigma_2(\tau'),\tau')$$
$$= -\frac{\sigma_1(\tau)-\sigma_1(\tau')-[\sigma_2(\tau)-\sigma_2(\tau')]}{4\sqrt{\pi}[\delta(\tau-\tau')]^{3/2}}\exp[-\frac{(\sigma_1(\tau)-\sigma_1(\tau'))^2}{4\delta(\tau-\tau')}]$$
$$-\frac{\sigma_2(\tau)-\sigma_2(\tau')}{4\sqrt{\pi}[\delta(\tau-\tau')]^{3/2}}\{\exp[-\frac{(\sigma_1(\tau)-\sigma_1(\tau'))^2}{4\delta(\tau-\tau')}] - \exp[-\frac{(\sigma_2(\tau)-\sigma_2(\tau'))^2}{4\delta(\tau-\tau')}]\}.$$

Writing

$$|\sigma_1(\tau) - \sigma_1(\tau') - [\sigma_2(\tau) - \sigma_2(\tau')]| = |\dot{\sigma}_1(\bar{\tau}) - \dot{\sigma}_2(\hat{\tau})|(\tau - \tau')$$

with $\bar{\tau}, \hat{\tau} \in (\tau', \tau)$, and

$$|\exp[-\frac{(\sigma_1(\tau) - \sigma_1(\tau'))^2}{4\delta(\tau - \tau')}] - \exp[-\frac{(\sigma_2(\tau) - \sigma_2(\tau'))^2}{4\delta(\tau - \tau')}]| \leq \frac{A}{2\delta}|\dot{\sigma}_1(\bar{\tau}) - \dot{\sigma}_2(\hat{\tau})|(\tau - \tau'),$$

we obtain the estimate

$$|\Omega(\tau, \tau')| \leq K\frac{||\dot{\sigma}_1 - \dot{\sigma}_2||_\tau}{\sqrt{\tau - \tau'}} \tag{4.28}$$

where K is a constant independent of the choice of σ_1, σ_2 in Σ.

Similar computations can be performed for the other terms involved, leading to the desired estimate (4.17). Coupling (4.16) and (4.17) leads to the conclusion that the mapping $\sigma \to \tilde{\sigma}$ is contractive for $\tilde{\tau}$ sufficiently small in the selected topology.

By means of standard arguments we can infer existence and uniqueness (any solution has to belong to Σ) up to the first time G vanishes.

5. Analysis of Stage 2: a priori results

First we prove that mass balance is expressed by an equation similar to (4.5)

Proposition 4. *For all $\tau \in (\tau_1, \tau_2)$ we have*

$$\sigma(\tau) + \int_{\sigma(\tau)}^{s(\tau)} [G(\xi, \tau) + c(\xi, \tau)]d\xi + \int_{s(\tau)}^{1} c(\xi, \tau)d\xi = c^*, \quad \text{where } c = c_S(1 - G). \tag{5.1}$$

Proof. Take the mass balance of the solvent separately in the domains $\sigma(\tau') < \xi < s(\tau')$, $\tau_1 < \tau' < \tau$; $s(\tau') < \xi < 1$, $\tau_1 < \tau' < \tau$, with $\tau \in (\tau_1, \tau_2)$.

Remembering that in the non-dimensional form J_γ is expressed by $J_\gamma = \delta(1 - c_s)G_\xi + (1 - G)c_s'$ in the first domain and simply by $J_\gamma = \dfrac{\partial c}{\partial \xi}$ in the second domain, and using $J_\gamma = \gamma\dot{\sigma}$ on the deposition front, $J_\gamma = 0$ on $\xi = 1$, $G = 0$, $[\gamma] = [J_\gamma] = 0$ on the desaturation front, (5.1) easily follows by integration of $\dfrac{\partial \gamma}{\partial r} + \dfrac{\partial J_\gamma}{\partial \xi} = 0$. □

Since Stage 2 is characterized by the presence of a saturated region, the same argument used in the proof of Prop. 1 leads to an analogous conclusion, i.e.,

Proposition 5. *The extinction time τ_2 of Stage 2 is finite. Moreover $\tau_2 < \tau^*$ defined by (4.8).*

Likewise we can say that Proposition 2 ($G < 1$) is still valid. It is enough to recall that $G(\xi, \tau_1) < 1$ and that $\dfrac{\partial G}{\partial \xi} \leq 0$ on the desaturation front, owing to (3.17).

Clearly we can also extend Proposition 3 ($\frac{\partial G}{\partial \xi} < 0$), implying that the saturated region remains connected during Stage 2.

A peculiar feature of Stage 2 is that there cannot be the analog of a "mushy region", in the following sense

Proposition 6. *During Stage* 2 *the complement of the set* $\{G > 0\}$ *cannot contain an open set where* $c \equiv c_s$.

Proof. The differential equation to be satisfied in such a set should be (3.4), with $\frac{\partial c}{\partial \tau} = 0$. Thus the presence of such a region is compatible only with $c_s'' = 0$. Because of the analyticity with respect to ξ of the solution of (3.4), the unique continuation of c up to $\xi = 1$ is a function constant in time, linear and increasing in ξ, thus contradicting the boundary condition $\frac{\partial c}{\partial \xi} = 0$. □

6. Analysis of Stage 2: weak formulation and existence

Clearly the nature of the free boundary conditions on the desaturation front, namely (3.14), (3.15), (3.16), is quite different from the conditions on the deposition front, which involve the free boundary velocity in an explicit way.

In order to prove existence the most convenient approach is to introduce a weak formulation, in which the desaturation front plays the role of a level set (the set of discontinuity of some coefficients).

The natural approach to a weak formulation seems to re-write the problem in terms of the solvent concentration γ.

We can identify the desaturation front with the level curve $\gamma = 1 - c_s(s)$.

We know that in non-dimensional variables the current density of the solvent has the expression

$$j_\gamma = -\delta \frac{\partial \gamma}{\partial \xi} + (1-\delta)\frac{\gamma c_s'}{1-c_s}, \quad \textit{for} \ \gamma < 1 - c_s \tag{6.1}$$

$$j_\gamma = -\frac{\partial \gamma}{\partial \xi} \ \textit{for} \ \gamma > 1 - c_s. \tag{6.2}$$

If we set

$$v = 1 - c_s - \gamma \tag{6.3}$$

and we define

$$A(v) = \{ {}^{\delta \ for v>0 \ (where \ v \equiv G(1-c_s))}_{1 \ for v<0 \ (where \ v \equiv c - c_s)} \tag{6.4}$$

then the balance equation

$$\frac{\partial \gamma}{\partial \tau} + \frac{\partial j_\gamma}{\partial \xi} = 0 \tag{6.5}$$

can be written in the distributional sense in the whole domain $D_\sigma = \{(\xi, r) : \sigma(\tau) < \xi < 1,\ \tau_1 < \tau < \bar{\tau}\}$:

$$\frac{\partial v}{\partial r} - \frac{\partial}{\partial \xi}\{A(v)\frac{\partial v}{\partial \xi} - [1 - A(v)]v\frac{c_s'}{1 - c_s}\} = c_s'' \tag{6.6}$$

Here $\bar{\tau}$ is a time instant sufficiently close to τ_1, still to be specified.

Equation (6.6) includes the free boundary conditions, that in the classical statements are

$$v = 0 \ \ on\ \ both\ \ sides\ \ of\ \ x = s(t) \tag{6.7}$$

$$[j_\gamma] = 0. \tag{6.8}$$

The latter condition, taking into account (6.7), reduces to

$$[A(v)\frac{\partial v}{\partial \xi}] = 0. \tag{6.9}$$

Thus, regarding the boundary $\xi = \sigma(\tau)$ as known, which is true for $\eta = 1$, the weak formulation of the problem for v is: find $v \in V^{1,0}(D_\sigma)$ such that

$$\int_{D_\sigma} \{[A(v)\frac{\partial v}{\partial \xi} - (1 - A(v))v\frac{c_s'}{1 - c_s} + c_s']\frac{\partial \phi}{\partial \xi} - v\frac{\partial \phi}{\partial \tau}\}d\xi d\tau \tag{6.10}$$

$$-\int_0^{\prime} v(\xi, \tau_1)\phi(\xi, \tau_1)d\xi + \int_{\tau_1}^{\bar{\tau}} \phi(\sigma(\tau), \tau)\chi c_s' \frac{1 - v}{1 - \eta v}(1 - c_s)|_{\xi=\sigma(\tau)}dr$$

$\forall \phi \in W_2^{1,1}(D_\sigma)$ such that $\phi = 0$ for $\tau = \bar{\tau}$. The notation of functional spaces is taken from [4].

(For the formulation of a similar problem in a cylindrical domain see [4], Chap. 3, Sect. 5). Existence and uniqueness can be established as in Theorem 5.1, p. 170 of [4].

At this point we can use Theorem 10.1, p. 204, of [4], ensuring that v is Hölder continuous, uniformly with respect to σ in the same class Σ used in the fixed point argument of Section 4, in a closed domain separated from σ and including an interval $[\xi_0, \xi_1] \subset (0, 1)$ for $\tau = \tau_1$, where we know that v is separated from zero.

Let $\Xi \in (\xi_0, \xi_1)$. On the basis of the above Hölder estimate we can find $\tilde{\tau}$ such that $v(\Xi, \tau) \geq \frac{1}{2}v(\Xi, \tau_1)$ for $\tau \in [\tau_1, \tilde{\tau}]$, for all $\sigma \in \Sigma$.

Thus, for a given σ we can solve the problem for $G(\xi, \tau)$ in the classical way in the domain $D_{\sigma,\Xi} = \{\sigma(\tau) < \xi < \Xi,\ \tau_1 < \tau < \tilde{\tau}\}$ with the boundary condition $G(\Xi, \tau) = \dfrac{v(\Xi, \tau)}{1 - c_s(\Xi)}$.

The function $v = G(1 - c_s)$ will necessarily be the restriction of the weak solution (i.e., the solution of (6.10)) to $D_{\sigma,\Xi}$.

Therefore, for each σ we know a domain $D_{\sigma,\Xi}$, such that $\Xi - \sigma(\tilde{\tau})$ remains positive for $\sigma \in \Sigma$, in which $A(v) = \delta$.

This is enough to apply the machinery of Section 4 to obtain a similar existence and uniqueness result in the interval $(\tau_1, \tilde{\tau})$. An additional information we have to provide is the continuous dependence of $v(\Xi, \tilde{\tau})$ on σ. Using the stability theorem on p. 166 of [4] in connection with the already quoted th. 10.1, p. 204, we can see that if $\sigma_1, \sigma_2 \to 0$ in the C^1 norm, then the corresponding difference $v_1(\Xi,\tau) - v_2(\Xi,\tau)$ tends to zero in the Hölder norm. What mainly matters, however, is the dependence of G_σ on σ. It is well known that $\dfrac{\partial G}{\partial \xi}$ can be estimated uniformly w.r.t. $\sigma \in \Sigma$ in a domain $D_{\sigma,\Xi'}$, for some $\Xi' < \Xi$. In practice it is possible to identify Ξ' with Ξ, by possibly reducing $\tilde{\tau}$, thanks to the arbitrariness of Ξ. In turn, writing the equation for the difference $G_1 - G_2$ after having performed the transformation which maps $D_{\sigma_i,\Xi}$ into the rectangle $(0,\Xi) \times (0,\tilde{\tau})$, it is easy to realize that $|G_{\sigma_1}(\tau) - G_{\sigma_2}(\tau)|$ can be estimated by a linear combination of $\sup_{\tau' \in (\tau_1,\tau)} |\sigma_1(\tau') - \sigma_2(\tau')|$ and $\int_{\tau_1}^{\tau} \dfrac{|\dot{\sigma}_1(\tau') - \dot{\sigma}_2(\tau')|}{\sqrt{\tau' - \tau}} d\tau'$.

This is the basic estimate in the fixed point argument already used in Stage 1 to obtain existence and uniqueness.

Precisely the same argument can be iterated (thanks to the a priori properties illustrated in the previous section) up to the extinction of the saturated zone.

We summarize the above results in the following statement

Theorem 2. *During Stage 2 the weak formulation of Problem 2 has one unique solution (σ, v) with $\sigma \in C^1$ and $v \in V^{1,0}$. The functions G and c can be easily deduced from v in the sets $\{v > 0\}$, $\{v < 0\}$, where they satisfy their respective differential equations in the classical sense. The set $\{v = 0\}$ must have zero measure.*

We conclude the paper by just remarking that the analysis of Stage 3 follows the pattern of the analysis of Stage 1 and the problem of existence and uniqueness Theorem for Problem 3 is in fact a simplification of the parallel result for Problem 1.

References

[1] A. Fasano, L. Fusi, S. Correra, *Mathematical models for waxy crude oils.* To appear in Meccanica (2004).

[2] A. Fasano, M. Primicerio, *Heat and mass transport in non-isothermal partially saturated oil-wax solutions.* To appear in *New Trends in Mathematical Physics*, P. Fergola et al., editors (2004).

[3] A. Friedman, *Partial differential equations of parabolic type.* Prentice-Hall, Englewood Cliffd, N.J. (1964).

[4] O.A. Ladyzenskaya, V.A. Solonnikov, N.N. Uralceva, *Linear and quasilinear equations of parabolic type.* AMS Translations of Mathematical Monographs 27, Providence R.I. (1968).

Antonio Fasano and Mario Primicerio
Dipartimento di Matematica "U.Dini"
Università di Firenze
Viale Morgagni 67/A
I-50134 Firenze, Italia

Progress in Nonlinear Differential Equations
and Their Applications, Vol. 61, 109–124

Solvability of a Free Boundary Problem for the Navier-Stokes Equations Describing the Motion of Viscous Incompressible Nonhomogeneous Fluid

Elena Frolova

In honor of the jubilee of Vsevolod A. Solonnikov

Abstract. We consider a time-dependent problem for a viscous incompressible nonhomogeneous fluid bounded by a free surface on which surface tension forces act. We prove the local in time solvability theorem for this problem in Sobolev function spaces. In the nonhomogeneous model the density of the fluid is unknown. Going over to Lagrange coordinates connected with the velocity vector field, we pass from the free boundary problem to the problem in the fixed boundary domain. Due to the continuity equation, in Lagrange coordinates the density is the same as at the initial moment of time. It gives us the possibility to apply the methods developed by V.A. Solonnikov for the case of incompressible fluid with constant density.

Mathematics Subject Classification (2000). Primary 35Q30,76D05; Secondary 35R35.

Keywords. Free boundary problem, Navier-Stokes equations, nonhomogeneous fluid.

1. Statement of the problem

The motion of incompressible fluids with varying densities in domains with fixed boundaries is considered in [1]–[8]. For the first time, the global existence of a weak solution was established by A.V. Kajikov [1]. In 1975 O.A. Ladizhenskaya and V.A. Solonnikov [3] proved local existence of a strong regular solution, and, for the sufficiently small data, global existence. Later on J. Simon [5] obtained the

This work is parially supported by US Civilian Research and Development Foundation, grant RU-M1-2596-ST04.

global existence theorem for this problem for less regular data than in [3], he also allows the initial density to have some zeros. In the present paper we consider a free boundary problem for a viscous incompressible nonhomogeneous fluid and prove local in time unique solvability of this problem in Sobolev-Slobodetskii spaces. The proof is based on the methods developed by V.A. Solonnikov for the case of incompressible fluid with constant density [9], [10].

We consider a horizontal layer of incompressible fluid with nonconstant density bounded below by a rigid plane and above by a free surface. Let x_1, x_2 are axes on the rigid plane, and the axis x_3 directed towards the upper surface Γ_t. In this case the gravitational force per unity of mass can be written in the form $\mathbf{f} = -g\nabla x_3$. We consider the periodic problem with respect to the variables $x' = (x_1, x_2)$ and denote the periodicity cell by Σ. We assume that the free surface Γ_t can be set by the explicit equation $x_3 = \eta(x', t)$ and has no joint points with the bottom. The problem is to find the domain $\Omega^t = \{(x', x_3)\,/x' \in \Sigma,\ \ 0 < x_3 < \eta(x', t)\}$, $t > 0$ filled by the fluid, the velocity vector field $\mathbf{u}$, the density ρ, and the pressure p which are the solution of the following problem:

$$\begin{aligned}
&\rho\Big(\frac{\partial \boldsymbol{u}}{\partial t} + (\boldsymbol{u}\cdot\nabla)\boldsymbol{u}\Big) - \nu\Delta\boldsymbol{u} = -\nabla p - \rho g \nabla x_3 \quad \text{in} \quad \Omega^t,\\
&\nabla\cdot\boldsymbol{u} = 0 \quad \text{in} \quad \Omega^t,\\
&\frac{d\rho}{dt} = \frac{\partial\rho}{\partial t} + \boldsymbol{u}\cdot\nabla\rho = 0 \quad \text{in} \quad \Omega^t,\\
&\boldsymbol{u}(x', 0, t) = 0,\\
&\boldsymbol{T}\boldsymbol{n} = (\alpha H - p_0)\boldsymbol{n} \quad \text{on} \quad \Gamma_t,\\
&V_n = \boldsymbol{u}\cdot\boldsymbol{n} \quad \text{on} \quad \Gamma_t,
\end{aligned} \tag{1.1}$$

where $\boldsymbol{T} = -p\boldsymbol{I} + \nu\boldsymbol{S}$ is the stress tensor, $\boldsymbol{S} = \frac{\partial u_i}{\partial x_j} + \frac{\partial u_j}{\partial x_i}$ is the strain velocity tensor, ν and α are positive constants (the coefficient of viscosity and the surface tension), H is the double mean curvature of Γ, which is negative for convex domains, $\boldsymbol{I}$ is the unit matrix, $\boldsymbol{n}|_{\Gamma_t} = \boldsymbol{n}(x', \eta(x', t), t)$ is the unit outward normal to the free boundary, V_n is the velocity of the free boundary in the direction of $\boldsymbol{n}$. The positive constants g and p_0 are the gravitational constant and the external pressure respectively.

The initial position of the free boundary $\Gamma_0 = \Gamma$ is known and set by the equation $x_3 = \eta_0(x')$, where $\eta_0(x')$ is a given function. In the domain $\Omega^0 = \{x \in \mathbb{R}^3 \mid 0 < x_3 < \eta(x')\}$ the initial conditions

$$\boldsymbol{u}(x, 0) = \boldsymbol{u}_0(x), \qquad \rho(x, 0) = \rho_0(x) \tag{1.2}$$

hold.

In [11] we investigate stability for problem (1.1) as t tends to infinity. In particular, we obtain control of perturbations in the L_2 norm by initial data (stability in the mean).

In the present paper we prove the local solvability theorem for problem (1.1). Our proof is essentially based on the investigation of a linear non-coercive initial boundary value problem for Stokes system, which arises in the linearization of the free boundary problem. As a consequence of boundary condition $(1.1)_5$, which includes the surface tension on the unknown boundary, in the linear problem we have at the part of the boundary the condition which contains two terms of different orders, neither of which can be considered dominant with respect to the other. The corresponding half-space model problem is studied by V.A. Solonnikov. Namely, the unique solvability of this model problem is proved in weighted Sobolev-Slobodetskii spaces [9]. As at the part of the boundary in the linearized problem we have also Dirichlet boundary condition, we need the solvability theorem for the model half-space problem with Dirichlet boundary condition in the same scale of function spaces. We prove this theorem in Section 5. Then, in Section 6 we obtain the a priori estimates for a solution of a linearized problem. The proof of the existence theorem for linear problem can be done by constructing a regularizer in the same manner as in [9] for incompressible fluid and in [12] for compressible fluid, and we omit the details. Then we use successive approximations and pass to the nonlinear problem. The scheme of this passage is described in details in [10], where the motion of a finite volume of incompressible fluid was studied, and then used in [13] for the case of two liquids.

2. Function spaces

We suppose that all the given functions are periodic with respect to the variables x' and are looking for a solution in the spaces of periodic functions. Let us introduce Sobolev-Slobodetskii spaces of periodic functions.

Let $\Omega \subset \mathbb{R}^3$ be a periodic domain bounded by non-contacting surfaces S_1 and S_2, $\Omega_T = \Omega \times [0,T)$, Σ_a – the periodicity cell connected with the point $a \in \mathbb{R}^2$. For example, a can be a center of the periodicity cell, then

$$\Sigma_a = \left\{x' \in \mathbb{R}^2 \ / \ |x_i - a_i| < \varepsilon_i, \ i = 1,2\right\}.$$

By $\Pi(a)$ we mean $\Sigma_a \times \mathbb{R}$; $\Omega(a) = \Omega \cap \Pi(a)$. By $\tilde{W}_2^r(\Omega)$ we denote the space of periodic functions $u(x)$ such that $u|_{x\in\Omega(a)} \in W_2^r(\Omega(a))$, for any $a \in \mathbb{R}^2$ with the finite norm

$$\|u\|^2_{\tilde{W}_2^r(\Omega)} = \|u\|^2_{W_2^r(\Omega(a))} = \sum_{|\alpha|<r} \|D^\alpha u\|^2_{\Omega(a)} + \|u\|^2_{\dot{W}_2^r(\Omega(a))}.$$

Here we denote by $\|\cdot\|_{\Omega(a)}$ the L_2 norm:

$$\|u\|^2_{\Omega(a)} = \int\limits_{\Omega(a)} |u(x)|^2 dx,$$

and by $\|\cdot\|^2_{\dot{W}_2^r(\Omega(a))}$ the leading term which is equal to

$$\|u\|^2_{\dot{W}_2^r(\Omega(a))} = \sum_{|\alpha|=r} \|D^\alpha u\|^2_{\Omega(a)}$$

for integer r, and to

$$\|u\|^2_{\dot{W}_2^r(\Omega(a))} = \sum_{|\alpha|=[r]} \int\limits_{\Omega(a)} \int\limits_{\Omega(a)} |D^\alpha u(x) - D^\alpha u(y)|^2 \frac{dx\,dy}{|x-y|^{3+2(r-[r])}}$$

for non-integer r. Here $[r]$ is the integral part of r, $D^\alpha u = \dfrac{\partial^{|\alpha|}}{\partial x_1^{\alpha_1} \partial x_2^{\alpha_2} \partial x_3^{\alpha_3}}$ is the generalized derivative of order $|\alpha| = \alpha_1 + \alpha_2 + \alpha_3$.

By $\tilde{W}_2^{r,\,r/2}(\Omega_T)$ we denote the space of periodic functions $u(x,t)$ defined in $\Omega \times [0,T)$ which belong to $W_2^{r,\,r/2}(\Omega_T(a))$ for any $a \in \mathbb{R}^2$ with the finite norm

$$\begin{aligned}
\|u\|^{(r,\,r/2)}_{\Omega_T} &= \|u\|_{\tilde{W}_2^{r,r/2}(\Omega_T)} = \|u\|_{W_2^{r,r/2}(\Omega_T(a))} \\
&= \Big(\|u\|^2_{W_2^{r,0}(\Omega_T(a))} + \|u\|^2_{W_2^{0,\,r/2}(\Omega_T(a))} \Big)^{1/2} \\
&= \Big(\int\limits_0^T \|u\|^2_{W_2^r(\Omega(a))} dt + \int\limits_{\Omega(a)} \|u\|^2_{W_2^{r/2}(0,T)} dx \Big)^{1/2},
\end{aligned}$$

where

$$\|u\|^2_{W_2^r(0,T)} = \sum_{j=0}^{[r]} \Big\| \frac{d^j u}{dt^j} \Big\|^2_{L_2(0,T)} + \int\limits_0^T dt \int\limits_0^t \Big| \frac{d^{[r]} u(t)}{dt^{[r]}} - \frac{d^{[r]} u(\tau)}{d\tau^{[r]}} \Big|^2 \frac{d\tau}{|t-\tau|^{1+2(r-[r])}}.$$

Periodic spaces $\tilde{W}_2^r(S)$ and $\tilde{W}_2^{r,\,r/2}(S_T)$ of functions defined on the periodic surface $S \subset \mathbb{R}^3$ are introduced in the similar way. To analyze problems with zero initial conditions, we use Sobolev-Slobodetskii spaces $H_\gamma^{r,r/2}(\Omega_T)$ with exponential weight, which elements admit zero extension to the domain $t < 0$ without loss of smoothness. These spaces were introduced by M.S. Agranovich and M.I. Vishik [16]. We denote by $\tilde{H}_\gamma^{r,r/2}(\Omega_T)$ the space of periodic functions $u(x,t)$ satisfying the zero initial conditions

$$\frac{\partial^j u}{\partial t^j}\Big|_{t=0} = 0, \qquad for \quad j = 0, \dots, [r/2] - 1$$

and belonging to $H_\gamma^{r,r/2}(\Omega_T(a))$ for any $a \in \mathbb{R}^2$, with the finite norm

$$\|u\|^2_{\tilde{H}_\gamma^{r,r/2}(\Omega_T)} = \|u\|^2_{H_\gamma^{r,r/2}(\Omega_T(a))} = \|u\|^2_{H_\gamma^{r,0}(\Omega_T(a))} + \|u\|^2_{H_\gamma^{0,r/2}(\Omega_T(a))},$$

where

$$\|u\|^2_{H_\gamma^{r,0}(\Omega_T(a))} = \int\limits_0^T e^{-2\gamma t} \Big(\|u\|^2_{\dot{W}_2^r(\Omega(a))} + \gamma^r \|u\|^2_{\Omega(a)} \Big) dt,$$

$$\|u\|^2_{H_\gamma^{0,r/2}(\Omega_T(a))} = \int\limits_0^T e^{-2\gamma t} \Big\| \frac{\partial^{r/2} u}{\partial t^{r/2}} \Big\|^2_{\Omega(a)} dt$$

for integer $r/2$, and

$$\|u\|^2_{H^{0,r/2}_\gamma(\Omega_T(a))} = \int_0^T e^{-2\gamma t} \int_0^\infty \left\| \frac{\partial^{[r/2]} u^0(\cdot, t-\tau)}{\partial t^{[r/2]}} - \frac{\partial^{[r/2]} u^0(\cdot, t)}{\partial t^{[r/2]}} \right\|^2_{\Omega(a)} \frac{d\tau}{\tau^{1+r-[r]}},$$

for non-integer $r/2$. Here

$$u^0(\cdot, t) = u(\cdot, t) \quad for \quad t > 0, \qquad u^0(\cdot, t) = 0 \quad for \quad t \le 0.$$

The detailed list of properties of weighted Sobolev-Slobodetskii spaces can be found in [9].

The norms in the spaces of vector fields $\boldsymbol{u}$ having components from $\tilde{W}_2^r(\Omega)$ $\left(\tilde{W}_2^{r,\,r/2}(\Omega_T)\right)$ are defined by the same formulas with replacement of the function u by the vector field $\boldsymbol{u}$.

3. Lagrange coordinates

Due to conditions $(1.1)_6$, $(1.1)_4$, domain Ω^t, $t > 0$ can be determined on the base of the given domain Ω^0 as the set of points $x = x(\xi, t)$ such that

$$\frac{\partial x}{\partial t} = \boldsymbol{u}(x, t), \qquad x(\xi, 0) = \xi \in \Omega^0.$$

Consequently, connection between Euler and Lagrange coordinates of a particle being at $t = 0$ in the point ξ takes the form

$$x = \xi + \int_0^t \boldsymbol{v}(\xi, \tau) d\tau \equiv X_v(\xi), \tag{3.1}$$

where $\boldsymbol{v}(\xi, t)$ is the velocity vector field expressed in Lagrange coordinates.

We denote by $J_{\boldsymbol{v}}(\xi, t)$ the Jacobi matrix of transform (3.1) with the elements $a_{ij} = \delta_i^j + \int_0^t \frac{\partial v_i}{\partial \xi_j} d\tau$. Then the Jacobi matrix of the inverse transform has the elements

$$(\det J_{\boldsymbol{v}}(\xi, t))^{-1} A_{ij}, \quad i, j = 1, 2, 3,$$

where A_{ij} is the algebraic complement of the element a_{ij}, and gradient in the variables ξ connected with the vector field $\boldsymbol{v}$ can be expressed by the following formula

$$\nabla_{\boldsymbol{v}} = A\nabla = \left(\sum_{m=1}^3 A_{1m} \frac{\partial}{\partial \xi_m}, \sum_{m=1}^3 A_{2m} \frac{\partial}{\partial \xi_m}, \sum_{m=1}^3 A_{3m} \frac{\partial}{\partial \xi_m} \right),$$

here $A = \{A_{ij}\}_{i,j=1}^3$ is the cofactor matrix for the matrix $J_{\boldsymbol{v}}(\xi, t)$.

Continuity equation $\frac{d\rho}{dt} = \frac{\partial \rho}{\partial t} + \boldsymbol{u} \cdot \nabla \rho = 0$ means that in Lagrange coordinates determined by (3.1), the density ρ is the same as at the initial moment of time, which is known. Going over to Lagrange coordinates, we pass from the free boundary problem (1.1) to the following problem in a domain with a fixed boundary

$$
\begin{aligned}
&\rho_0 \boldsymbol{v}_t - \nu \nabla^2_{\boldsymbol{v}} \boldsymbol{v} + \nabla_{\boldsymbol{v}} q = \rho_0 g \nabla \xi_3 \quad \text{in} \quad \Omega^0,\ t>0,\\
&\nabla_{\boldsymbol{v}} \cdot \boldsymbol{v} = 0 \quad \text{in} \quad \Omega^0, \quad t>0,\\
&\boldsymbol{v}\Big|_{t=0} = \boldsymbol{u}_0, \qquad \boldsymbol{v}\Big|_{\xi\in S} = 0,\\
&\boldsymbol{T}_{\boldsymbol{v}} \boldsymbol{n} - \sigma \Delta(t) X_{\boldsymbol{v}}\Big|_{\xi\in\Gamma_0} = p_0 \boldsymbol{n},
\end{aligned} \tag{3.2}
$$

where q is the pressure expressed in Lagrange coordinates, $S = \{\xi \in \mathbb{R}^3 | \xi_3 = 0\}$, $\boldsymbol{n} = (A\boldsymbol{n}_0)/|A\boldsymbol{n}_0|$, $\mathbf{n_0}(\xi)$ is the unit vector of the outer normal to Γ_0 at the point ξ, $\Delta(t)$ is the Laplace-Beltrami operator on $\Gamma_t = \{x = X_{\boldsymbol{v}}(\xi,t), \quad \xi \in \Gamma_0\}$,

$$
\begin{aligned}
&\boldsymbol{T}_{\boldsymbol{v}} = -q\boldsymbol{I} + \nu \boldsymbol{S}_{\boldsymbol{v}}(\boldsymbol{v}),\\
&(\boldsymbol{S}_{\boldsymbol{v}})_{ij} = \sum_{m=1}^{3} \left(A_{im} \frac{\partial v_j}{\partial \xi_m} + A_{jm} \frac{\partial v_i}{\partial \xi_m} \right).
\end{aligned}
$$

As it is suggested in [9], [10], we introduce for an arbitrary vector field $\boldsymbol{w}$ defined on Γ the projections Π_0 and Π onto the tangent planes to Γ and Γ_t, we have

$$
\Pi_0 \boldsymbol{w} = \boldsymbol{w} - \boldsymbol{n}_0(\boldsymbol{w}\cdot\boldsymbol{n}_0), \quad \Pi \boldsymbol{w} = \boldsymbol{w} - \boldsymbol{n}(\boldsymbol{w}\cdot\boldsymbol{n}).
$$

Then we separate condition $(3.2)_5$ on tangential and normal components. For $\boldsymbol{n}\cdot\boldsymbol{n}_0 > 0$, we have

$$
\nu \Pi_0 \Pi S_{\boldsymbol{v}}(\boldsymbol{v})\boldsymbol{n} = 0, \quad \boldsymbol{n}_0 \cdot \boldsymbol{T}_{\boldsymbol{v}}(\boldsymbol{v},q)\boldsymbol{n} - \sigma \boldsymbol{n}_0 \cdot \Delta(t) X_{\boldsymbol{n}} = 0 \quad \text{on} \quad \Gamma_0. \tag{3.3}
$$

Solvability theorem for problem (3.2) is formulated as follows:

Theorem 3.1. *Let $\Gamma_0 \in \tilde{W}_2^{5/2+l}(\mathbb{R}^2)$, $\boldsymbol{u}_0 \in \tilde{W}_2^{1+l}(\Omega^0)$ with some $l \in (1/2, 1)$, $\rho_0 \in \tilde{W}_2^{1+l}(\Omega^0)$, $\rho_0(x) \geq \beta > 0$, and the compatibility conditions*

$$
\nabla\cdot\boldsymbol{u}_0 = 0, \qquad \Pi_0 \boldsymbol{S}(\boldsymbol{u}_0)\boldsymbol{n}_0\Big|_{\xi\in\Gamma_0} = 0, \qquad \boldsymbol{u}_0\Big|_{x_3=0} = 0
$$

hold.

Then there exists a unique solution $(\boldsymbol{v}, q)$ to problem (3.2) defined on some interval of time $(0,T)$ and having the following regularity properties:

$$
\begin{aligned}
&\boldsymbol{v} \in \tilde{W}_2^{l+2,l/2+1}(\Omega^0_T), \quad q \in \tilde{W}_2^{l,\,l/2}(\Omega^0_T), \quad \nabla q \in \tilde{W}_2^{l,l/2}(\Omega^0_T),\\
&q\Big|_{\Gamma_T} \in \tilde{W}_2^{l+1/2,l/2+1/4}(\Gamma_T), \quad \Omega^0_T = \Omega^0 \times [0,T), \quad \Gamma_T = \Gamma \times [0,T).
\end{aligned}
$$

This solution satisfies the following estimate:

$$
\begin{aligned}
&\|\boldsymbol{v}\|^{(l+2,l/2+1)}_{\Omega^0_T} + \|q\|^{(l,l/2)}_{\Omega^0_T} + \|\nabla q\|^{(l,l/2)}_{\Omega^0_T} + \|q\|^{(l+1/2,l/2+1/4)}_{\Gamma_T}\\
&\leq C\left(\|\eta_0\|_{\tilde{W}_2^{5/2+l}(\mathbb{R}^2)}, \|\rho_0\|_{\tilde{W}_2^{1+l}(\Omega^0)} \right) \|\boldsymbol{u}_0\|_{\tilde{W}_2^{l+1}(\Omega^0)}.
\end{aligned} \tag{3.4}
$$

This theorem is proved by successive approximations, similar to the case of incompressible fluid with constant density [10]. Namely, we set $\boldsymbol{v}^{(0)}=0$, $q^{(0)}=0$ and for $m\in\mathbb{N}\cup\{0\}$ determine $\boldsymbol{v}^{(m+1)}$, $q^{(m+1)}$ as a solution of the following linear problem

$$\begin{aligned}
&\rho_0\boldsymbol{v}_t^{(m+1)}-\nu\nabla_m^2\boldsymbol{v}^{(m+1)}+\nabla_m q^{(m+1)}=\rho_0 g\nabla\xi_3, \quad \xi\in\Omega^0,\\
&\nabla_m\cdot\boldsymbol{v}^{(m+1)}=0, \quad \xi\in\Omega^0,\\
&\boldsymbol{v}^{(m+1)}\Big|_{t=0}=\boldsymbol{u}_0,\\
&\boldsymbol{v}^{(m+1)}\Big|_{\xi\in S}=0,\\
&\nu\Pi_0\Pi_m\boldsymbol{S}_m(\boldsymbol{v}^{(m+1)})n_m\Big|_{\xi\in\Gamma}=0, \qquad (3.5)\\
&\boldsymbol{n}_0\cdot\boldsymbol{T}_m(\boldsymbol{v}^{(m+1)},q^{(m+1)})\boldsymbol{n}_m-\sigma\boldsymbol{n}_0\cdot\Delta_m(t)\int_0^t\boldsymbol{v}^{(m+1)}d\tau\Big|_{\xi\in\Gamma}\\
&=\sigma H_0(\xi)+\sigma\int_0^t\boldsymbol{n}_0\cdot\Delta_m(\tau)\xi d\tau\Big|_{\xi\in\Gamma},
\end{aligned}$$

here by ∇_m we mean $\nabla_{\boldsymbol{v}^{(m)}}$, $\boldsymbol{S}_m=\boldsymbol{S}_{\boldsymbol{v}^{(m)}}$, $\boldsymbol{T}_m=\boldsymbol{T}_{\boldsymbol{v}^{(m)}}$, $\boldsymbol{n}_m$ is the outward normal to $\Gamma_m(t)=\{x=X_m(\xi,t),\ \xi\in\Gamma\}$, $X_m=X_{\boldsymbol{v}^{(m)}}$, $\Delta_m(t)$ is the Laplace-Beltrami operator on $\Gamma_m(t)$.

4. Auxiliary linear problem

Existence of a solution to problem (3.5) for $m\geq 0$ follows from the solvability of the linear problem:

$$\begin{aligned}
&\rho_0\boldsymbol{w}_t-\nu\nabla_{\boldsymbol{v}}^2\boldsymbol{w}+\nabla_{\boldsymbol{v}}q=\boldsymbol{f} \quad \text{in} \quad \Omega^0, \quad t>0,\\
&\nabla_{\boldsymbol{v}}\cdot\boldsymbol{w}=\varphi(\xi,t) \quad \text{in} \quad \Omega^0,\ t>0,\\
&\boldsymbol{w}\Big|_{t=0}=\boldsymbol{w}_0,\\
&\boldsymbol{w}\Big|_{\xi\in S}=0, \qquad (4.1)\\
&\nu\Pi_0\Pi S_{\boldsymbol{v}}(\boldsymbol{w})\boldsymbol{n}\Big|_{\xi\in\Gamma}=\Pi_0 d,\\
&\boldsymbol{n}_0\cdot\boldsymbol{T}_{\boldsymbol{v}}(\boldsymbol{w},q)\boldsymbol{n}-\sigma\boldsymbol{n}_0\cdot\Delta(t)\int_0^t\boldsymbol{w}d\tau|_{\xi\in\Gamma}=\boldsymbol{b}+\sigma\int_0^t\boldsymbol{B}d\tau,
\end{aligned}$$

where $\boldsymbol{v}\in\tilde{W}_2^{l+2,l/2+1}(\Omega_T^0)$ is a given vector field, $\Delta(t)$ is the Laplace-Beltrami operator on the surface $\Gamma_t=\{x=X_{\boldsymbol{v}}(\xi,t),\ \xi\in\Gamma\}$.

Following the scheme suggested in [9], at first, we consider problem (4.1) with $\boldsymbol{v} = 0$ and zero initial conditions, it is reduced to the problem

$$\begin{aligned}
&\boldsymbol{w}_t - \frac{\nu}{\rho_0(x)}\nabla^2 \boldsymbol{w} + \frac{1}{\rho_0(x)}\nabla q = \boldsymbol{f} \quad \text{in} \quad \Omega^0, \quad t > 0,\\
&\nabla \cdot \boldsymbol{w} = \varphi \qquad \text{in} \quad \Omega^0,\\
&\boldsymbol{w}\big|_{t=0} = 0 \qquad \text{in} \quad \Omega^0,\\
&\boldsymbol{w}\big|_{\xi \in S} = 0, \qquad\qquad (4.2)\\
&\nu \Pi_0 \Pi S(\boldsymbol{w})\boldsymbol{n}\big|_{\xi\in\Gamma} = \Pi_0 d,\\
&\Big(\boldsymbol{n}_0 \cdot \boldsymbol{T}(\boldsymbol{w}, q)\boldsymbol{n} - \sigma \boldsymbol{n}_0 \cdot \Delta \int_0^t \boldsymbol{w} d\tau\Big)\big|_{\xi\in\Gamma} = \boldsymbol{b} + \sigma \int_0^t \boldsymbol{B} d\tau.
\end{aligned}$$

The solvability theorem to problem (4.2) in weighted Sobolev-Slobodetskii spaces is formulated as follows.

Theorem 4.1. *Let $\eta_0 \in \tilde{W}_2^{3/2+l}(\mathbb{R}^2)$, and $\rho_0 \in \tilde{W}_2^{1+l}(\Omega_0)$ with some $l > 1/2$ and satisfies the inequality $0 < \beta \leq \rho_0(x)$. Then for any $\boldsymbol{f} \in \tilde{H}_\gamma^{l,l/2}(\Omega_T^0)$, $\varphi \in \tilde{H}_\gamma^{1+l,1/2+l/2}(\Omega_T^0)$ such that $\varphi = \nabla \cdot \Phi$, $\Phi \in \tilde{H}_\gamma^{0,1/2+l}(\Omega_T^0)$, $d \in \tilde{H}_\gamma^{1/2+l,1/4+l/2}(\Gamma_T)$, $\boldsymbol{b} \in \tilde{H}_\gamma^{l+1/2,l/2+1/4}(\Gamma_T)$, $\boldsymbol{B} \in \tilde{H}_\gamma^{l-1/2,l/2-1/4}(\Gamma_T)$, where $T \leq +\infty$ and the number γ is supposed to be sufficiently large ($\gamma \geq \gamma_0 > 1$), problem (4.2) has a unique solution $(\boldsymbol{w}, q)$ with the following properties:*

$$\boldsymbol{w} \in \tilde{H}_\gamma^{l+2,l/2+1}(\Omega_T^0), \quad q \in \tilde{H}_\gamma^{l,l/2}(\Omega_T^0), \quad \nabla q \in \tilde{H}_\gamma^{l,l/2}(\Omega_T^0),$$

and

$$\begin{aligned}
&\|\boldsymbol{w}\|_{\tilde{H}_\gamma^{l+2,l/2+1}(\Omega_T^0)} + \|q\|_{\tilde{H}_\gamma^{l,l/2}(\Omega_T^0)} + \|\nabla q\|_{\tilde{H}_\gamma^{l,l/2}(\Omega_T^0)}\\
&\leq C\Big(\|\boldsymbol{f}\|_{\tilde{H}_\gamma^{l,l/2}(\Omega_T^0)} + \|\varphi\|_{\tilde{H}_\gamma^{l+1,l/2+1/2}(\Omega_T^0)} + \|\Phi\|_{\tilde{H}_\gamma^{l+1,l/2+1/2}(\Omega_T^0)} \qquad (4.3)\\
&+\|\boldsymbol{d}\|_{\tilde{H}_\gamma^{l+1/2,l/2+1/4}(\Gamma_T)} + \|\boldsymbol{b}\|_{\tilde{H}_\gamma^{l+1/2,l/2+1/4}(\Gamma_T)} + \sigma\|\boldsymbol{B}\|_{\tilde{H}_\gamma^{l-1/2,l/2-1/4}(\Gamma_T)}\Big),
\end{aligned}$$

where the constant C depends on ρ_0, η_0 and can be chosen not depending on T.

To prove Theorem 4.1, we need to analyze the auxiliary model half-space problems.

5. Model problems

The non-coercive model problem arising as a consequence of the fact that condition $(1.1)_5$ includes the surface tension on the free boundary has the form:

$$\begin{aligned}
&\boldsymbol{v}_t - \frac{\nu}{\rho}\nabla^2\boldsymbol{v} + \frac{1}{\rho}\nabla p = \boldsymbol{f} \qquad \text{in} \quad \mathbb{R}^3_{+,T} = \mathbb{R}^3_+ \times [0,T),\\
&\nabla\cdot\boldsymbol{v} = r \qquad\qquad \text{in} \quad \mathbb{R}^3_{+,T},\\
&\boldsymbol{v}\Big|_{t=0} = 0 \qquad\qquad \text{in} \quad \mathbb{R}^3_+,\\
&\nu\Big(\frac{\partial v_3}{\partial y_j} + \frac{\partial v_j}{\partial y_3}\Big)\Big|_{y_3=0} = b_j(y',t), \qquad j=1,2,\\
&\Big(-p + 2\nu\frac{\partial v_3}{\partial y_3} + \sigma\int_0^t \nabla'^2 v_3 d\tau\Big)\Big|_{y_3=0} = b_3(y',t) + \sigma\int_0^t \boldsymbol{B}d\tau,
\end{aligned} \tag{5.1}$$

here ν, ρ, σ are positive constants.

This problem contains in the boundary condition two terms of different orders neither of which can be regarded as a dominant term with respect to the other and so being nonstandard. Problem (5.1) (with $\rho = 1$) is carefully analyzed by V.A. Solonnikov [9]. In particular, the solvability theorem for this problem in Sobolev-Slobodetskii spaces is proved. It is clear that in the case when the positive constant ρ is not equal to 1, the solvability result is just the same while at the estimates we have a constant depending on ρ.

Theorem 5.1. *Let $l > 1/2$, $\gamma > 0$. We assume that $\boldsymbol{f} \in H^{l,l/2}_\gamma(\mathbb{R}^3_{+,T})$, $b_j \in H^{l+1/2,l/2+1/4}_\gamma(\mathbb{R}^2_T)$, $j = 1,2$, $b_3 \in H^{l+1/2,1/2,l/2}_\gamma(\mathbb{R}^2_T)$, $\boldsymbol{B} \in H^{l-1/2,l/2-1/4}_\gamma(\mathbb{R}^2_T)$, $r \in H^{l+1,l/2+1/2}_\gamma(\mathbb{R}^3_{+,T})$, and $r = \nabla\cdot R$, $R \in H^{0,l/2+1}_\gamma(\mathbb{R}^3_{+,T})$. Problem (5.1) has a unique solution $(\boldsymbol{v},p)$ such that $\boldsymbol{v} \in H^{l+2,l/2+1}_\gamma(\mathbb{R}^3_{+,T})$, $\nabla p \in H^{l,l/2}_\gamma(\mathbb{R}^3_{+,T})$, and the following estimate*

$$\begin{aligned}
&\|\boldsymbol{v}\|^2_{H^{l+2,l/2+1}_\gamma(\mathbb{R}^3_{+,T})} + \|\nabla p\|^2_{H^{l,l/2}_\gamma(\mathbb{R}^3_{+,T})} \le c(\gamma,\rho)\Big(\|\boldsymbol{f}\|^2_{H^{l,l/2}_\gamma(\mathbb{R}^3_{+,T})}\\
&+\|r\|^2_{H^{l+1,l/2+1/2}_\gamma(\mathbb{R}^3_{+,T})} + \|\boldsymbol{R}\|^2_{H^{0,l/2+1}_\gamma(\mathbb{R}^3_{+,T})} + \|b_1\|^2_{H^{l+1/2,l/2+1/4}_\gamma(\mathbb{R}^2_T)}\\
&+\|b_2\|^2_{H^{l+1/2,l/2+1/4}_\gamma(\mathbb{R}^2_T)} + \sigma^2\|B\|^2_{H^{l-1/2,l/2-1/4}_\gamma(\mathbb{R}^2_T)} + \|b_3\|^2_{H^{l+1/2,l/2,l/2}_\gamma(\mathbb{R}^2_T)}\Big)
\end{aligned} \tag{5.2}$$

holds true.

The model problem with the Dirichlet boundary condition has the form

$$\begin{aligned}
&\boldsymbol{v}_t - \frac{\nu}{\rho}\nabla^2\boldsymbol{v} + \frac{1}{\rho}\nabla p = \boldsymbol{f} \quad \text{in} \quad \mathbb{R}^3_{+,T},\\
&\nabla\cdot\boldsymbol{v} = g \quad \text{in} \quad \mathbb{R}^3_{+,T},\\
&\boldsymbol{v}|_{t=0} = 0 \quad \text{in} \quad \mathbb{R}^3_T, \qquad \boldsymbol{v}|_{x_3=0} = \boldsymbol{\psi},
\end{aligned} \tag{5.3}$$

where ν and ρ are positive constants.

Theorem 5.2. *Let $l > \frac{1}{2}$, $\gamma > 0$. We assume that*

$$\boldsymbol{f} \in H_\gamma^{l,l/2}(\mathbb{R}^3_{+,T}), \qquad \boldsymbol{\psi} \in H_\gamma^{l+3/2,l/2+3/4}(\mathbb{R}^2_T), \quad \psi_3 = 0,$$
$$g \in H_\gamma^{l+1,l/2+1/2}(\mathbb{R}^3_{+,T}), \qquad g = \nabla \cdot \boldsymbol{R}, \quad \text{and} \quad \boldsymbol{R} \in H_\gamma^{0,l/2+1}(\mathbb{R}^3_{+,T}).$$

Problem (5.3) *has a unique solution* $\boldsymbol{v} \in H_\gamma^{l+2,l/2+1}(\mathbb{R}^3_{+,T})$, $\nabla p \in H_\gamma^{l,l/2}(\mathbb{R}^3_{+,T})$, *and the following estimate*

$$\|\boldsymbol{v}\|^2_{H_\gamma^{l+2,l/2+1}(\mathbb{R}^3_{+,T})} + \|\nabla p\|^2_{H_\gamma^{l,l/2}(\mathbb{R}^3_{+,T})} \leq c(\gamma,\rho)\Big(\|\boldsymbol{f}\|^2_{H_\gamma^{l,l/2}(\mathbb{R}^3_{+,T})} \tag{5.4}$$
$$+\|g\|^2_{H_\gamma^{l+2,l/2+1}(\mathbb{R}^3_{+,T})} + \|\boldsymbol{R}\|^2_{H_\gamma^{0,l/2+1}(\mathbb{R}^3_{+,T})} + \|\boldsymbol{\psi}\|^2_{H_\gamma^{l+3/2,l/2+3/4}(\mathbb{R}^2_T)}\Big)$$

holds true.

At first we consider problem (5.3) with homogeneous equations. In this case we extend all the given functions for $t \in [0, +\infty)$ with preservation of class and use the Laplace-Fourier transform

$$\tilde{u}(\xi_1, \xi_2, x_3, s) = \int_0^{+\infty} e^{-st}dt \int_{\mathbb{R}^2} u(x', x_3, t)e^{-i(x_1\xi_1 + x_2\xi_2)}dx'.$$

We arrive at

$$\nu\left(-\frac{d^2}{dx_3^2} + r^2\right)\tilde{v}_j + i\xi_j\tilde{p} = 0, \quad j = 1, 2,$$
$$\nu\left(-\frac{d^2}{dx_3^2} + r^2\right)\tilde{v}_3 + \frac{d\tilde{p}}{dx_3} = 0, \tag{5.5}$$

where $r^2 = \frac{s\rho}{\nu} + \xi^2$, $\xi^2 = \xi_1^2 + \xi_2^2$, $\arg(r) \in (-\frac{\pi}{4}, \frac{\pi}{4})$,

$$i\xi_1\tilde{v}_1 + i\xi_2\tilde{v}_2 + \frac{d\tilde{v}_3}{dx_3} = 0,$$
$$\tilde{v}_j|_{x_3=0} = \tilde{\psi}_j, \quad j = 1, 2, 3. \quad \tilde{v}_j \to 0, \quad p \to 0 \quad \text{for} \quad x_3 \to +\infty.$$

Problem (5.5) can be solved in the explicit form in the same way as it is done in [14].

$$\tilde{v}_1 = \tilde{\psi}_1 e^{-rx_3} + \frac{(\xi_1^2\tilde{\psi}_1 + \xi_1\xi_2\tilde{\psi}_2)}{|\xi|}\frac{e^{-rx_3} - e^{-|\xi|x_3}}{r - |\xi|}, \tag{5.6}$$
$$\tilde{v}_2 = \tilde{\psi}_2 e^{-rx_3} + \frac{(\xi_2^2\tilde{\psi}_2 + \xi_1\xi_2\tilde{\psi}_1)}{|\xi|}\frac{e^{-rx_3} - e^{-|\xi|x_3}}{r - |\xi|},$$
$$\tilde{v}_3 = \left(i\xi_1\tilde{\psi}_1 + i\xi_2\tilde{\psi}_2\right)\frac{e^{-rx_3} - e^{-|\xi|x_3}}{r - |\xi|},$$
$$p = -\frac{\rho s\left(i\xi_1\tilde{\psi}_1 + i\xi_2\tilde{\psi}_2\right)}{|\xi|(r - |\xi|)}e^{-|\xi|x_3}.$$

The equivalent norm in $H_\gamma^{l,l/2}(\mathbb{R}^2_\infty)$ connected with the Laplace-Fourier transform has the form

$$|||\, u\, |||^2_{l,\gamma,\mathbb{R}^2_\infty} = \int\limits_{\mathbb{R}^2} d\xi \int\limits_{-\infty}^{+\infty} |\tilde{u}(\xi,\gamma+i\xi_0)|^2(|s|+\xi^2)^l d\xi_0, \quad s=\gamma+i\xi_0,$$

and in the space $H_\gamma^{l,l/2}(\mathbb{R}^3_{+,\infty})$

$$|||\, u\, |||^2_{l,\gamma,\mathbb{R}^3_{+,\infty}} = \sum_{j<l} \int\limits_{\mathbb{R}^2} d\xi \int\limits_{-\infty}^{+\infty} \left\| \frac{\partial^j}{\partial x_3} \tilde{u}(\xi,\cdot,s) \right\|^2_{\mathbb{R}_+} (\xi^2+|s|)^{l-j} d\xi_0$$
$$+ \int\limits_{\mathbb{R}^2} d\xi \int\limits_{-\infty}^{+\infty} \|\, \tilde{u}(\xi,\cdot,s)\, \|^2_{\dot{W}^l_2(\mathbb{R}_+)}\, d\xi_0.$$

Theorem 5.3. *Let in problem* (5.3) $\boldsymbol{f}=0$, $\boldsymbol{g}=0$. *For any* $\boldsymbol{\psi} \in H_\gamma^{l+3/2,l/2+3/4}(\mathbb{R}^2_\infty)$ *with* $\psi_3 = 0$, *solution* (5.6) *of problem* (5.3) *satisfies the following estimate*

$$\sum_{j=1}^{3} |||\, v_k\, |||^2_{l+2,\gamma,\mathbb{R}^3_{+,\infty}} + |||\, \nabla p\, |||^2_{l,\gamma,\mathbb{R}^3_{+,\infty}} \le c(\gamma,\rho) \sum_{k=1}^{2} |||\, \psi_k\, |||^2_{l+3/2,\mathbb{R}^2_\infty},$$

where the constant $c(\gamma,\rho)$ *remains bounded for* $\gamma \ge \gamma_0 > 0$.

Proof. The proof of Theorem 5.3 is based on the estimates of the functions $e_0(x_3) = e^{-rx_3}$ and $e_1(x_3) = \frac{e^{-rx_3}-e^{-|\xi|x_3}}{r-|\xi|}$ proved in [9].

Lemma 5.4. *(Lemma 3.1 in* [9]*) For any* $\xi \in \mathbb{R}^2$, $s=\gamma+i\xi_0$, $\gamma>0$ *the functions* $e_0(x_3)$, $e_1(x_3)$ *satisfy the following estimates*

$$\begin{aligned}
&\left\| \frac{d^j e_0(x_3)}{dx_3^j} \right\|^2_{L_2(\mathbb{R}_+)} \le \frac{1}{\sqrt{2}} |r|^{2j-1},\\
&\left\| \frac{d^j e_1(x_3)}{dx_3^j} \right\|^2_{L_2(\mathbb{R}_+)} \le \frac{c|r|^{2j-1}+|\xi|^{2j-1}}{|r|^2},\\
&\int\limits_0^{+\infty}\int\limits_0^{+\infty} \left| \frac{d^j e_0(x_3+z)}{dx_3^j} - \frac{d^j e_0(x_3)}{dx_3^j} \right|^2 \frac{dx_3 dz}{z^{1+2\varsigma}} \le c|r|^{2(j+\varsigma)-1},\\
&\int\limits_0^{+\infty}\int\limits_0^{+\infty} \left| \frac{d^j e_1(x_3+z)}{dx_3^j} - \frac{d^j e_1(x_3)}{dx_3^j} \right|^2 \frac{dx_3 dz}{z^{1+2\varsigma}} \le c\frac{|r|^{2(j+\varsigma)-1}+|\xi|^{2(j+\varsigma)-1}}{|r|^2}.
\end{aligned} \tag{5.7}$$

Here $j \ge 0$, $\varsigma \in (0,1)$. *Constants in* (5.7) *are independent on* r, $|\xi|$.

Taking into account that $|r|^2 \le c(|s|+\xi^2)$ and using (5.7), we have for the function $v_1(x',x_3,t)$ the estimate:

$$
\begin{aligned}
&|||\, v_1 \,|||^2_{l+2,\gamma,\mathbb{R}^3_{+,\infty}} \\
&\le \int\limits_{\mathbb{R}^2} d\xi \int\limits_{-\infty}^{+\infty} \Bigg(|\tilde{\psi}_1|^2 \Big(\sum_{j<l+2} \Big\| \frac{d^j e_0(x_3)}{dx_3^j} \Big\|^2_{L_2(\mathbb{R}_+)} (|s|+\xi^2)^{l+2-j} + \|e_0(x_3)\|_{\dot{W}_2^{l+2}(\mathbb{R}_+)} \Big) \\
&\quad + \frac{|\xi_1^2\tilde{\psi}_1 + \xi_1\xi_2\tilde{\psi}_2|^2}{|\xi|^2} \Big(\sum_{j<l+2} \Big\| \frac{d^j e_1(x_3)}{dx_3^j} \Big\|^2_{L_2(\mathbb{R}_+)} \\
&\qquad \times (|s|+\xi^2)^{l+2-j} + \|e_1(x_3)\|_{\dot{W}_2^{l+2}(\mathbb{R}_+)} \Big) \Bigg) d\xi_0 \\
&\le c \int\limits_{\mathbb{R}^2} d\xi \int\limits_{-\infty}^{+\infty} \sum_{k=1}^{2} |\tilde{\psi}_k|^2 \Big(\sum_{j<l+2} |r|^{2j-1}(|s|+\xi^2)^{l+2-j} + |r|^{2(l+2)-1} + |\xi|^{2(l+2)-1} \Big) d\xi_0 \\
&\le c \sum_{k=1}^{2} \int\limits_{\mathbb{R}^2} d\xi \int\limits_{-\infty}^{+\infty} |\tilde{\psi}_k|^2 (|s|+\xi^2)^{l+3/2} d\xi_0 = c \sum_{k=1}^{2} |||\, \psi_k \,|||^2_{l+3/2,\mathbb{R}^2_\infty} .
\end{aligned}
$$

Similar estimates are evidently valid for $|||\, v_k \,|||^2_{l+2,\gamma,\mathbb{R}^3_{+,\infty}}$, $k=2,3$. To estimate $|||\, \nabla p \,|||^2_{l,\gamma,\mathbb{R}^3_{+,\infty}}$, we use estimates for the function $e^{-|\xi|x_3}$, which can be obtained from estimates (5.7) for $e_0(x_3)$ by replacement r to $|\xi|$. We have:

$$
\begin{aligned}
|||\, \nabla p \,|||^2_{l,\gamma,\mathbb{R}^3_{+,\infty}} &\le \rho^2 \int\limits_{\mathbb{R}^2} d\xi \int\limits_{-\infty}^{+\infty} \frac{|s|^2|\xi|^2(|\tilde{\psi}_1|^2 + |\tilde{\psi}_2|^2)}{|r-|\xi||^2} \\
&\Bigg(\sum_{j<l} \Big\| \frac{d^j}{dx_3^j}(e^{-|\xi|x_3}) \Big\|^2_{L_2(\mathbb{R}_+)} (|s|+\xi^2)^{l-j} + \|e^{-|\xi|x_3}\|_{\dot{W}_2^{l}(\mathbb{R}_+)} \Bigg) d\xi_0 \\
&\le c\rho \int\limits_{\mathbb{R}^2} d\xi \int\limits_{-\infty}^{+\infty} |s||\xi|^2 \sum_{k=1}^{2} |\tilde{\psi}_k|^2 \Bigg(\sum_{j<l} |\xi|^{2j-1}(|s|+\xi^2)^{l-j} + |\xi|^{2l-1} \Bigg) d\xi_0 \\
&\le c \sum_{k=1}^{2} \int\limits_{\mathbb{R}^2} d\xi \int\limits_{-\infty}^{+\infty} |\tilde{\psi}_k|^2 (|s|+\xi^2)^{l+3/2} d\xi_0 = c \sum_{k=1}^{2} |||\, \psi_k \,|||^2_{l+3/2,\mathbb{R}^2_\infty} .
\end{aligned}
$$

□

Theorem 5.3 is proved. Theorem 5.2 is deduced from Theorem 5.3 on the base of the same arguments as in [15] by construction the auxiliary vector field. As it is clearly described in [15] (Section 9), under this construction, the assumption that $v_3 = 0$ on the plane $x_3 = 0$ can be preserved.

6. Proof of Theorem 4.1

Existence of a solution to linear problem (4.2) can be proved by construction of a regularizer in the same way as it is done in [9] for the linear problem corresponding to the case of incompressible homogeneous fluid and in [12] for the linear problem corresponding to the case of compressible fluid. An example of construction of a regularizer in a periodic layer is described in [17]. Here we concentrate on estimate (4.3) and give the scheme of checking its validity as an a priori one by the Schauder method. We introduce the partition of unity

$$1 = \sum_{j=1}^{N} \zeta_j(x), \qquad x \in \overline{\Omega},$$

where $\zeta_j(x)$ are smooth periodic functions such that

$$\mathrm{diam}(\mathrm{supp}\zeta_j \cap \Pi(a)) < \delta,$$

for any periodicity cell $\Sigma_a \subset \mathbb{R}^2$. Here δ is a sufficiently small positive number. For every $j = 1, \dots, N$ we choose the periodicity cell Σ_{a_j} in such a way that $\zeta_j\Big|_{\Pi(a_j)}$ is a finite function in $\Pi(a_j)$. We consider the zero extension of the function $\zeta_j\Big|_{\Pi(a_j)}$ to the space $\mathbb{R}^3$ and denote this extension by $\eta_j(x)$. We fix points $\lambda_j \in \mathrm{supp}(\eta_j)$, $j = 1, \dots, N$. Then we multiply all the relations in problem (4.2) by ζ_j and arrive at a periodic problem for the unknown functions $\boldsymbol{w}\zeta_j$, $q\zeta_j$. In accordance with the definition of norms in periodic spaces, to estimate $\boldsymbol{w}\zeta_j$, $q\zeta_j$, we are to estimate the functions $\boldsymbol{w}^j = \boldsymbol{w}\eta_j$, $q^j = q\eta_j$.

1. In the case when $\mathrm{supp}(\eta_j) \subset \Omega$, we have for the functions $\boldsymbol{w}^j$, q^j the Cauchy problem.
2. In the case when $\mathrm{supp}(\eta_j) \cap S \neq \emptyset$, we have for the functions $\boldsymbol{w}^j$, q^j the half-space problem with the Dirichlet boundary condition.
3. In the case when $\mathrm{supp}(\eta_j) \cap \Gamma \neq \emptyset$, we make the coordinate transformation straighten the boundary and obtain for the functions $\boldsymbol{w}^j$, q^j the half- space problem with the boundary conditions in the same form as in problem (5.1).

In the case of homogeneous incompressible fluid a detailed proof of estimates for the functions $\boldsymbol{w}^j$, q^j on the base of the estimates for solutions to corresponding model problems is given in [9] for problems of types 1, 3. For a problem of type 2 these estimates are proved in the same way, on the base of Theorem 5.2. In comparison with the homogeneous case, we have nonconstant coefficients at the equation in problem (4.2). So, we have to freeze the coefficients at the selected points λ_j, it generates additional terms at the right-hand sides of the equations. Precisely, we have

$$\boldsymbol{w}^j_t - \frac{\nu}{\rho_0(\lambda_j)}\Delta \boldsymbol{w}^j + \frac{1}{\rho_0(\lambda_j)}\nabla q^j = \boldsymbol{f}\eta_j + Q_1 + Q_2 + Q_3 + Q_4,$$

where

$$Q_1 = \nu\left(\frac{1}{\rho_0(x)} - \frac{1}{\rho_0(\lambda_j)}\right)\Delta \boldsymbol{w}^j, \qquad Q_2 = \left(\frac{1}{\rho_0(\lambda_j)} - \frac{1}{\rho_0(x)}\right)\nabla q^j,$$

$$Q_3 = \frac{1}{\rho_0(x)} q\nabla\eta_j - \frac{\nu}{\rho_0(x)}\left(\boldsymbol{w}\Delta\eta_j + 2\nabla\eta_j\cdot\nabla\boldsymbol{w}\right),$$

the term Q_4 is arising only if we make a coordinate transform (in problem of type 3) and has the form

$$Q_4 = \frac{1}{\rho_0(x)}\left(\nabla - \nabla_1\right) q^j + \frac{\nu}{\rho_0(x)}\left(\nabla_1^2 - \nabla^2\right)\boldsymbol{w}^j,$$

where $\nabla_1 = F^*\nabla$, F is the Jacobi matrix of the inverse coordinate transform.

It is clear that to obtain the estimates for $\boldsymbol{w}^j$, q^j in our case, we need to supplement the reasoning given in [9] by consideration of the terms Q_1, Q_2. To estimate these terms, we use the product lemma.

Lemma 6.1. (*Corollary from Lemma* 4.1 *in* [9]) *For any functions* $a \in W_2^{1+l}(\Omega)$, $f \in H_\gamma^{l,l/2}(\Omega\times[0,T))$, *where* $\Omega\subset\mathbb{R}^3$, *the following estimate*

$$\|af\|_{H_\gamma^{l,l/2}(\Omega_T)} \le \|f\|_{H_\gamma^{l,l/2}(\Omega_T)}\left(c_1\sup_\Omega|a(x)| + (\varepsilon + c_2(\varepsilon)\gamma^{-1/2})\|a\|_{W_2^{1+l}}\right)$$

holds true.

By Lemma 6.1 we have

$$\|Q_1\|_{H_\gamma^{l,l/2}(\mathbb{R}^3_T)} + \|Q_2\|_{H_\gamma^{l,l/2}(\mathbb{R}^3_T)} \le c_3\left(\|\boldsymbol{w}^j\|_{H_\gamma^{l+2,l/2+1}(\mathbb{R}^3_T)} + \|\nabla q^j\|_{H_\gamma^{l,l/2}(\mathbb{R}^3_T)}\right)$$
$$\left(\sup_{\mathrm{supp}(\eta_j)}|\rho_0(\lambda_j) - \rho_0(x)| + (\varepsilon + c_4(\varepsilon)\gamma^{-1/2})\|\rho_0\|_{\tilde{W}_2^{1+l}}\right). \tag{6.1}$$

Because of the fact that $\mathrm{diam}(\mathrm{supp}(\eta_j)) < \delta$, by the embedding theorem, we conclude that [9], [12]

$$\sup_{\mathrm{supp}(\eta_j)}|\rho_0(\lambda_j) - \rho_0(x)| \le c_5\|\rho_0\|_{W_2^{l+1}}\delta^\alpha, \quad \alpha\in(0,1), \quad \alpha < l - 1/2. \tag{6.2}$$

If we choose parameter ε sufficiently small and parameter γ sufficiently large, then (6.1), (6.2) imply

$$\|Q_1\|_{H_\gamma^{l,l/2}(\mathbb{R}^3_T)} + \|Q_2\|_{H_\gamma^{l,l/2}(\mathbb{R}^3_T)} \le \frac{1}{4}\left(\|\boldsymbol{w}^j\|_{H_\gamma^{l+2,l/2+1}(\mathbb{R}^3_T)} + \|\nabla q^j\|_{H_\gamma^{l,l/2}(\mathbb{R}^3_T)}\right). \tag{6.3}$$

Inequality (6.3) gives us the possibility to use calculations which was done in [9] for the case of homogeneous fluid and arrive at the estimate

$$\|\boldsymbol{w}\zeta_j\|^2_{\tilde{H}_\gamma^{l+2,l/2+1}(\Omega_T^0)} + \|\nabla(q\zeta_j)\|^2_{\tilde{H}_\gamma^{l,l/2}(\Omega_T^0)}$$
$$\le c_6\Big(\|\boldsymbol{w}\|^2_{\tilde{H}_\gamma^{l+3/2,l/2+3/4}(\Omega_T^0)} + \|q\|^2_{\tilde{H}_\gamma^{l,l/2}(\Omega_T^0)} + \|q\|^2_{\tilde{H}_\gamma^{0,l/2}(\Gamma_T)} + \|\boldsymbol{f}\|^2_{\tilde{H}_\gamma^{l,l/2}(\Omega_T^0)}$$
$$+\|\varphi\|^2_{\tilde{H}_\gamma^{l+1,l/2+1/2}(\Omega_T^0)} + \|\Phi\|_{\tilde{H}_\gamma^{l+1,l/2+1/2}(\Omega_T^0)} + K\Big), \tag{6.4}$$

where

$$K = \|\boldsymbol{d}\|_{\tilde{H}_\gamma^{l+1/2,l/2+1/4}(\Gamma_T)} + \|\boldsymbol{b}\|_{\tilde{H}_\gamma^{l+1/2,l/2+1/4}(\Gamma_T)} + \sigma\|\boldsymbol{B}\|_{\tilde{H}_\gamma^{l-1/2,l/2-1/4}(\Gamma_T)}$$

and arising in the case when we have for $\boldsymbol{w}\zeta^j$, $q\zeta^j$ problem of type 3.

Then we summarize estimates (6.4) on $j = 1,\dots,N$, use interpolation inequality for the norm $\|\boldsymbol{w}\|^2_{\tilde{H}_\gamma^{l+3/2,l/2+3/4}(\Omega_T^0)}$, estimates for pressure obtained in [9], [13] and arrive at (4.3).

It is clear that for any $T < +\infty$ we can pass from weighted spaces $\tilde{H}_\gamma^{l,l/2}(\Omega_T^0)$ to the spaces $\tilde{W}_2^{l,l/2}(\Omega_T^0)$, because they are equivalent for finite values of T. The solvability theorem to problem (4.1) with $\boldsymbol{v} = 0$ and nonhomogeneous initial conditions is deduced from Theorem 4.1 with the help of construction of the auxiliary vector field $\boldsymbol{V} \in \tilde{W}_2^{2+l,1+l/2}(\Omega_T^0)$ such that $\boldsymbol{V}\Big|_{t=0} = \boldsymbol{w}_0$.

7. Nonlinear problem

We have at linear problem (4.1) the same known function $\rho_0(x)$ as at nonlinear problem (3.1), consequently, the proof of Theorem 3.1 is similar to the case when $\rho = 1$ and can be done by the same scheme. The solvability theorem for problem (4.1) with $\boldsymbol{v} \in \tilde{W}_2^{l+2,l/2+1}(\Omega_T^0)$ satisfying the condition

$$T^{1/2}\|\boldsymbol{v}\|^{(l+2,l/2+1)}_{\Omega_T^0} \le \delta$$

with a sufficiently small δ is proved by successive approximations. Based on this theorem, Theorem 3.1 is proved also by successive approximations, solving on every step problem (3.4). As on every step we have at the equations coefficients depending on one and the same function $\rho_0(x)$, this passage is just the same as in [10], [12].

Returning from Lagrangian coordinates ξ to Eulerian coordinates, we find the density $\rho(x)$ by the following formula $\rho(x,t) = \rho_0(\xi(x,t))$.

References

[1] A.V. Kajikov, Resolution of boundary value problems for nonhomogeneous viscous fluids, *Doklady Akad. Nauk,* **216** (1974), No. 5, 1008–1010.

[2] S.N. Antonsev and A.V. Kajikov, The mathematical problems of the dynamics of nonhomogeneous fluids, Novosibirsk, Lectures of the University, 1973.

[3] O.A. Ladyzhenskaya and V.A. Solonnikov, On unique solvability of an initial-boundary value problem for viscous incompressible nonhomogeneous liquids, *Zap. Nauchn. Semin. LOMI* **52** (1975), 52–109 (English transl. in *J. Soviet Math.*, **9** (1978), 697–749.)

[4] A.A. Arkhipova and O.A. Ladyzhenskaya, On inhomogeneous incompressible fluids and reverse Holder inequalities, *Ann. Scuola Norm. Sup. Pisa Cl. Sci,* **25** (1997), no. 4, 51–67.

[5] J. Simon, Sur les fluides visqueux incompressibles et non-homogènes, C.R.A.S. Paris, **309** (1989), sér. I, 447–452.

[6] J.L. Lions, On some problems connected with Navier-Stokes equations, Nonlinear evolution equations, (1978) Academic Press, 59–84.

[7] N.N. Frolov, Solvability of a boundary problem of motion of an inhomogeneous fluid, *Mat. Zametki* **53** (1993), no. 6, 130–140.

[8] M.M. Santos, Stationary solution of the Navier-Stokes equations in 2d bounded domain for incompressible flow with discontinuous density, *Z. angew. Math. Phys.*, **53** (2002), 661–675.

[9] V.A. Solonnikov, On an initial-boundary value problem for the Stokes system arising in the study of a problem with a free boundary, *Trudy Mat. Inst. Steklov.*, **188** (1990), 150–188, (English transl. in *Proc. Steklov Inst. Math.* **3** (1991) 191–239.)

[10] V.A. Solonnikov, Solvability of the problem of evolution of a viscous incompressible fluid bounded by a free surface on a finite time interval, *Algebra i Analiz* **3** (1), (1991) 222–257, (English transl. in *St. Petersburg Math. J.* **3** (1), (1992) 189–220.)

[11] E.V. Frolova and M. Padula, Free boundary problem for a layer of inhomogeneous fluid, *European J. of Mechanics / B Fluids* **23** (2004) 665–679.

[12] V.A. Solonnikov and A. Tani, Free boundary problem for a viscous compressible flow with surface tension, *Constantin Caratheodory: An International Tribute* (1991), World Scientific, 1270–1303.

[13] I.V. Denisova, Evolution of compressible and incompressible fluids separated by a closed interface, *Interfaces and Free Boundaries* **2**, (2000) 283–312.

[14] V.A. Solonnikov, Estimates for the solution of an initial-boundary value problem for the linear nonstationary Navier-Stokes system, *Zap. Nauchn. semin. LOMI* **59** (1976), 178–254.

[15] V.A. Solonnikov, Estimates of solutions to nonstationary linearised Navier-Stokes system, *Trudy Mat. Inst. Steklov,* **70** (1964), 213–317.

[16] M.S. Agranovich and M.I. Vishik, Elliptic problems with a parameter and parabolic problems of general type, *Usp. Mat. Nauk* **19**(3), (1964) 53–152.

[17] E. Frolova, Solvability in Sobolev spaces of a problem for a second-order parabolic equation with time derivative in the boundary condition, *Portugaliae Mathematica* **56** (1999), no. 4, 419–441.

Elena Frolova
St.Petersburg Electrotechnical University,
Prof. Popov st., 5,
St.Petersburg, 197376 Russia
e-mail: `elenafr@mail.ru`

Progress in Nonlinear Differential Equations
and Their Applications, Vol. 61, 125–136

Duality Principles for Fully Nonlinear Elliptic Equations

Diogo Aguiar Gomes

Abstract. In this paper we use duality theory to associate certain measures to fully-nonlinear elliptic equations. These measures are the natural extension of the Mather measures to controlled stochastic processes and associated second-order elliptic equations. We apply these ideas to prove new a priori estimates for smooth solutions of fully nonlinear elliptic equations.

1. Introduction

This paper builds upon the connections between variational principles in classical mechanics, such as Aubry-Mather theory, and viscosity solutions of Hamilton-Jacobi equations and tries to illustrate how similar techniques can be used to study fully nonlinear elliptic equations and associated controlled Markov processes.

The variational principle in Classical Mechanics asserts that the trajectories $x(\cdot)$ of a mechanical system are critical points of the *action*:

$$\int_0^T L(x(t), \dot{x}(t))dt.$$

Of particular interest are the minimizers of the action. In Mather's theory, the problem of determining minimizers is relaxed and, instead, one looks for probability measures $\mu(x, v)$ which are generalized curves, that is,

$$\int vD_x\phi(x)d\mu = 0,$$

for all C^1 functions $\phi(x)$ and minimize the action, which is:

$$\int L(x, v)d\mu. \tag{1}$$

Supported in part by FCT/POCTI/FEDER.

The support of such minimizing measures, the *Mather set*, is invariant under the *Euler-Lagrange equations*

$$\frac{d}{dt}D_vL(x,\dot{x}) - D_xL(x,\dot{x}) = 0.$$

Since the work of J. Mather [Mat91] this area of research has been extremely active. Several authors [E99], [Fat97a, Fat97b, Fat98a, Fat98b], [EG01a, EG01b], among others, have studied the connection between Mather's theory and Hamilton-Jacobi partial differential equations. The minimization problem (1) is a infinite-dimensional linear programming problem. The dual problem is related with the stationary Hamilton-Jacobi equation

$$H(D_xu,x) = \overline{H}. \tag{2}$$

As in finite-dimensional linear programming, the dual problem yields important information about the primal and vice versa. For instance, if μ is a minimizing measure and u a solution of (2) then μ is supported on a graph $(x, v(x))$, with $v(x)$ is defined through

$$v(x) = -D_pH(D_xu,x).$$

In the other direction, one can use the measure to prove partial regularity [EG01a, Gom03] for the solutions to (2), and, in fact, one has

$$\int |D^2_{xx}u|^2 d\mu \leq C,$$

which is a weaker version of the Lipschitz graph theorem for Mather sets [Mat91].

In this paper we focus our attention at the class of nonlinear elliptic operators which have the form:

$$H(D^2_{xx}u, D_xu, x) = \sup_{\omega\in U}\left[-A_\omega u - L(\omega,x)\right].$$

The set $U \subset \mathbb{R}^m$ is the control space. We assume U to be a closed and convex set. The linear operator $A_\omega u$ is, for each ω, a (possibly degenerate) second-order elliptic operator whose zeroth-order coefficient vanishes, that is,

$$A_\omega u = a_\omega(x) : D^2_{xx}u + b_\omega(x)\cdot D_xu,$$

in which $a : b = \mathrm{Tr}(a^Tb)$. We assume that the Lagrangian $L(\omega,x)$ is a convex and superlinear (if U is unbounded) function in ω. Observe that $H(M,p,x)$ is jointly convex in (M,p) and monotone in M, that is,

$$H(M+B,p,x) \geq H(M,p,x)$$

for all non-negative matrices B. This class of operators arises in controlled Stochastic Dynamics and has been studied extensively, see, for instance, [CC95], [FS93] and the references therein for an introduction to fully nonlinear elliptic equations and stochastic optimal control theory.

We are particularly interested in periodic solutions to the stationary Hamilton-Jacobi equation

$$H(D^2_{xx}u, D_xu, x) = \overline{H}.$$

We associate to this equation a variational problem in a space of measures. The dual of this variational problem is closely related with the Hamilton-Jacobi equation. We apply these methods to study regularity of second-order Hamilton-Jacobi equations, extending some of the results from [Gom02b] concerning generalizations of Aubry-Mather theory to a stochastic setting.

The outline of the paper is as follows: in Section 2 we prove a representation formula for $\overline{H}$, and study its connections with generalized Mather measures. In Section 3 we study some applications to prove a priori regularity results for smooth solutions.

2. Duality

Proposition 1. *There is at most one value $\overline{H}$ for which*

$$H(D^2_{xx}u, D_xu, x) = \overline{H}. \tag{3}$$

has a periodic viscosity solution.

Proof. Suppose, by contradiction, $\overline{H}_1 > \overline{H}_2$ are such that (3) admits a viscosity solutions u_1 and u_2 for $\overline{H} = \overline{H}_1, \overline{H}_2$. We may assume $v_1 \equiv u_1 + C > u_2$, for a sufficiently large positive constant C. For ϵ sufficiently small

$$\epsilon v_1 + H(D^2_{xx}v_1, D_xv_1, x) \geq \epsilon u_2 + H(D^2_{xx}u_2, D_xu_2, x).$$

in the viscosity sense. The comparison principle for viscosity solutions implies $v_1 \leq u_2$, which is a contradiction. □

Proposition 2. *Suppose that there exists a viscosity solution u of* (3) *then*

$$\overline{H} = \inf_{\phi \in C^2_{per}} \sup_x H(D^2_{xx}\phi, D_x\phi, x). \tag{4}$$

in which the infimum is taken over all C^2 periodic functions.

Remark. In the case of first-order Hamiltonians this was proved in [CIPP98]. A proof for a special class of second-order equations can be found in [Gom02b]
Proof. Let

$$\overline{H}^* = \inf_{\phi \in C^2_{per}} \sup_x H(D^2_{xx}\phi, D_x\phi, x).$$

At some point x_0, $u - \phi$ has a local minimum. By the viscosity property

$$H(D^2_{xx}\phi, D_x\phi, x_0) \geq \overline{H}.$$

which implies $\overline{H}^* \geq \overline{H}$.

To prove the reverse inequality we need to recall a few facts concerning the sup convolution whose proof can be found in [FS93].

Lemma 1. *Suppose u is a viscosity of* (3). *Define*

$$u_\epsilon(x) = \sup_y \left[u(y) - \frac{|x-y|^2}{\epsilon} \right].$$

Then

1. $u_\epsilon \to u$ *uniformly as* $\epsilon \to 0$,
2. u_ϵ *is semiconvex,*
3. u_ϵ *satisfies*
$$H(D^2_{xx}u_\epsilon, D_x u_\epsilon, x) \le \overline{H} + O(\epsilon),$$
in the viscosity sense and almost everywhere.

Set $v_\epsilon = u_\epsilon * \eta_\epsilon$. Then
$$H(D^2_{xx}v_\epsilon, D_x v_\epsilon, x) \le \overline{H} + O(\epsilon),$$
thus $\overline{H}^* \le \overline{H}$. □

This representation formula can be best understood in light of a dual problem that involves generalized Mather measures. Choose a function $\gamma : \mathbb{T}^n \times U \to \mathbb{R}$, $\gamma \ge 1$, that satisfies
$$\lim_{|\omega|\to\infty} \frac{L(x,\omega)}{\gamma(\omega)} = +\infty \qquad \lim_{|\omega|\to\infty} \frac{|\omega|}{\gamma(\omega)} = 0,$$
we use the convention that if U is bounded then the previous identities are trivially satisfied.

Let $\mathcal{M}$ be the set of Radon measures on $\mathbb{T}^n \times U$ that satisfy
$$\int_{\mathbb{T}^n \times U} \gamma d\mu \le \infty.$$
Note that $\mathcal{M}$ can be identified with the dual space of $C^0_\gamma(\mathbb{T}^n \times U)$, that is, the set of continuous functions ϕ such that
$$\|\phi\|_\gamma = \sup_{\mathbb{T}^n \times U} \left|\frac{\phi}{\gamma}\right| \qquad \lim_{|\omega|\to\infty} \frac{\phi}{\gamma} = 0.$$
Define $\mathcal{M}_0$ to be the set of all measures in $\mathcal{M}$ that satisfy the constraint
$$\int_{\mathbb{T}^n \times U} A_\omega \phi d\mu = 0$$
for all $\phi \in C^2(\mathbb{T}^n)$. Let $\mathcal{M}_1 \subset \mathcal{M}$ be the set of all positive probability measures that belong to $\mathcal{M}$.

We look for measures in $\mathcal{M}_0 \cap \mathcal{M}_1$ that minimize the action
$$\int_{\mathbb{T}^n \times U} L(\omega, x) d\mu.$$

If L is strictly convex in ω and A_ω is linear in ω then $\omega = \omega(x)$ almost everywhere in the support of μ. However, we do not make this assumption and will work in a more general framework.

This variational problem is, in fact, a linear programming problem, in an infinite-dimensional space, and by Fenchel-Rockafellar duality theory [Roc66] it admits a dual problem. This is close in spirit to the papers [VL78a], [VL78b], [LV80], [FV89], [FV88] and [Fle89], in which Fenchel-Rockafellar duality theory is used to analyze optimal control problems. In the first order case this dual problem has been identified and studied [CIPP98] and involves a Hamilton-Jacobi equation.

Before proceeding we need to recall some facts concerning convex duality. Let E be a Banach space with dual E'. The pairing between E and E' is denoted by $(\cdot,\cdot)$. Suppose $h_1 : E \to (-\infty,+\infty]$ is a convex, lower semicontinuous function. The Legendre-Fenchel transform $h_1^* : E' \to [-\infty,+\infty]$ of h_1 is defined by

$$h_1^*(y) = \sup_{x\in E} \left(-(x,y) - h_1(x)\right),$$

for $y \in E'$. Similarly, for concave, upper semicontinuous functions $h_2 : E \to (-\infty,+\infty]$ we define

$$h_2^*(y) = \inf_{x\in E} \left(-(x,y) - h_2(x)\right).$$

Theorem 1 (Rockafellar [Roc66]**).** *Let E be a locally convex Hausdorff topological vector space over $\mathbb{R}$ with dual E^*. Suppose $h_1 : E \to (-\infty,+\infty]$ is convex and lower semicontinuous, $h_2 : E \to [-\infty,+\infty)$ is concave and upper semicontinuous. Then*

$$\sup_x h_2(x) - h_1(x) = \inf_y h_1^*(y) - h_2^*(y), \tag{5}$$

provided that either h_1 or h_2 is continuous at some point where both functions are finite.

To apply this theorem we define two functions h_1 and h_2 on $C^0_\gamma(\mathbb{T}^n \times U)$ and consider the dual problem of

$$\sup_{\phi\in C^0_\gamma(\mathbb{T}^n\times U)} h_2(\phi) - h_1(\phi).$$

The first function is defined by

$$h_1(\phi) = \sup_{(x,\omega)\in\mathbb{T}^n\times U} \left[-\phi(x,\omega) - L(x,\omega)\right].$$

Let

$$\mathcal{C} = \mathrm{cl}\left\{\phi : \phi = A_\omega\varphi, \varphi\in C^2(\mathbb{T}^n)\right\},$$

and set

$$h_2(\phi) = \begin{cases} 0 & \text{if } \phi\in\mathcal{C} \\ -\infty & \text{otherwise.}\end{cases}$$

Proposition 3. *We have*

$$h_1^*(\mu) = \begin{cases} \int L d\mu & \text{if } \mu\in\mathcal{M}_1 \\ +\infty & \text{otherwise,}\end{cases}$$

and

$$h_2^*(\mu) = \begin{cases} 0 & \text{if } \mu\in\mathcal{M}_0 \\ -\infty & \text{otherwise.}\end{cases}$$

Proof. The Legendre-Fenchel transform h_1^* of h_1 is

$$h_1^*(\mu) = \sup_{\phi \in C^0_\gamma(\mathbb{T}^n \times U)} \left(-\int \phi d\mu - h_1(\phi) \right).$$

We claim that for all non-positive measures μ, $h_1^*(\mu) = \infty$.

Lemma 2. *If $\mu \not\geq 0$ then $h_1^*(\mu) = +\infty$.*

Proof. If $\mu \not\geq 0$ there is a sequence of non-negative functions $\phi_n \in C^0_\gamma(\mathbb{T}^n \times U)$ such that

$$\int -\phi_n d\mu \to +\infty.$$

Thus, since $L \geq 0$,

$$\sup_{\mathbb{T}^n \times U} -\phi_n - L \leq 0.$$

Thus, if $\mu \not\geq 0$, then $h_1^*(\mu) = +\infty$. □

Lemma 3. *If $\mu \geq 0$ then*

$$h_1^*(\mu) \geq \int L d\mu + \sup_{\psi \in C^0_\gamma(\mathbb{T}^n \times U)} \left(\int \psi d\mu - \sup \psi \right).$$

Proof. Let L_n be a sequence of functions in $C^0_\gamma(\mathbb{T}^n \times U)$ increasing pointwise to L. Any function ϕ in $C^0_\gamma(\mathbb{T}^n \times U)$ can be written as $\phi = -L_n - \psi$, for some ψ also in $C^0_\gamma(\mathbb{T}^n \times U)$. Therefore

$$\sup_{\phi \in C^0_\gamma(\mathbb{T}^n \times U)} \left(-\int \phi d\mu - h_1(\phi) \right)$$
$$= \sup_{\psi \in C^0_\gamma(\mathbb{T}^n \times U)} \left(\int L_n d\mu + \int \psi d\mu - \sup(L_n + \psi - L) \right).$$

Note that $L_n - L \leq 0$ implies

$$\sup_{\mathbb{T}^n \times U} L_n - L \leq 0,$$

thus

$$\sup_{\mathbb{T}^n \times U} (L_n + \psi - L) \leq \sup_{\mathbb{T}^n \times U} \psi.$$

Thus

$$\sup_{\phi \in C^0_\gamma(\mathbb{T}^n \times U)} \left(-\int \phi d\mu - h_1(\phi) \right)$$
$$\geq \sup_{\psi \in C^0_\gamma(\mathbb{T}^n \times U)} \left(\int L_n d\mu + \int \psi d\mu - \sup(\psi) \right).$$

By the monotone convergence theorem $\int L_n d\mu \to \int L d\mu$. Therefore

$$\sup_{\phi\in C^0_\gamma(\mathbb{T}^n\times U)}\left(-\int \phi d\mu - h_1(\phi)\right) \geq \int L d\mu + \sup_{\psi\in C^0_\gamma(\mathbb{T}^n\times U)}\left(\int \psi d\mu - \sup(\psi)\right),$$

as required. □

If $\int L d\mu = +\infty$ then $h_1^*(\mu) = +\infty$. Also if $\int d\mu \neq 1$ then

$$\sup_{\psi\in C^0_\gamma(\mathcal{D})}\left(\int \psi d\mu - \sup\psi\right) \geq \sup_{\alpha\in\mathbb{R}} \alpha(\int d\mu - 1) = +\infty,$$

by taking $\psi \equiv \alpha$, constant. So, $h_1^*(\mu) = +\infty$, and therefore a finite value of h_1^* is only possible if $\int d\mu = 1$.

If $\int d\mu = 1$ we have, from the previous lemma,

$$h_1^*(\mu) \geq \int L d\mu,$$

by taking $\psi \equiv 0$.

Also, for any function ϕ

$$\int (-\phi - L) d\mu \leq \sup_{\mathbb{T}^n\times U}(-\phi - L),$$

if $\int d\mu = 1$. Hence

$$\sup_{\phi\in C^0_\gamma(\mathbb{T}^n\times U)}\left(-\int \phi d\mu - h_1(\phi)\right) \leq \int L d\mu.$$

Thus

$$h_1^*(\mu) = \begin{cases} \int L d\mu & \text{if} \quad \mu \in \mathcal{M}_1 \\ +\infty & \text{otherwise.} \end{cases}$$

Now we compute h_2^*. Note that if $\mu \notin \mathcal{M}_0$ there exists $\hat{\phi} \in \mathcal{C}$ such that

$$\int \hat{\phi} d\mu \neq 0.$$

and so

$$\inf_{\phi\in\mathcal{C}} -\int \phi d\mu \leq \inf_{\alpha\in\mathbb{R}} \alpha \int \hat{\phi} d\mu = -\infty.$$

If $\mu \in \mathcal{M}_0$ then $\int \phi d\mu = 0$, for all $\phi \in \mathcal{C}$. Therefore

$$h_2^*(\mu) = \inf_{\phi\in\mathcal{C}} -\int \phi d\mu = \begin{cases} 0 & \text{if } \mu \in \mathcal{M}_0 \\ -\infty & \text{otherwise.} \end{cases}$$ □

Theorem 1 yields then

$$\sup_{\phi\in C^0_\gamma(\mathbb{T}^n\times U)} h_2(\phi) - h_1(\phi) = \inf_{\mu\in\mathcal{M}} h_1^*(\mu) - h_2^*(\mu),$$

provided we prove that h_1 is continuous on the set $h_2 > -\infty$. This is the content of the next lemma.

Lemma 4. *h_1 is continuous.*

Proof. Suppose $\phi_n \to \phi$ in C^0_γ. Then $\|\phi_n\|_\gamma$ and $\|\phi\|_\gamma$ are bounded uniformly by some constant C. The growth condition on L implies that there exists $R > 0$ such that

$$\sup_{\mathbb{T}^n \times U} -\hat{\phi} - L = \sup_{T^n \times (B_R \cap U)} -\hat{\phi} - L,$$

for all $\hat{\phi}$ in $C^0_\gamma(\mathbb{T}^n \times U)$ with $\|\hat{\phi}\|_\gamma < C$. On $B_R \cap U$, $\phi_n \to \phi$ uniformly and so

$$\sup_{\mathbb{T}^n \times U} -\phi_n - L \to \sup_{\mathbb{T}^n \times U} -\phi - L.$$ □

The next theorem summarizes the main result of this section.

Theorem 2.

$$\overline{H} = -\inf_{\mu \in \mathcal{M}_0 \cap \mathcal{M}_1} \int L d\mu. \tag{6}$$

Proof. This is a corollary to Proposition 2 and the duality result proved above. □

3. A priori estimates

In this section we apply the ideas from the previous section to prove a-priori bounds for smooth solutions of second-order nonlinear equations such as the maximal eigenvalue operator, streamline diffusion controlled dynamics, and mean curvature flow.

Proposition 4. *Let u be a smooth periodic solution to*

$$H(D^2_{xx}u, D_x u, x) = \overline{H},$$

and μ a corresponding Mather measure. Then

$$\omega \in \operatorname{argmin}\left[A_\omega u + L(x, \omega)\right]$$

μ almost everywhere.

Proof. Since

$$-\overline{H} = -H(D^2_{xx}u, D_x u, x) \leq A_\omega u + L(x, \omega),$$

with equality if and only if $\omega \in \operatorname{argmin}\left[A_\omega u + L(x, \omega)\right]$, integrating with respect to μ yields

$$-\overline{H} \leq \int L(x, \omega) d\mu,$$

and, unless

$$\omega \in \operatorname{argmin}\left[A_\omega u + L(x, \omega)\right],$$

μ-a.e. this would yield a contradiction. □

Proposition 5. *Suppose u is a smooth periodic solution to*

$$H(D^2_{xx}u, D_x u, x) = \overline{H}.$$

Then, μ almost everywhere,

$$A_\omega \varphi = a_\omega : D^2_{xx}\varphi + b_\omega \cdot D_x \varphi,$$

for any $\phi \in C^2(\mathbb{T}^n)$, with

$$a_\omega = -D_M H(D^2_{xx}u, D_x u, x),$$

and

$$b_\omega = -D_p H(D^2_{xx}u, D_x u, x).$$

Proof. It suffices to observe that almost everywhere in the support of μ one has

$$A_\omega u + L(x,\omega) = -H(D^2_{xx}u, D_x u, x),$$

and, for any $\varphi \in C^2(\mathbb{T}^n)$

$$A_\omega(u+\epsilon\varphi) + L(x,\omega) \geq -H(D^2_{xx}(u+\epsilon\varphi), D_x(u+\epsilon\varphi), x).$$

Thus, by subtracting the last two equations and sending $\epsilon \to 0$ one gets the theorem. □

Theorem 3. *Suppose u is a smooth periodic solution to*

$$H(D^2_{xx}u, D_x u, x) = \overline{H}.$$

Then u satisfies the following a priori identity: let $\xi \in \mathbb{R}^n$ be arbitrary, then

$$\begin{aligned}\int \Big[& H_{M_{ij}M_{lm}} \xi_k D^3_{x_i x_j x_k} u \xi_{k'} D^3_{x_i x_j x_{k'}} u + 2 H_{M_{ij} p_m} \xi_k D^3_{x_i x_j x_k} u \xi_{k'} D^2_{x_m x_{k'}} u \\ & + H_{p_i p_j} \xi_k D^2_{x_i x_k} u \xi_{k'} D^2_{x_j x_{k'}} u \Big] d\mu \\ = & -\int \Big[2 H_{M_{ij} x_m} \xi_k D^3_{x_i x_j x_k} u + \xi_{k'} 2 H_{p_i x_{k'}} \xi_k D^2_{x_i x_k} u + \xi_{k'} H_{x_k x_{k'}} \Big] d\mu.\end{aligned}$$

Remark. Note that the left-hand side of this estimate is a non-negative quadratic form on $D^2(D_\xi u)$ and $D(D_\xi u)$ since $H(M,p,x)$ is jointly convex in M and p and the right-hand side depends on lower-order terms. Therefore this identity yields an a priori estimate for second and third derivatives.

Proof. By differentiating the equation

$$H(D^2_{xx}u, D_x u, x) = \overline{H}$$

with respect to x_k and multiplying by ξ_k we obtain

$$H_{M_{ij}} \xi_k D^3_{x_i x_j x_k} u + H_{p_i} \xi_k D^2_{x_i x_k} u + \xi_k H_{x_k} = 0.$$

Differentiating this last expression with respect to $x_{k'}$ and multiplying by $\xi_{k'}$ we obtain

$$
\begin{aligned}
&H_{M_{ij}M_{lm}}\xi_k D^3_{x_i x_j x_k}u\xi_{k'}D^3_{x_i x_j x_{k'}}u + 2H_{M_{ij}p_m}\xi_k D^3_{x_i x_j x_k}u\xi_{k'}D^2_{x_m x_{k'}}u\\
&+ H_{p_i p_j}\xi_k D^2_{x_i x_k}u\xi_{k'}D^2_{x_j x_{k'}}u + 2H_{M_{ij}x_m}\xi_k D^3_{x_i x_j x_k}u\\
&+ 2\xi_{k'}H_{p_i x_{k'}}\xi_k D^2_{x_i x_k}u + \xi_k\xi_{k'}H_{x_k x_{k'}} + H_{M_{ij}}\xi_k\xi_{k'}D^4_{x_i x_j x_k x_{k'}}u\\
&+ \xi_k\xi_{k'}H_{p_i}D^3_{x_i x_k x_{k'}}u = 0.
\end{aligned}
$$

Integrating with respect to μ, and observing that the last two terms integrate to 0 since

$$
\begin{aligned}
H_{M_{ij}}\xi_k\xi_{k'}D^4_{x_i x_j x_k x_{k'}}u &+ \xi_k\xi_{k'}H_{p_i}D^3_{x_i x_k x_{k'}}u\\
&= H_{M_{ij}}D^2_{x_i x_j}(D^2_{\xi\xi}u) + H_{p_i}D_{x_i}(D^2_{\xi\xi}u),
\end{aligned}
$$

we obtain the result. □

Next, we briefly illustrate the estimates discussed above for a fully nonlinear second-order equation. Let u be a periodic solution to the one-dimensional equation

$$e^{u_{xx}} + \frac{u_x^2}{2} + V(x) = \overline{H}.$$

The projection $\theta(x)$ of a minimizing measure in the x-axis satisfies

$$(\theta(x)e^{u_{xx}})_{xx} - (\theta u_x)_x = 0,$$

weakly as a measure. Furthermore

$$\int \left[e^{u_{xx}}u_{xxx}^2 + u_{xx}^2\right]\theta dx \leq C.$$

There are several important examples for which these estimates apply, two of them, the Stochastic Mather problem [Gom02b]

$$A_\omega u + L(x,\omega) = \Delta u + \omega\cdot D_x u + \frac{|\omega|^2}{2} - V(x),$$

and the vakonomic mechanics operator [Gom02a]

$$A_\omega u + L(x,\omega) = \omega f(x)\cdot D_x u + \frac{|\omega|^2}{2} - V(x),$$

have been studied in detail. However, important cases such as the maximal eigenvalue operator

$$A_\omega u = \frac{\omega\otimes\omega}{|\omega|^2} : D^2_{xx}u,$$

the related streamline-diffusion controlled dynamics problem

$$A_\omega u + L(x,\omega) = \frac{\omega\otimes\omega}{|\omega|^2} : D^2_{xx}u + \omega\cdot D_x u + \frac{|\omega|^2}{2} + V(x),$$

and the mean curvature flow (see [ST02] for a control theory formulation of the mean curvature flow) with drift

$$A_\omega u + L(x,\omega) = (I - \frac{\omega \otimes \omega}{|\omega|^2}) : D^2_{xx}u + b(x)D_x u + V(x),$$

have not been studied using these techniques. We believe that our estimates and ideas may give important insight on these problems.

References

[CC95] Luis A. Caffarelli and Xavier Cabré. *Fully nonlinear elliptic equations.* American Mathematical Society, Providence, RI, 1995.

[CIPP98] G. Contreras, R. Iturriaga, G.P. Paternain, and M. Paternain. Lagrangian graphs, minimizing measures and Mañé's critical values. *Geom. Funct. Anal.*, 8(5):788–809, 1998.

[E99] Weinan E. Aubry-Mather theory and periodic solutions of the forced Burgers equation. *Comm. Pure Appl. Math.*, 52(7):811–828, 1999.

[EG01a] L.C. Evans and D. Gomes. Effective Hamiltonians and averaging for Hamiltonian dynamics. I. *Arch. Ration. Mech. Anal.*, 157(1):1–33, 2001.

[EG01b] L.C. Evans and D. Gomes. Effective Hamiltonians and averaging for Hamiltonian dynamics II. *Preprint*, 2001.

[Fat97a] Albert Fathi. Solutions KAM faibles conjuguées et barrières de Peierls. *C. R. Acad. Sci. Paris Sér. I Math.*, 325(6):649–652, 1997.

[Fat97b] Albert Fathi. Théorème KAM faible et théorie de Mather sur les systèmes lagrangiens. *C. R. Acad. Sci. Paris Sér. I Math.*, 324(9):1043–1046, 1997.

[Fat98a] Albert Fathi. Orbite hétéroclines et ensemble de Peierls. *C. R. Acad. Sci. Paris Sér. I Math.*, 326:1213–1216, 1998.

[Fat98b] Albert Fathi. Sur la convergence du semi-groupe de Lax-Oleinik. *C. R. Acad. Sci. Paris Sér. I Math.*, 327:267–270, 1998.

[Fle89] Wendell H. Fleming. Generalized solutions and convex duality in optimal control. In *Partial differential equations and the calculus of variations, Vol. I*, pages 461–471. Birkhäuser Boston, Boston, MA, 1989.

[FS93] Wendell H. Fleming and H. Mete Soner. *Controlled Markov processes and viscosity solutions.* Springer-Verlag, New York, 1993.

[FV88] Wendell H. Fleming and Domokos Vermes. Generalized solutions in the optimal control of diffusions. In *Stochastic differential systems, stochastic control theory and applications (Minneapolis, Minn., 1986)*, pages 119–127. Springer, New York, 1988.

[FV89] Wendell H. Fleming and Domokos Vermes. Convex duality approach to the optimal control of diffusions. *SIAM J. Control Optim.*, 27(5):1136–1155, 1989.

[Gom02a] D. Gomes. Hamilton-Jacobi methods for vakonomic mechanics. *preprint*, 2002.

[Gom02b] Diogo Aguiar Gomes. A stochastic analogue of Aubry-Mather theory. *Nonlinearity*, 15(3):581–603, 2002.

[Gom03] Diogo Aguiar Gomes. Regularity theory for Hamilton-Jacobi equations. *J. Differential Equations*, 187(2):359–374, 2003.

[LV80] R.M. Lewis and R.B. Vinter. Relaxation of optimal control problems to equivalent convex programs. *J. Math. Anal. Appl.*, 74(2):475–493, 1980.

[Mat91] John N. Mather. Action minimizing invariant measures for positive definite Lagrangian systems. *Math. Z.*, 207(2):169–207, 1991.

[Roc66] R.T. Rockafellar. Extension of Fenchel's duality theorem for convex functions. *Duke Math. J.*, 33:81–89, 1966.

[ST02] H. Mete Soner and Nizar Touzi. Dynamic programming for stochastic target problems and geometric flows. *J. Eur. Math. Soc. (JEMS)*, 4(3):201–236, 2002.

[VL78a] Richard B. Vinter and Richard M. Lewis. The equivalence of strong and weak formulations for certain problems in optimal control. *SIAM J. Control Optim.*, 16(4):546–570, 1978.

[VL78b] Richard B. Vinter and Richard M. Lewis. A necessary and sufficient condition for optimality of dynamic programming type, making no a priori assumptions on the controls. *SIAM J. Control Optim.*, 16(4):571–583, 1978.

Diogo Aguiar Gomes
Departamento de Matemática,
Instituto Superior Técnico
Av. Rovisco Pais
1049-001 Lisboa, Portugal
e-mail: dgomes@math.ist.utl.pt

Progress in Nonlinear Differential Equations
and Their Applications, Vol. 61, 137–148

On the Bénard Problem

Giovanna Guidoboni and Mariarosaria Padula

Abstract. In literature there is no mathematical proof of the experimentally trivial stability of the rest state for a layer of compressible fluid heated from above. In the case of layer heated from below it is known that the system shows a threshold in the temperature gradient below which the fluid is not sensible to the imposed difference of temperature. Only semi-empirical justifications are available for this phenomenon, see [6].

Neglecting the thermal conductivity, we are able to prove that for a layer of compressible fluid between two rigid planes kept at constant temperature, the rest state is linearly stable for every values of the parameters involved in two cases: *a)* the layer is heated from above; *b)* the layer is heated from below and the imposed gradient of temperature is less than a precise quantity, namely g/c_p, where g is the gravity constant, and c_p is the specific heat at constant pressure, known as *adiabatic gradient.*

1. Introduction

This represents part of a research on the Bénard problem shared with professor T. Nishida, who should be considered morally as coauthor.

In the literature the *Bénard problem* refers to investigation of the onset of convection in a horizontal layer of fluid heated from below. If we do not consider surface effects (e.g., fluid between rigid planes), buoyancy rules the instability: hotter particles are lighter and tend to rise as colder particles tend to sink according to the action of the gravity force. To take this effect into account rigorously, the density should be variable and then the compressible scheme should be used. Since this approach is very complicated, to model the problem it is adopted the *Boussinesq approximation* where the density is considered as a constant in all the terms of the equations except for the gravity term in which it is assumed to vary linearly with the temperature (see [3]).

Actually this approximation is reasonable only if the thickness of the layer is small and then a large class of physical phenomena are left aside, as, e.g., convection in stars.

This justifies the large extent of literature concerning a comparison between the equations governing the real convection and those of the Boussinesq approximation. Here we wish to give small remarks on this subject considering the case of a layer between rigid planes kept at constant temperature: mainly our observations concern the most trivial case of a fluid heated from above.

Let us indicate with Θ_0 and Θ_1 the temperatures of the lower and the upper plane respectively. On one hand, if $\Theta_1 > \Theta_0$ (layer heated from above) it is expected that the rest state is always *nonlinearly stable*, however, stability has been proven only in the Boussinesq approximation, (see [3]). On the other hand, if $\Theta_1 < \Theta_0$ (layer heated from below) experimentally the rest state is stable only if the Rayleigh number is sufficiently small. By linear methods, for the general compressible scheme, it is possible to compute only critical Rayleigh number R_c below which there holds linear stability, above which also nonlinear instability holds (see [1]). It is worth of notice that in the Boussinesq scheme the critical number R_c ensures nonlinear stability too! This fact explains the reason of the success of the Boussinesq method.

Our work begins just by noticing that in literature there is no mathematical proof of the experimentally trivial stability for a layer of compressible fluid heated from above. Moreover, in the case of $\Theta_1 < \Theta_0$ it is known that the system shows a threshold in the temperature gradient below which the fluid is not sensible to the imposed difference of temperature. Only semi-empirical justifications are available for this phenomenon, see, e.g., Jeffreys [6].

Object of this note is to give rigorous mathematical proofs of facts that have experimental evidence. Unfortunately we give only partial answers, precisely, for zero thermal conductivity.

Actually, neglecting thermal conductivity, we are able to prove that for a layer of compressible fluid between two rigid planes kept at constant temperature, the rest state is *linearly stable* for every values of the parameters involved if

- the layer is heated from above; (see Section 3)
- the layer is heated from below and the imposed gradient of temperature is less than a precise quantity, namely g/c_p, where g is the gravity constant, and c_p is the specific heat at constant pressure, known as *adiabatic gradient*, (see Section 4) the same that we find in Jeffreys' paper.

Since the method employed is the energy method, the proof could infer also nonlinear stability for small initial data, only if the kinematic and thermal diffusivity are non zero, which is not our case!

We are conscious that the action of thermal conductivity cannot be neglected because it acts in favor of stability, therefore our note should be considered only as starting point, and we hope that these results could throw more light in the comprehension of these phenomena. We expect that the terms related to the thermal conductivity are dissipative but we were not able to prove their sign. We are performing numerical computations to check if these terms can in some cases have a destabilizing effect.

The plane of the work is the following: in Section 2 we remind the Boussinesq approximation; then, we study linear stability criteria for a compressible fluid heated from above (in Section 3) and from below (in Section 4). The stability criteria are deduced by use of the classical Liapunov method [2], [10], [9], however for inviscid fluids, we have been not able to extend these results to the rigorous nonlinear scheme.

2. Boussinesq approximation

Let us consider a horizontal layer of fluid bounded by two rigid planes. Let us consider a cartesian coordinates system upward directed whose origin lies on the lower plane. We consider a periodicity cell Σ in the horizontal direction and then the domain is $\Omega = \Sigma \times [z_0, z_0 + d]$. In the Boussinesq approximation (see [3]) the equations of motion are:

$$\begin{cases} \nabla \cdot \mathbf{u} = 0 \\ \varrho_0[\partial_t \mathbf{u} + (\mathbf{u} \cdot \nabla)\mathbf{u}] = -\nabla p + \mu \Delta \mathbf{u} - g\varrho_0[1 - \alpha(\Theta - \Theta_0)]\mathbf{e}_3 \\ c_v \varrho_0[\partial_t \Theta + (\mathbf{u} \cdot \nabla)\Theta] = \chi \Delta \Theta \end{cases} \tag{2.1}$$

where $\mathbf{u}$ is the velocity field, p the pressure, Θ the temperature, ϱ_0 the density, α the coefficient of volume expansion, μ the dynamic viscosity, g the acceleration of gravity, c_v the specific heat at constant volume, χ the coefficient of termal conductivity. We add the following boundary conditions:

$$\begin{cases} \mathbf{u} = \mathbf{0} & \text{at } z = z_0, z_0 + d \\ \Theta = \Theta_0 & \text{at } z = z_0 \\ \Theta = \Theta_1 & \text{at } z = z_0 + d. \end{cases} \tag{2.2}$$

A stationary solution of the previous problem is the rest state:

$$\mathbf{u}_r = \mathbf{0}, \qquad \Theta_r = \beta_0(z - z_0) + \Theta_0 \tag{2.3}$$

where $\beta_0 = \Theta_1 - \Theta_0/d$ is the imposed gradient of temperature. It is positive if the layer is heated from above, negative otherwise.

We now have to write the problem in a dimensionless form. We introduce the *Rayleigh number* and the *Prandtl number* defined as follows:

$$R = \sqrt{\frac{g\alpha|\beta_0|d^4\varrho_0^2 c_v}{\chi\mu}}, \qquad Pr = \frac{\mu c_v}{\chi}\,. \tag{2.4}$$

We remark that R is positive both in the case of layer heated from above and below. Taking d, $\varrho d^2/\mu$, $\sqrt{g\alpha|\beta_0|d^2\chi/c_v\mu}$, $|\beta_0|d$, $\sqrt{g\alpha|\beta_0|\chi\mu/c_v}$ as units of length, time, velocity, temperature and pressure, we can write the dimensionless equations for the perturbations:

$$\begin{cases} \nabla\cdot\mathbf{u}=0 \\ \partial_t\mathbf{u}+\dfrac{R}{Pr}\mathbf{u}\cdot\nabla\mathbf{u}=-\nabla\pi+\Delta\mathbf{u}+R\theta\mathbf{e}_3 \\ Pr\partial_t\theta+R\mathbf{u}\cdot\nabla\theta=\Delta\theta-R\mathbf{u}\cdot\nabla\widetilde{\Theta_r} \end{cases} \tag{2.5}$$

where $\mathbf{u}$, π, θ are the dimensionless perturbations to the velocity, the pressure and the temperature respectively, and $\nabla\widetilde{\Theta_r}$ is the dimensionless temperature gradient at the rest state. It is useful to remark that:

$$\begin{array}{lcl} \text{layer heated from above} & \longrightarrow & \nabla\widetilde{\Theta_r}=\mathbf{e}_3 \\ \text{layer heated from below} & \longrightarrow & \nabla\widetilde{\Theta_r}=-\mathbf{e}_3\,. \end{array}$$

The boundary conditions now read:

$$\begin{cases} \mathbf{u}=\mathbf{0} & \text{at } z=Z_0, Z_0+1 \\ \theta=0 & \text{at } z=Z_0, Z_0+1 \end{cases} \tag{2.6}$$

where $Z_0=z_0/d$.

Multiplying scalarly in L^2 $(2.5)_2$ by $\mathbf{u}$ and $(2.5)_3$ by θ we get:

$$\frac{1}{2}\frac{d}{dt}\mathcal{E}=\frac{1}{2}\frac{d}{dt}(\|\mathbf{u}\|^2+Pr\|\theta\|^2)=-\|\nabla\mathbf{u}\|^2-\|\nabla\theta\|^2+R\int_\Omega\theta u_3-R\int_\Omega\theta\mathbf{u}\cdot\nabla\widetilde{\Theta_r} \tag{2.7}$$

where $\|\cdot\|$ indicates the L^2-norm on Ω.

In the case of layer heated from above

$$\frac{1}{2}\frac{d}{dt}\mathcal{E}=-\|\nabla\mathbf{u}\|^2-\|\nabla\theta\|^2 \tag{2.8}$$

which means that the rest state is nonlinearly stable for every R.

In the case of layer heated from below

$$\frac{1}{2}\frac{d}{dt}\mathcal{E}=-\|\nabla\mathbf{u}\|^2-\|\nabla\theta\|^2+2R\int_\Omega\theta u_3 \tag{2.9}$$

the rest state is stable only if R is small enough.

3. Compressible scheme: layer heated from above

The nonlinear stability of the rest state for a layer of compressible fluid heated from below has been studied by Padula and Coscia [4], Padula and Benabidallah [2]. Actually the investigation of the stability of the rest state in the case of layer heated from above is not immediate as in the previous section. This is the reason why in the following we are dealing with the linearized equations for the perturbations: we try to obtain some information from the simplified problem.

We consider separately the case of layer heated from above and below. We begin with the first case.

Let us consider a horizontal layer of fluid in the same frame of reference described in the previous section. Let us consider a perfect gas

$$p = R_* \varrho \Theta \tag{3.1}$$

where $R_* = c_p - c_v$, and c_p is the specific heat at constant pressure. The Navier-Stokes equations for a viscous newtonian compressible fluid are:

$$\begin{cases} \partial_t \varrho + \nabla \cdot (\varrho \mathbf{u}) = 0 \\ \varrho[\partial_t \mathbf{u} + (\mathbf{u} \cdot \nabla)\mathbf{u}] = -R_* \nabla(\varrho\Theta) + \mu\Delta\mathbf{u} + (\mu + \lambda)\nabla(\nabla \cdot \mathbf{u}) - g\varrho\mathbf{e}_3 \\ c_v \varrho[\partial_t \Theta + (\mathbf{u} \cdot \nabla)\Theta] = \chi\Delta\Theta - R_* \varrho\Theta\nabla \cdot \mathbf{u} + \mathbf{V} : \mathbf{D} \end{cases} \tag{3.2}$$

where $\mathbf{D} = (\nabla\mathbf{u} + \nabla^T\mathbf{u})/2$ is the rate-of-strain tensor and $\mathbf{V} = \lambda\nabla \cdot \mathbf{u} + 2\mu\mathbf{D}$. The domain is $\Omega = \Sigma \times (z_0, z_0 + d)$, where Σ is the periodicity cell in the horizontal plane. The boundary conditions:

$$\begin{cases} \mathbf{u} = \mathbf{0} & \text{at } z = z_0, z_0 + d \\ \Theta = \Theta_0 & \text{at } z = z_0 \\ \Theta = \Theta_1 & \text{at } z = z_0 + d. \end{cases} \tag{3.3}$$

Let us introduce a new dimensionless variable for the height

$$\zeta = \frac{z - z_0}{d} + \frac{\Theta_0}{\beta_0 d}, \tag{3.4}$$

where we recall the definition of the imposed temperature gradient

$$\beta_0 = \frac{\Theta_1 - \Theta_0}{d} > 0. \tag{3.5}$$

The rest state $S_r = (\mathbf{u}_r, \Theta_r, \varrho_r)$ is then

$$\mathbf{u}_r = \mathbf{0}, \qquad \Theta_r = \beta_0 d\zeta, \qquad \varrho_r = \varrho_0 \zeta^{-m} \tag{3.6}$$

where

$$m = 1 + \frac{g}{R_* \beta_0}. \tag{3.7}$$

Since ζ is dimensionless, $\beta_0 d$ and ϱ_0 have the dimensions of temperature and density respectively. They will be taken as units of scale of the corresponding quantities.

Let us consider a perturbed motion $S = (\mathbf{u}, \Theta, \varrho)$ with

$$\mathbf{u}, \qquad \Theta = \Theta_r + \theta, \qquad \varrho = \varrho_r + \sigma \tag{3.8}$$

where $\mathbf{u}$, θ and σ are the disturbances in the velocity, temperature and density respectively.

We write the linear equations for the disturbances:

$$\begin{cases} \partial_t \sigma + \nabla \cdot (\varrho_r \mathbf{u}) = 0 \\ \varrho_r \partial_t \mathbf{u} = -R_* \nabla(\varrho_r \theta) - R_* \nabla(\sigma\Theta_r) + \mu\Delta\mathbf{u} + (\mu + \lambda)\nabla(\nabla \cdot \mathbf{u}) - g\sigma\mathbf{e}_3 \\ c_v \varrho_r[\partial_t \theta + (\mathbf{u} \cdot \nabla)\Theta_r] = \chi\Delta\theta - R_* \varrho_r \Theta_r \nabla \cdot \mathbf{u}. \end{cases} \tag{3.9}$$

Let introduce the following dimensionless groups:

$$
\begin{aligned}
&A = d/tV \\
&B = V/R_* \beta_0 t \\
&D_1 = \mu V / R_* \varrho_0 \beta_0 d^2 \\
&D_2 = \chi / R_* \varrho_0 d V \\
&l = 1 + \lambda/\mu \\
&n = c_v / R_*
\end{aligned}
\tag{3.10}
$$

where d is the distance between the planes, t and V are the units of time and velocity respectively.

Using (3.6) the dimensionless equations become:

$$
\begin{cases}
A\partial_t \sigma = -\nabla \cdot (\zeta^{-m}\mathbf{u}) \\
B\zeta^{-m}\partial_t \mathbf{u} = D_1 \Delta \mathbf{u} + lD_1 \nabla(\nabla \cdot \mathbf{u}) - \nabla(\zeta^{-m}\theta) - \nabla(\zeta\sigma) + (1-m)\sigma\nabla\zeta \\
nA\zeta^{-m}\partial_t \theta = D_2 \Delta\theta - n\zeta^{-m}\mathbf{u}\cdot\nabla\zeta - \zeta^{-m+1}\nabla\cdot\mathbf{u}.
\end{cases}
\tag{3.11}
$$

This is not the usual dimensional analysis but it turns out to be useful from the view point of mathematics since it puts the right-hand side of the equations in a handling form.

Let us consider $(3.11)_2$. We can see that:

$$
-\nabla(\zeta\sigma) + (1-m)\sigma\nabla\zeta = -\zeta\nabla\sigma - m\sigma\nabla\zeta \tag{3.12}
$$

but since

$$
\nabla(\zeta^m \sigma) = m\zeta^{m-1}\sigma\nabla\zeta + \zeta^m \nabla\sigma \tag{3.13}
$$

the $(3.11)_2$ reads

$$
B\zeta^{-m}\partial_t \mathbf{u} = D_1\Delta\mathbf{u} + lD_1\nabla(\nabla\cdot\mathbf{u}) - \nabla(\zeta^{-m}\theta) - \zeta^{-m+1}\nabla(\zeta^m\sigma)\,. \tag{3.14}
$$

Let us now consider $(3.11)_3$. Noticing that

$$
-n\zeta^{-m}\mathbf{u}\cdot\nabla\zeta - \zeta^{-m+1}\nabla\cdot\mathbf{u} = -\zeta^{-m-n+1}\nabla\cdot(\zeta^n\mathbf{u}) \tag{3.15}
$$

the $(3.11)_3$ reads

$$
nA\zeta^{-m}\partial_t\theta = D_2\Delta\theta - \zeta^{-m-n+1}\nabla\cdot(\zeta^n\mathbf{u})\,. \tag{3.16}
$$

Then the equations (3.11) become

$$
\begin{cases}
A\partial_t \sigma = -\nabla \cdot (\zeta^{-m}\mathbf{u}) \\
B\zeta^{-m}\partial_t \mathbf{u} = D_1 \Delta \mathbf{u} + lD_1 \nabla(\nabla \cdot \mathbf{u}) - \nabla(\zeta^{-m}\theta) - \zeta^{-m+1}\nabla(\zeta^m\sigma) \\
nA\zeta^{-m}\partial_t \theta = D_2 \Delta\theta - \zeta^{-m-n+1}\nabla\cdot(\zeta^n\mathbf{u})\,.
\end{cases}
\tag{3.17}
$$

3.1. Case $D_2 = 0$

We consider here a fluid whose thermal conductivity is equal to zero. The equations are simply

$$\begin{cases} A\partial_t\sigma = -\nabla\cdot(\zeta^{-m}\mathbf{u}) \\ B\zeta^{-m}\partial_t\mathbf{u} = D_1\Delta\mathbf{u} + lD_1\nabla(\nabla\cdot\mathbf{u}) - \nabla(\zeta^{-m}\theta) - \zeta^{-m+1}\nabla(\zeta^m\sigma) \\ nA\zeta^{-m}\partial_t\theta = -\zeta^{-m-n+1}\nabla\cdot(\zeta^n\mathbf{u})\,. \end{cases} \tag{3.18}$$

Remark 1. *In this case temperature profile at rest state is not determined and can be taken arbitrarily. We consider a linear profile in analogy with the classical case.*

Multiplying scalarly in L^2 $(3.18)_1$ by $\zeta^{m+1}\sigma$ we obtain:

$$\frac{A}{2}\frac{d}{dt}\int_\Omega \zeta^{m+1}\sigma^2 = -\int_\Omega \zeta^{m+1}\sigma\nabla\cdot(\zeta^{-m}\mathbf{u})\,. \tag{3.19}$$

Multiplying scalarly in L^2 $(3.18)_2$ by $\mathbf{u}$ we obtain:

$$\frac{B}{2}\frac{d}{dt}\int_\Omega \zeta^{-m}\mathbf{u}^2 = -D_1\|\nabla\mathbf{u}\|^2 - lD_1\|\nabla\cdot\mathbf{u}\|^2 + \int_\Omega \zeta^{-m}\theta\nabla\cdot\mathbf{u} + \int_\Omega \zeta^m\sigma\nabla\cdot(\zeta^{-m+1}\mathbf{u})\,. \tag{3.20}$$

Multiplying scalarly in L^2 $(3.18)_3$ by $\zeta^{-1}\theta$ we obtain:

$$n\frac{A}{2}\frac{d}{dt}\int_\Omega \zeta^{-m-1}\theta^2 = -\int_\Omega \zeta^{-m-n}\theta\nabla\cdot(\zeta^n\mathbf{u})\,. \tag{3.21}$$

Multiplying scalarly in L^2 $(3.18)_1$ by $n\theta$, $(3.18)_3$ by $\zeta^m\sigma$ and adding the resulting equations we obtain:

$$nA\frac{d}{dt}\int_\Omega \sigma\theta = -n\int_\Omega \theta\nabla\cdot(\zeta^{-m}\mathbf{u}) - \int_\Omega \zeta^{-n+1}\sigma\nabla\cdot(\zeta^n\mathbf{u})\,. \tag{3.22}$$

Adding to (3.20) the (3.19), (3.21) and (3.22) multiplied respectively by the coupling constants λ_1, λ_2 and λ_3 respectively, we obtain:

$$\frac{d\mathcal{E}}{dt} = -D_1\|\nabla\mathbf{u}\|^2 - lD_1\|\nabla\cdot\mathbf{u}\|^2 + \mathcal{F} \tag{3.23}$$

where

$$\mathcal{E} = \int_\Omega \left\{\frac{B}{2}\zeta^{-m}\mathbf{u}^2 + \lambda_1\frac{A}{2}\zeta^{m+1}\sigma^2 + \lambda_2 n\frac{A}{2}\zeta^{-m-1}\theta^2 + \lambda_3 nA\sigma\theta\right\}$$

$$\begin{aligned} \mathcal{F} = {} & [1-\lambda_2-\lambda_3 n]\int_\Omega \zeta^{-m}\theta\nabla\cdot\mathbf{u} + [1-\lambda_1-\lambda_3]\int_\Omega \zeta\sigma\nabla\cdot\mathbf{u} \\ & + [-m+1+m\lambda_1-n\lambda_3]\int_\Omega \sigma w + [-n\lambda_2+mn\lambda_3]\int_\Omega \zeta^{-m-1}\theta w \end{aligned}$$

where w is the vertical component of the velocity. It is possible to choose the coupling constants in such a way that $\mathcal{F} = 0$:

$$\lambda_1 = 1 - \frac{1}{m+n} \qquad \lambda_2 = \frac{m}{m+n} \qquad \lambda_3 = \frac{1}{m+n}\,. \tag{3.24}$$

The energy functional now reads:

$$\mathcal{E} = \int_\Omega \left\{ \frac{B}{2}\zeta^{-m}\mathbf{u}^2 + \left(1 - \frac{1}{m+n}\right)\frac{A}{2}\zeta^{m+1}\sigma^2 + \frac{mn}{m+n}\frac{A}{2}\zeta^{-m-1}\theta^2 + \frac{n}{m+n}A\theta\sigma \right\} \tag{3.25}$$

and we ask it is positive definite. A sufficient condition is that

$$\left(1 - \frac{1}{m+n}\right)\frac{mn}{m+n} > \frac{n^2}{(m+n)^2} \tag{3.26}$$

which gives

$$(m+n)(m-1) > 0\,. \tag{3.27}$$

We remind that

$$m = 1 + \frac{g}{R_*\beta_0} \qquad n = \frac{c_v}{R_*} \qquad R_* = c_p - c_v \tag{3.28}$$

and β_0 is positive since the layer is heated from the top. In this case (3.27) is always satisfied and the linear stability of the rest state follows from the inequality:

$$\frac{d\mathcal{E}}{dt} = -D_1\|\nabla\mathbf{u}\|^2 - lD_1\|\nabla\cdot\mathbf{u}\|^2 \leq 0\,. \tag{3.29}$$

Remark 2. Case $D_2 \neq 0$

If we consider $D_2 \neq 0$ *we obtain:*

$$\frac{d}{dt}\mathcal{E} = -D_1\|\nabla\mathbf{u}\|^2 - lD_1\|\nabla\cdot\mathbf{u}\|^2 + \lambda_2 D_2\int_\Omega \Delta\theta\frac{\theta}{\zeta} + \lambda_3 D_2\int_\Omega \Delta\theta\zeta^m\sigma \tag{3.30}$$

Now we have the last two terms more whose sign is not defined.

4. Compressible scheme: layer heated from below

In this case we cannot anymore use the variable ζ in the same way, since it would be negative! Then we take another frame of reference, with the origin on the upper plane, downward directed. We rewrite the problem and redo the computations in this new scenario:

$$\begin{cases} \partial_t\varrho + \nabla\cdot(\varrho\mathbf{u}) = 0 \\ \varrho[\partial_t\mathbf{u} + (\mathbf{u}\cdot\nabla)\mathbf{u}] = -R_*\nabla(\varrho\Theta) + \mu\Delta\mathbf{u} + (\mu+\lambda)\nabla(\nabla\cdot\mathbf{u}) + g\varrho\mathbf{e}_3 \\ c_v\varrho[\partial_t\Theta + (\mathbf{u}\cdot\nabla)\Theta] = \chi\Delta\Theta - R_*\varrho\Theta\nabla\cdot\mathbf{u} + \mathbf{V}:\mathbf{D} \end{cases} \tag{4.1}$$

where $\mathbf{D} = (\nabla\mathbf{u} + \nabla^T\mathbf{u})/2$ is the rate-of-strain tensor and $\mathbf{V} = \lambda\nabla\cdot\mathbf{u} + 2\mu\mathbf{D}$. The domain is $\Omega = \Sigma \times (z_0, z_0 + d)$, where Σ is the periodicity cell in the horizontal plane. The boundary conditions:

$$\begin{cases} \mathbf{u} = \mathbf{0} & \text{at } z = z_0, z_0 + d \\ \Theta = \Theta_1 & \text{at } z = z_0 \\ \Theta = \Theta_0 & \text{at } z = z_0 + d \end{cases} \tag{4.2}$$

Let us define the imposed temperature gradient

$$\beta_0 = \frac{\Theta_0 - \Theta_1}{d} > 0. \tag{4.3}$$

Let us introduce a new dimensionless variable for the height

$$\zeta = \frac{z - z_0}{d} + \frac{\Theta_1}{\beta_0 d}. \tag{4.4}$$

The rest state $S_r = (\mathbf{u}_r, \Theta_r, \varrho_r)$ is then

$$\mathbf{u}_r = \mathbf{0}, \qquad \Theta_r = \beta_0 d\zeta, \qquad \varrho_r = \varrho_0 \zeta^m \tag{4.5}$$

where now

$$m = \frac{g}{R_* \beta_0} - 1 \tag{4.6}$$

is slightly different than the previous case. The one defined in the previous section is always positive, while the parameter defined in (4.6) is positive if $\beta_0 < g/R_*$, and negative if $\beta_0 > g/R_*$.

Following the line of the previous section, we write the linear equations for the disturbances:

$$\begin{cases} \partial_t \sigma + \nabla \cdot (\varrho_r \mathbf{u}) = 0 \\ \varrho_r \partial_t \mathbf{u} = -R_* \nabla(\varrho_r \theta) - R_* \nabla(\sigma \Theta_r) + \mu \Delta \mathbf{u} + (\mu + \lambda)\nabla(\nabla \cdot \mathbf{u}) + g\sigma \mathbf{e}_3 \\ c_v \varrho_r [\partial_t \theta + (\mathbf{u} \cdot \nabla)\Theta_r] = \chi \Delta \theta - R_* \varrho_r \Theta_r \nabla \cdot \mathbf{u}. \end{cases} \tag{4.7}$$

Introducing the dimensionless groups already defined, we write the dimensionless equations:

$$\begin{cases} A\partial_t \sigma = -\nabla \cdot (\zeta^m \mathbf{u}) \\ B\zeta^m \partial_t \mathbf{u} = D_1 \Delta \mathbf{u} + lD_1 \nabla(\nabla \cdot \mathbf{u}) - \nabla(\zeta^m \theta) - \nabla(\zeta\sigma) + (1+m)\sigma\nabla\zeta \\ nA\zeta^m \partial_t \theta = D_2 \Delta \theta - n\zeta^m \mathbf{u} \cdot \nabla\zeta - \zeta^{m+1} \nabla \cdot \mathbf{u}. \end{cases} \tag{4.8}$$

Let us consider $(4.8)_2$. We can see that:

$$-\nabla(\zeta\sigma) + (1+m)\sigma\nabla\zeta = -\zeta\nabla\sigma + m\sigma\nabla\zeta \tag{4.9}$$

but since

$$\nabla(\zeta^{-m}\sigma) = -m\zeta^{-m-1}\sigma\nabla\zeta + \zeta^{-m}\nabla\sigma \tag{4.10}$$

the $(4.8)_2$ reads

$$B\zeta^m \partial_t \mathbf{u} = D_1 \Delta \mathbf{u} + lD_1 \nabla(\nabla \cdot \mathbf{u}) - \nabla(\zeta^m \theta) - \zeta^{m+1}\nabla(\zeta^{-m}\sigma). \tag{4.11}$$

Let us now consider $(4.8)_3$. Noticing that

$$-n\zeta^m \mathbf{u} \cdot \nabla\zeta - \zeta^{m+1}\nabla \cdot \mathbf{u} = -\zeta^{m-n+1}\nabla \cdot (\zeta^n \mathbf{u}) \tag{4.12}$$

the $(4.8)_3$ reads

$$nA\zeta^m \partial_t \theta = D_2 \Delta \theta - \zeta^{m-n+1}\nabla \cdot (\zeta^n \mathbf{u}). \tag{4.13}$$

Then the equations (4.8) become

$$\begin{cases} A\partial_t\sigma = -\nabla\cdot(\zeta^m\mathbf{u}) \\ B\zeta^m\partial_t\mathbf{u} = D_1\Delta\mathbf{u} + lD_1\nabla(\nabla\cdot\mathbf{u}) - \nabla(\zeta^m\theta) - \zeta^{m+1}\nabla(\zeta^{-m}\sigma) \\ nA\zeta^m\partial_t\theta = D_2\Delta\theta - \zeta^{m-n+1}\nabla\cdot(\zeta^n\mathbf{u})\,. \end{cases} \tag{4.14}$$

4.1. Case $D_2 = 0$

The equations are simply

$$\begin{cases} A\partial_t\sigma = -\nabla\cdot(\zeta^m\mathbf{u}) \\ B\zeta^m\partial_t\mathbf{u} = D_1\Delta\mathbf{u} + lD_1\nabla(\nabla\cdot\mathbf{u}) - \nabla(\zeta^m\theta) - \zeta^{m+1}\nabla(\zeta^{-m}\sigma) \\ nA\zeta^m\partial_t\theta = -\zeta^{m-n+1}\nabla\cdot(\zeta^n\mathbf{u})\,. \end{cases} \tag{4.15}$$

Multiplying scalarly in L^2 $(4.15)_1$ by $\zeta^{1-m}\sigma$ we obtain:

$$\frac{A}{2}\frac{d}{dt}\int_\Omega \zeta^{1-m}\sigma^2 = -\int_\Omega \zeta^{1-m}\sigma\nabla\cdot(\zeta^m\mathbf{u})\,. \tag{4.16}$$

Multiplying scalarly in L^2 $(4.15)_2$ by $\mathbf{u}$ we obtain:

$$\frac{B}{2}\frac{d}{dt}\int_\Omega \zeta^m\mathbf{u}^2 = -D_1\|\nabla\mathbf{u}\|^2 - lD_1\|\nabla\cdot\mathbf{u}\|^2 + \int_\Omega \zeta^m\theta\nabla\cdot\mathbf{u} + \int_\Omega \zeta^{-m}\sigma\nabla\cdot(\zeta^{m+1}\mathbf{u})\,. \tag{4.17}$$

Multiplying scalarly in L^2 $(4.15)_3$ by $\zeta^{-1}\theta$ we obtain:

$$n\frac{A}{2}\frac{d}{dt}\int_\Omega \zeta^{m-1}\theta^2 = -\int_\Omega \zeta^{m-n}\theta\nabla\cdot(\zeta^n\mathbf{u})\,. \tag{4.18}$$

Multiplying scalarly in L^2 $(4.15)_1$ by $n\theta$, $(4.15)_3$ by $\zeta^{-m}\sigma$ and adding the resulting equations we obtain:

$$nA\frac{d}{dt}\int_\Omega \sigma\theta = -n\int_\Omega \theta\nabla\cdot(\zeta^m\mathbf{u}) - \int_\Omega \zeta^{-n+1}\sigma\nabla\cdot(\zeta^n\mathbf{u})\,. \tag{4.19}$$

Adding to (4.17) the (4.16), (4.18) and (4.19) multiplied respectively by the coupling constants λ_1, λ_2 and λ_3 respectively, we obtain:

$$\frac{d\mathcal{E}}{dt} = -D_1\|\nabla\mathbf{u}\|^2 - lD_1\|\nabla\cdot\mathbf{u}\|^2 + \mathcal{F} \tag{4.20}$$

where

$$\mathcal{E} = \int_\Omega \left\{\frac{B}{2}\zeta^m\mathbf{u}^2 + \lambda_1\frac{A}{2}\zeta^{1-m}\sigma^2 + \lambda_2 n\frac{A}{2}\zeta^{m-1}\theta^2 + \lambda_3 nA\sigma\theta\right\}$$

$$\begin{aligned} \mathcal{F} = {} & [1-\lambda_2-\lambda_3 n]\int_\Omega \zeta^m\theta\nabla\cdot\mathbf{u} + [1-\lambda_1-\lambda_3]\int_\Omega \zeta\sigma\nabla\cdot\mathbf{u} \\ & + [m+1-m\lambda_1-n\lambda_3]\int_\Omega \sigma w + [-n\lambda_2 - mn\lambda_3]\int_\Omega \zeta^{m-1}\theta w \end{aligned}$$

where w is the vertical component of the velocity. It is possible to choose the coupling constants in such a way that $\mathcal{F} = 0$:

$$\lambda_1 = 1 - \frac{1}{n-m} \qquad \lambda_2 = \frac{m}{m-n} \qquad \lambda_3 = \frac{1}{n-m}\,. \tag{4.21}$$

The energy functional now reads:

$$\mathcal{E} = \int_\Omega \left\{ \frac{B}{2}\zeta^m \mathbf{u}^2 + \left(1 - \frac{1}{n-m}\right)\frac{A}{2}\zeta^{1-m}\sigma^2 + \frac{mn}{m-n}\frac{A}{2}\zeta^{m-1}\theta^2 + \frac{n}{n-m}A\theta\sigma \right\} \tag{4.22}$$

and we ask it is positive definite. A sufficient condition is that

$$\left(1 - \frac{1}{n-m}\right)\frac{mn}{m-n} > \frac{n^2}{(m-n)^2} \tag{4.23}$$

which gives

$$(m+1)(m-n) > 0\,. \tag{4.24}$$

which is satisfied if $m - n > 0$, alias if $\beta_0 < g/c_p$.

Acknowledgments

The authors wish to thank professor Nishida for helpful and stimulating discussions. The authors also acknowledge the Italian GNFM-INDAM and 60% and 40% Murst.

References

[1] P. Aye & T. Nishida, *Heat convection of compressible fluid*, GAKUTO International Series, Mathematical Sciences and Applications, *Recent developments in domain decomposition methods and flow problems* **11**, (1998), 107–115.

[2] R. Benabidallah & M. Padula, *Sulla stabilità di un fluido politropico, viscoso, pesante in un contenitore limitato da pareti perfette conduttrici di calore*, Annali Univ. Ferrara, sez VII, Sc. Mat. vol XLV, (1999), 127–161.

[3] Chandrasekhar, S., *Hydrodynamic and Hydromagnetic Stability*, University Press, Oxford (1961).

[4] V. Coscia & M. Padula, *Nonlinear energy stability in compressible atmosphere*, Geophys. Astrophys. Fluid Dynamics **54** (1990) 49–83.

[5] G. Guidoboni & B.J. Jin, *On the nonlinear stability of Marangoni-Bénard problem with free surface in Boussinesq approximation*, to appear.

[6] H. Jeffreys , *The instability of a compressible fluid heated from below*, Proc. Cam. Phil. Soc., **26**, (1930), 170–172.

[7] B.J. Jin & M. Padula,*On existence of nonsteady compressible viscous flows in a horizontal layer with free upper surface*, Comm. Pure Appl. Anal. 1(3) (2002), 370–415.

[8] M. Padula, *On the exponential decay to the rest state for a viscous isothermal fluid*, J. Math. Fluid Mech., **1** (1999), 1–16.

[9] M. Padula & M. Pokorny, *Stability and decay to zero of the L^2 norms of perturbations to a viscous compressible heat conductive fluid motion exterior to a ball*, J. Math. Fluid Mech., **3**, (2001), 317–408.

[10] M. Padula, *On direct Ljapunov method in continuum theories*, Nonlinear problems in mathematical physics and related topics I, in honor of Prof. Ladyzhenskaya, eds. Sh. Birman, S. Hildebrandt, V.A. Solonnikov, O.A. Uralsteva, Kluwer Academic/Plenum Publisher, New York, Boston, Dordrecht, London, Moscow, 289–302, Novosibirsk, 2002, 271–283.

Giovanna Guidoboni and Mariarosaria Padula
University of Ferrara
Via Machiavelli 35
I-44100 Ferrara, Italy
e-mail: gio@dm.unife.it
e-mail: pad@unife.ii

Progress in Nonlinear Differential Equations
and Their Applications, Vol. 61, 149–160

Exact Boundary Controllability for Quasilinear Wave Equations

Li Tatsien

First of all, I would like to give all my congratulations and best wishes to our friend, Prof. V.A. Solonnikov, for his 70th birthday, for his great contribution to partial differential equations and for his health, success and happiness.

1. Introduction and main results

There are many publications concerning the exact controllability for linear hyperbolic systems (see [9], [10] and the references therein). Using the HUM method suggested by J.-L. Lions [9] and Schauder's fixed point theorem, E. Zuazua [12] proved the global (resp. local) exact boundary controllability for semilinear wave equations in the asymptotically linear case (resp. the super-linear case with suitable growth conditions).

Moreover, in the one-dimensional case, E. Zuazua [13] obtained the global exact controllability for semilinear wave equations. On the other hand, using a global inversion theorem, I. Lasiecka and R. Triggiani [2] gave the global exact boundary controllability for semilinear wave equations in the asymptotically linear case. However, even in the one-dimensional case, only a few results are known for quasilinear wave equations.

Consider the following quasilinear wave equation

$$\frac{\partial^2 u}{\partial t^2} - \frac{\partial}{\partial x}(K(u, \frac{\partial u}{\partial x})) = F(u, \frac{\partial u}{\partial x}, \frac{\partial u}{\partial t}), \tag{1.1}$$

where $K = K(u, v)$ is a given C^2 function of u and v, such that

$$K_v(u, v) > 0, \tag{1.2}$$

and $F = F(u, v, w)$ is a given C^1 function of u, v and w, satisfying

$$F(0, 0, 0) = 0. \tag{1.3}$$

On one end $x = 0$, we prescribe any one of the following boundary conditions:

$$u = h(t) \qquad \text{(of Dirichlet type)}, \tag{1.4.1}$$

$$u_x = h(t) \qquad \text{(of Neumann type)}, \tag{1.4.2}$$

$$u_x - \alpha u = h(t) \qquad \text{(of the third type)} \tag{1.4.3}$$

or

$$u_x - \bar{\alpha} u_t = h(t) \qquad \text{(of the dissipative type)}, \tag{1.4.4}$$

where α and $\bar{\alpha}$ are given positive constants, $h(t)$ is a C^2 function (in case (1.4.1)) or a C^1 function (in cases (1.4.2)–(1.4.4)).

Similarly, on another end $x = 1$, the boundary condition is

$$u = \bar{h}(t) \qquad \text{(of Dirichlet type)}, \tag{1.5.1}$$

$$u_x = \bar{h}(t) \qquad \text{(of Neumann type)}, \tag{1.5.2}$$

$$u_x + \beta u = \bar{h}(t) \qquad \text{(of the third type)} \tag{1.5.3}$$

or

$$u_x + \bar{\beta} u_t = \bar{h}(t) \qquad \text{(of the dissipative type)}, \tag{1.5.4}$$

where β and $\bar{\beta}$ are given positive constants, $\bar{h}(t)$ is a C^2 function (in case (1.5.1)) or a C^1 function (in cases (1.5.2)–(1.5.4)).

In the case that K and F in equation (1.1) are independent of u, if on one end, say, on $x = 0$, the boundary condition is of Dirichlet type, by means of the theory on the semi-global C^1 solution and the local exact boundary controllability for quasilinear hyperbolic systems without zero eigenvalues, Li Tatsien and Rao Bopeng [5] established the corresponding local exact boundary controllability with the boundary control $\bar{h}(t)$ acting on another end $x = 1$; while, if on $x = 0$, the boundary condition is of the third type, by means of the theory on the semi-global C^1 solution and the local exact boundary controllability for quasilinear hyperbolic systems without zero eigenvalues together with a kind of nonlocal boundary conditions, Li Tatsien and Xu Yulan [7] established the corresponding local exact boundary controllability with the boundary control $\bar{h}(t)$ acting on another end $x = 1$. However, the method used in [5] and [7] cannot be applied to the following cases: (1). K and F in equation (1.1) depend explicitly on u; (2). The boundary condition on $x = 0$ is of Neumann type or of the dissipative type; (3). Boundary controls are simultaneously given on both ends $x = 0$ and $x = 1$.

In this talk, we will present a unified method to get the local exact boundary controllability with boundary controls acting on one end or on two ends for equation (1.1) with boundary conditions (1.4) and (1.5) of different types. The main results are the following two theorems.

Theorem 1 (Exact boundary controllability with boundary controls acting on two ends). *Let*

$$T > \frac{1}{\sqrt{K_v(0,0)}}. \tag{1.6}$$

For any given initial data (φ, ψ) and final data (Φ, Ψ) with small norms

$$\|(\varphi, \psi)\|_{C^2[0,1]\times C^1[0,1]} \qquad \textit{and} \qquad \|(\Phi, \Psi)\|_{C^2[0,1]\times C^1[0,1]},$$

there exist boundary controls $h(t)$ *and* $\bar{h}(t)$ *with small norms*

$$\|h\|_{C^2[0,T]} \qquad \text{and} \qquad \|\bar{h}\|_{C^2[0,T]}$$

(for (1.4.1) *and* (1.5.1)*) or with small norms*

$$\|h\|_{C^1[0,T]} \qquad \text{and} \qquad \|\bar{h}\|_{C^1[0,T]}$$

(for (1.4.2)–(1.4.4) *and* (1.5.2)–(1.5.4)*), such that the mixed initial-boundary value problem for equation* (1.1) *with the initial condition*

$$t = 0: \quad u = \varphi(x), \quad u_t = \psi(x), \quad 0 \le x \le 1, \tag{1.7}$$

one of the boundary conditions (1.4) *on* $x = 0$ *and one of the boundary conditions* (1.5) *on* $x = 1$ *admits a unique* C^2 *solution* $u = u(t,x)$ *with small* C^2 *norm on the domain*

$$R(T) = \{(t,x) |\ 0 \le t \le T, \ \ 0 \le x \le 1\}, \tag{1.8}$$

which exactly satisfies the final condition

$$t = T: \quad u = \Phi(x), \quad u_t = \Psi(x), \quad 0 \le x \le 1. \tag{1.9}$$

□

Theorem 2 (Exact boundary controllability with boundary controls acting on one end). *Let*

$$T > \frac{2}{\sqrt{K_v(0,0)}}. \tag{1.10}$$

Suppose that

$$\bar{\alpha} \ne \frac{1}{\sqrt{K_v(0,0)}}, \tag{1.11}$$

where $\bar{\alpha}$ *is given in* (1.4.4)*. For any given initial data* (φ, ψ) *and final data* (Φ, Ψ) *with small norms* $\|(\varphi,\psi)\|_{C^2[0,1]\times C^1[0,1]}$ *and* $\|(\Phi,\Psi)\|_{C^2[0,1]\times C^1[0,1]}$ *and any given function* $h(t)$ *with small norm* $\|h\|_{C^2[0,T]}$ *(in case* (1.4.1)*) or with small norm* $\|h\|_{C^1[0,T]}$ *(in cases* (1.4.2)–(1.4.4)*), such that the usual conditions of* C^2 *compatibility are satisfied at the points* $(0,0)$ *and* $(T,0)$*, respectively, there exists a boundary control* $\bar{h}(t)$ *with small norm* $\|\bar{h}\|_{C^2[0,T]}$ *(in case* (1.5.1)*) or with small norm* $\|\bar{h}\|_{C^1[0,T]}$ *(in cases* (1.5.2)–(1.5.4)*), such that the mixed initial-boundary value problem for equation* (1.1) *with the initial condition* (1.7)*, one of the boundary conditions* (1.4) *on* $x = 0$ *and one of the boundary conditions* (1.5) *on* $x = 1$ *admits a unique* C^2 *solution* $u = u(t,x)$ *with small* C^2 *norm on the domain* $R(T) = \{(t,x)|\ 0 \le t \le T,\ 0 \le x \le 1\}$*, which exactly satisfies the final condition* (1.9). □

Remark 1. Comparing with Theorem 1, only one boundary control acting on one end is used in Theorem 2, however, the exact controllability time is doubled.

Remark 2. The exact controllability time given in Theorem 1 or Theorem 2 is optimal.

Remark 3. The boundary controls given in Theorem 1 or Theorem 2 are not unique.

2. Reduction of equation and boundary conditions

Setting

$$v=\frac{\partial u}{\partial x},\qquad w=\frac{\partial u}{\partial t},\tag{2.1}$$

equation (1.1) can be reduced to the following first-order quasilinear system

$$\begin{cases}\dfrac{\partial u}{\partial t}=w,\\ \dfrac{\partial v}{\partial t}-\dfrac{\partial w}{\partial x}=0,\\ \dfrac{\partial w}{\partial t}-K_v(u,v)\dfrac{\partial v}{\partial x}=F(u,v,w)+K_u(u,v)v\triangleq\tilde{F}(u,v,w),\end{cases}\tag{2.2}$$

where $\tilde{F}(u,v,w)$ is still a C^1 function of u, v and w, satisfying

$$\tilde{F}(0,0,0)=0.\tag{2.3}$$

Noting (1.2), (2.2) is a strictly hyperbolic system with three distinct real eigenvalues $\dfrac{dx}{dt}=\lambda_i(i=1,2,3)$:

$$\lambda_1=-\lambda<\lambda_2=0<\lambda_3=\lambda,\tag{2.4}$$

in which

$$\lambda=\sqrt{K_v(u,v)}.\tag{2.5}$$

The corresponding left eigenvectors can be taken as

$$l_1=(0,\sqrt{K_v},1),\quad l_2=(1,0,0),\quad l_3=(0,-\sqrt{K_v},1).\tag{2.6}$$

Let

$$U=(u,v,w)^T.\tag{2.7}$$

Setting

$$v_i=l_i(U)U\qquad (i=1,\ 2,\ 3),\tag{2.8}$$

namely,

$$v_1=\sqrt{K_v(u,v)}\,v+w,\quad v_2=u,\quad v_3=-\sqrt{K_v(u,v)}\,v+w,\tag{2.8$'$}$$

we have

$$\begin{cases}v_1+v_3=2w,\\ v_1-v_3=2\sqrt{K_v(u,v)}\,v.\end{cases}\tag{2.9}$$

Changing the order of t and x, similarly to (2.2), equation (1.1) can be also reduced to the following first-order quasilinear system

$$\begin{cases}\dfrac{\partial u}{\partial x}=v,\\ \dfrac{\partial v}{\partial x}-\dfrac{1}{K_v(u,v)}\dfrac{\partial w}{\partial t}=-\dfrac{\tilde{F}(u,v,w)}{K_v(u,v)}\triangleq\tilde{\tilde{F}}(u,v,w),\\ \dfrac{\partial w}{\partial x}-\dfrac{\partial v}{\partial t}=0,\end{cases}\tag{2.10}$$

where $\tilde{\tilde{F}}(u, v, w)$ is still a C^1 function of u, v and w, satisfying

$$\tilde{\tilde{F}}(0,0,0) = 0. \tag{2.11}$$

(2.10) is also a strictly hyperbolic system with three distinct real eigenvalue $\frac{dt}{dx} = \mu_i$ $(i = 1, 2, 3)$:

$$\mu_1 = -\frac{1}{\lambda} < \mu_2 = 0 < \mu_3 = \frac{1}{\lambda}, \tag{2.12}$$

where λ is still given by (2.5). The corresponding left eigenvectors can still be taken as (2.6), then we also have (2.8) and (2.9).

It is easy to get the following

Lemma 1.

(i) *If on the domain*

$$D = \{(t, x)|\ t_1 \le t \le t_2,\ \ x_1 \le x \le x_2\}, \tag{2.13}$$

$U = U(t, x) = (u(t, x), v(t, x), w(t, x))^T$ *is a* C^1 *solution to system* (2.2), *satisfying*

$$t = t_1\ (or\ t_2):\quad u_x(t, x) = v(t, x),\quad x_1 \le x \le x_2, \tag{2.14}$$

then

$$\frac{\partial u}{\partial x} = v\ \ on\ D, \tag{2.15}$$

hence, on the domain D, $U = U(t, x)$ *is a* C^1 *solution to system* (2.10) *and* $u = u(t, x)$ *is a* C^2 *solution to equation* (1.1).

(ii) *If on the domain* D, $U = U(t, x)$ *is a* C^1 *solution to system* (2.10), *satisfying*

$$x = x_1\ (or\ x_2):\quad u_t(t, x) = w(t, x),\quad t_1 \le t \le t_2, \tag{2.16}$$

then

$$\frac{\partial u}{\partial t} = w \qquad on\ D, \tag{2.17}$$

hence, on the domain D, $U = U(t, x)$ *is a* C^1 *solution to system* (2.2) *and* $u = u(t, x)$ *is a* C^2 *solution to equation* (1.1). □

When equation (1.1) is reduced to system (2.2), it is easy to see that the boundary condition (1.4) can be rewritten as

$$x = 0: \qquad v_1 + v_3 = 2h'(t), \tag{2.18.1}$$

$$x = 0: \qquad v_1 - v_3 = 2\sqrt{K_v(v_2, h(t))}h(t) \triangleq p_2(h(t), v_2) + \bar{p}_2(t), \tag{2.18.2}$$

$$x = 0: \qquad v_1 - v_3 = 2\sqrt{K_v(v_2, \alpha v_2 + h(t))}(\alpha v_2 + h(t)) \triangleq p_3(h(t), v_2) + \bar{p}_3(t) \tag{2.18.3}$$

or

$$x = 0: \qquad v_3 = p_4(h(t), v_1, v_2) + \bar{p}_4(t) \tag{2.18.4}$$

and respectively (noting (1.11))

$$x = 0: \qquad v_1 = q_4(h(t), v_2, v_3) + \bar{q}_4(t), \tag{2.18.4$'$}$$

where v_i $(i = 1, 2, 3)$ are defined by (2.8),

$$p_2(h(t), 0) \equiv p_3(h(t), 0) \equiv p_4(h(t), 0, 0) \equiv q_4(h(t), 0, 0) \equiv 0. \tag{2.19}$$

Moreover, when the C^1 norm of $h(t)$ is small enough, the C^1 norms of $\bar{p}_2(t)$, $\bar{p}_3(t)$, $\bar{p}_4(t)$ and $\bar{q}_4(t)$ are also small.

3. Semi-global C^1 solution for quasilinear hyperbolic systems with zero eigenvalues

Since the hyperbolic wave has a finite speed of propagation, the exact boundary controllability of a hyperbolic equation (system) requires that the controllability time T must be suitably large. In order to get the exact boundary controllability, we should first prove the existence and uniqueness of the semi-global classical solution, namely, the classical solution on the interval $0 \le t \le T_0$, where $T_0 > 0$ is a preassigned and possibly quite large number.

Noting that (2.2) or (2.10) is a hyperbolic system with one zero eigenvalue, in order to study the exact boundary controllability for equation (1.1), the results in [3] and [1] on the existence and uniqueness of semi-global C^1 solution to the mixed initial-boundary value problem for quasilinear hyperbolic systems without zero eigenvalues should be generalized to quasilinear hyperbolic systems with zero eigenvalues.

Consider the following first order quasilinear hyperbolic system

$$\frac{\partial u}{\partial t} + A(u)\frac{\partial u}{\partial x} = F(u), \tag{3.1}$$

where $u = (u_1, \dots, u_n)^T$ is a unknown vector function of (t, x), $A(u)$ is a $n \times n$ matrix with suitably smooth elements $a_{ij}(u)$ $(i, j = 1, \dots, n)$, $F : \mathbb{R}^n \to \mathbb{R}$ is a vector function with suitably smooth components $f_i(u)$ $(i = 1, \dots, n)$ and

$$F(0) = 0. \tag{3.2}$$

By the definition of hyperbolicity, for any given u on the domain under consideration the matrix $A(u)$ has n real eigenvalues $\lambda_i(u)$ $(i = 1, \dots, n)$ and a complete set of left eigenvectors $l_i(u) = (l_{i1}(u), \dots, l_{in}(u))$ $(i = 1, \dots, n)$:

$$l_i(u)A(u) = \lambda_i(u)l_i(u). \tag{3.3}$$

We have

$$\det |l_{ij}(u)| \neq 0. \tag{3.4}$$

Moreover, suppose that on the domain under consideration we have

$$\lambda_p(u) < \lambda_q(u) \equiv 0 < \lambda_r(u) \quad (p = 1, \dots, l;\ q = l+1, \dots, m;\ r = m+1, \dots, n). \tag{3.5}$$

For the mixed initial-boundary value problem of system (3.1) with the initial condition

$$t = 0: \qquad u = \varphi(x), \quad 0 \le x \le 1 \tag{3.6}$$

and the boundary conditions

$$x = 0: \qquad v_r = G_r(t, v_1, \ldots, v_l, v_{l+1}, \ldots, v_m) + H_r(t) \quad (r = m+1, \ldots, n), \tag{3.7}$$

$$x = 1: \qquad v_p = G_p(t, v_{l+1}, \ldots, v_m, v_{m+1}, \ldots, v_n) + H_p(t) \quad (p = 1, \ldots, l), \tag{3.8}$$

where

$$v_i = l_i(u)u \quad (i = 1, \ldots, n) \tag{3.9}$$

and without loss of generality we assume that

$$G_r(t, 0, \ldots, 0) \equiv G_p(t, 0, \ldots, 0) \equiv 0 \quad (r = m+1, \ldots, n;\ p = 1, \ldots, l), \tag{3.10}$$

we have the following lemma on the existence and uniqueness of semi-global C^1 solution (see [11]).

Lemma 2. *Under the previous assumptions, suppose that $l_{ij}(u)$, $\lambda_i(u)$, $f_i(u)$, $G_r(t, \cdot)$, $G_p(t, \cdot)$, $H_r(t)$, $H_p(t)$ $(i, j = 1, \ldots, n; r = m+1, \ldots, n; p = 1, \ldots, l)$ and $\varphi(x)$ are all C^1 functions with respect to their arguments. Assume furthermore that the conditions of C^1 compatibility are satisfied at the points $(0,0)$ and $(0,1)$ respectively. Then, for a given (possibly quite large) $T_0 > 0$, the mixed initial-boundary value problem (3.1) and (3.6)–(3.8) admits a unique C^1 solution $u = u(t,x)$ (called the semi-global C^1 solution) with small C^1 norm on the domain*

$$R(T_0) = \{(t,x)|\ \ 0 \le t \le T_0,\ \ 0 \le x \le 1\}, \tag{3.11}$$

provided that the C^1 norms $\|\varphi\|_{C^1[0,1]}$ and $\|(H_r, H_p)\|_{C^1[0,T_0]}$ $(r = m+1, \ldots, n;\ p = 1, \ldots, l)$ are small enough (depending on T_0). □

4. Proof of Theorems 1 and 2

In order to prove Theorem 1 it suffices to prove the following

Lemma 3. *Let $T > 0$ be defined by (1.6). For any given initial data (φ, ψ) and final data (Φ, Ψ) with small norms $\|(\varphi, \psi)\|_{C^2[0,1]\times C^1[0,1]}$ and $\|(\Phi, \Psi)\|_{C^2[0,1]\times C^1[0,1]}$, the quasilinear wave equation (1.1) admits a C^2 solution $u = u(t,x)$ with small C^2 norm on the domain $R(T) = \{(t,x)|\ \ 0 \le t \le T,\ 0 \le x \le 1\}$, which satisfies simultaneously the initial condition (1.7) and the final condition (1.9).* □

In fact, substituting a C^2 solution $u = u(t,x)$ given by Lemma 3 into the boundary conditions (1.4) and (1.5), we get the boundary controls

$$h(t) = \begin{cases} u|_{x=0}, \\ u_x|_{x=0}, \\ (u_x - \alpha u)|_{x=0} \\ \text{or } (u_x - \bar{\alpha} u_t)|_{x=0} \end{cases} \qquad \text{and} \qquad \bar{h}(t) = \begin{cases} u|_{x=1}, \\ u_x|_{x=1}, \\ (u_x + \beta u)|_{x=1} \\ \text{or } (u_x + \bar{\beta} u_t)|_{x=1}. \end{cases}$$

Then $u = u(t,x)$ is just the C^2 solution to the mixed initial-boundary value problem for equation (1.1) with the initial condition (1.7) and the corresponding boundary conditions (1.4) and (1.5), which verifies the final condition (1.9).

Similarly, in order to get Theorem 2, it suffices to prove the following

Lemma 4. *Let $T > 0$ be defined by* (1.10). *Under the assumptions of Theorem* 2, *the quasilinear wave equation* (1.1) *with the boundary condition* (1.4) *on $x = 0$ admits a C^2 solution $u = u(t,x)$ with small C^2 norm on the domain $R(T) = \{(t,x)|\ 0 \leq t \leq T,\ \ 0 \leq x \leq 1\}$, which satisfies simultaneously the initial condition* (1.7) *and the final condition* (1.9). □

We now give the main steps of the proof of Lemma 3. The proof of Lemma 4 is similar (cf. [5]).

Proof of Lemma 3. Since system (2.2) or (2.10) possesses one zero eigenvalue, the method used in [4] cannot be directly applied to the present situation and it should be modified according to the concrete character of system (2.2).

By (1.6), there exists an $\varepsilon_0 > 0$ so small that

$$T > \max_{|U|\leq\varepsilon_0} \frac{1}{\sqrt{K_v(u,v)}}, \tag{4.1}$$

where $U = (u,v,w)^T$. Let

$$T_1 = \frac{1}{2} \max_{|U|\leq\varepsilon_0} \frac{1}{\sqrt{K_v(u,v)}}. \tag{4.2}$$

We divide the proof into several steps.

(i) We first consider the following forward mixed initial-boundary value problem of system (2.2) with the initial condition

$$t = 0: \qquad U = U_0(x) \triangleq (\varphi(x), \varphi'(x), \psi(x))^T, \quad 0 \leq x \leq 1 \tag{4.3}$$

and the boundary conditions

$$x = 0: \qquad v_3 = f_3(t), \tag{4.4}$$

$$x = 1: \qquad v_1 = \bar{f}_1(t), \tag{4.5}$$

where v_i $(i = 1,2,3)$ are defined by (2.8), f_3 and $\bar{f}_1$ are any given C^1 functions of t with small $C^1[0,T_1]$ norm, such that the conditions of C^1 compatibility are satisfied at the points $(0,0)$ and $(0,1)$ respectively. By Lemma 2, on the domain

$$\{(t,x)|\ \ 0 \leq t \leq T_1,\ \ 0 \leq x \leq 1\}, \tag{4.6}$$

there exists a unique semi-global C^1 solution $U = U^{(1)}(t,x)$ with small C^1 norm and

$$|U^{(1)}(t,x)| \leq \varepsilon_0. \tag{4.7}$$

Thus, we can determine the value of $U^{(1)}(t,x)$ on $x = \frac{1}{2}$ as

$$x = \frac{1}{2}: \qquad U = a(t) \triangleq (a_1(t), a_2(t), a_3(t))^T, \quad 0 \leq t \leq T_1 \tag{4.8}$$

and the $C^1[0,T_1]$ norm of $a(t)$ is small. Moreover, noting the first equation of (2.2), we have

$$a_3(t) = a_1'(t), \quad 0 \leq t \leq T_1, \tag{4.9}$$

then $a_1(t)$ is a C^2 function with small $C^2[0,T_1]$ norm.

(ii) Similarly, we consider the following backward mixed initial-boundary value problem of system (2.2) with the initial condition

$$t = T: \qquad U = U_T(x) \triangleq (\Phi(x), \Phi'(x), \Psi(x))^T, \quad 0 \le x \le 1 \tag{4.10}$$

and the boundary conditions

$$x = 0: \qquad v_1 = g_1(t), \tag{4.11}$$

$$x = 1: \qquad v_3 = \bar{g}_3(t), \tag{4.12}$$

where v_i $(i = 1, 2, 3)$ are still defined by (2.8), g_1 and $\bar{g}_3$ are any given C^1 functions of t with small $C^1[T-T_1, T]$ norm, such that the conditions of C^1 compatibility are satisfied at the points $(T, 0)$ and $(T, 1)$ respectively. By Lemma 2, on the domain

$$\{(t,x)|\ T - T_1 \le t \le T, \quad 0 \le x \le 1\}, \tag{4.13}$$

there exists a unique semi-global C^1 solution $U = U^{(2)}(t, x)$ with small C^1 norm and

$$|U^{(2)}(t,x)| \le \varepsilon_0. \tag{4.14}$$

Thus, we can determine the value of $U^{(2)}(t, x)$ on $x = \frac{1}{2}$ as

$$x = \frac{1}{2}: \qquad U = b(t) \triangleq (b_1(t), b_2(t), b_3(t))^T, \quad T - T_1 \le t \le T \tag{4.15}$$

and the $C^1[T - T_1, T]$ norm of $b(t)$ is small. Similarly to (4.9), we have

$$b_3(t) = b_1'(t), \quad T - T_1 \le t \le T, \tag{4.16}$$

then $b_1(t)$ is a C^2 function with small $C^2[T - T_1, T]$ norm.

(iii) Noting (4.9) and (4.16), we can find $c(t) = (c_1(t), c_2(t), c_3(t))^T \in C^1[0, T]$ with small C^1 norm, such that

$$c(t) = \begin{cases} a(t), & 0 \le t \le T_1, \\ b(t), & T - T_1 \le t \le T \end{cases} \tag{4.17}$$

and

$$c_3(t) = c_1'(t), \quad 0 \le t \le T, \tag{4.18}$$

then $c_1(t) \in C^2[0, T]$ with small C^2 norm.

We now change the order of t and x, and on the domain

$$R_l(T) = \{(t,x)|\ 0 \le t \le T, \quad 0 \le x \le \frac{1}{2}\} \tag{4.19}$$

we consider the following leftward mixed initial-boundary value problem of system (2.10) with the initial condition

$$x = \frac{1}{2}: \qquad U = c(t), \quad 0 \le t \le T \tag{4.20}$$

and the boundary conditions

$$t = 0: \qquad v_1 = l_1(U_0(x))U_0(x), \qquad 0 \le x \le \frac{1}{2}, \tag{4.21}$$

$$t = T: \qquad v_3 = l_3(U_T(x))U_T(x), \qquad 0 \le x \le \frac{1}{2}, \tag{4.22}$$

where $l_i(U)$ $(i = 1, 2, 3)$ are defined by (2.6), $U_0(x)$ and $U_T(x)$ are given in (4.3) and (4.10) respectively.

By Lemma 1 (i), it is easy to see that the corresponding conditions of C^1 compatibility are satisfied at the points $(0, \frac{1}{2})$ and $(T, \frac{1}{2})$ respectively. Hence, by Lemma 2, there exists a unique semi-global C^1 solution $U = U_l(t, x)$ with small C^1 norm on the domain $R_l(T)$ and

$$|U_l(t,x)| \leq \varepsilon_0 \qquad \text{on } R_l(T), \tag{4.23}$$

provided that the C^1 norms of $U_0(x)$, $U_T(x)$, $f_3(t)$, $\bar{f}_1(t)$, $g_1(t)$ and $\bar{g}_3(t)$ are small enough.

(iv) Similarly, the rightward mixed initial-boundary value problem of system (2.10) with the initial condition (4.20) and the boundary conditions

$$t = 0: \qquad v_3 = l_3(U_0(x))U_0(x), \qquad \frac{1}{2} \leq x \leq 1, \tag{4.24}$$

$$t = T: \qquad v_1 = l_1(U_T(x))U_T(x), \qquad \frac{1}{2} \leq x \leq 1 \tag{4.25}$$

admits a unique semi-global C^1 solution $U = U_r(t, x)$ with small C^1 norm on the domain

$$R_r(T) = \{(t,x)|\ 0 \leq t \leq T,\ \frac{1}{2} \leq x \leq 1\} \tag{4.26}$$

and

$$|U_r(t,x)| \leq \varepsilon_0 \qquad \text{on } R_r(T). \tag{4.27}$$

(v) Let

$$U(t,x) = \begin{cases} U_l(t,x), & (t,x) \in R_l(T), \\ U_r(t,x), & (t,x) \in R_r(T). \end{cases} \tag{4.28}$$

We claim that

$$t = 0: \qquad U = U_0(x), \qquad 0 \leq x \leq 1, \tag{4.29}$$

$$t = T: \qquad U = U_T(x), \qquad 0 \leq x \leq 1. \tag{4.30}$$

In fact, by Lemma 1 (i), the C^1 solutions $U = U_l(t, x)$ [resp. $U = U_r(t, x)$] and $U = U^{(1)}(t, x)$ satisfy the system (2.10), the initial condition

$$x = \frac{1}{2}: \qquad U = a(t), \qquad 0 \leq t \leq T_1 \tag{4.31}$$

and the boundary condition (4.21) [resp. (4.24)]. By the uniqueness of C^1 solution (see [8]) and the choice of T_1 given in (4.2), the mixed initial-boundary value problem (2.10), (4.31) and (4.21) [resp. (4.24)] has a unique C^1 solution on the domain

$$\{(t,x)|\ 0 \leq t \leq 2T_1 x,\ \ 0 \leq x \leq \frac{1}{2}\} \tag{4.32}$$

$$[\text{resp. } \{(t,x)|\ 0 \leq t \leq 2T_1(1-x),\ \ \frac{1}{2} \leq x \leq 1\}].$$

Then

$$U(t,x) \equiv U^{(1)}(t,x) \tag{4.33}$$

on these domains and, in particular, we get (4.29).

In a similar manner we obtain (4.30).

Thus, by Lemma 1 (ii), it is easy to see that the first component $u = u(t,x)$ of $U = U(t,x)$ satisfies all the requirements of Lemma 3.

5. Remarks

If in the boundary conditions (1.4.2)–(1.4.4) on $x = 0$ and the boundary conditions (1.5.2)–(1.5.4) on $x = 1$, u_x is replaced by $K(u, u_x)$, the conclusions of Theorem 1 and Theorem 2 are still valid, provided that (1.11) is replaced by

$$\bar{\alpha} \neq \sqrt{K_v(0,0)}. \tag{5.1}$$

References

[1] M. Cirinà, *Nonlinear hyperbolic problems with solutions on preassigned sets*, Michigan Math. J., 17(1970), 193–209.

[2] I. Lasiecka, R. Triggiani, *Exact controllability of semilinear abstract systems with applications to waves and plates boundary control problems*, Appl. Math. Optim., 23 (1991), 109–154.

[3] Li Tatsien, Jin Yi, *Semi-global C^1 solution to the mixed initial-boundary value problem for quasilinear hyperbolic systems*, Chin. Ann. Math., 22B (2001), 325–336.

[4] Li Tatsien, Rao Bopeng, *Exact boundary controllability for quasilinear hyperbolic systems*, SIAM J. Control Optim., 41(2003), 1748–1755.

[5] Li Tatsien, Rao Bopeng, *Local exact boundary controllability for a class of quasilinear hyperbolic systems*, Chin. Ann. Math., 23B(2002), 209–218.

[6] Li Tatsien, Rao Bopeng, Jin Yi, *Solution C^1 semi-globale et contrôlabilité exacte frontière de systèmes hyperboliques quasilinéaires,*, C. R. Acad. Sci. Paris, t. 333, Série I (2001), 219–224.

[7] Li Tatsien, Xu Yulan, *Local exact boundary controllability for nonlinear vibrating string equations*, International Journal of Mordern Physics B, 17 (2003), 4062–4071.

[8] Li Tatsien, Yu Wenci, *Boundary Value Problems for Quasilinear Hyperbolic Systems*, Duck University Mathematics Series V, 1985.

[9] J.-L. Lions, *Contrôlabilité Exacte, Perturbations et Stabilisation de Systèmes Distribués*, Vol. I, Masson, 1988.

[10] D.L. Russell, *Controllability and stabilizability theory for linear partial differential equations, Recent progress and open questions*, SIAM Rev., 20 (1978), 639–739.

[11] Yu Lixin, *Semi-global C^1 solutions to the mixed initial-boundary value problem for a general class of quasilinear hyperbolic systems*, to appear in Chin. Ann. Math., 25A (2004).

[12] E. Zuazua, *Exact controllability for the semilinear wave equation*, J. Math. Pures et Appl., 69 (1990), 1–31.

[13] E. Zuazua, *Exact controllability for semilinear wave equations in one sapce dimension*, Annales de l'Institut Henri Poincaré, Analyse non linéaire, 10(1993), 109–129.

Li Tatsien
Department of Mathematics
Fudan University
Shanghai, 200433 China

Progress in Nonlinear Differential Equations
and Their Applications, Vol. 61, 161–185

Regularity of Euler Equations for a Class of Three-Dimensional Initial Data

A. Mahalov, B. Nicolaenko, C. Bardos and F. Golse

Dedicated to V.A. Solonnikov with admiration

Abstract. The 3D incompressible Euler equations with initial data characterized by uniformly large vorticity are investigated. We prove existence on long time intervals of regular solutions to the 3D incompressible Euler equations for a class of large initial data in bounded cylindrical domains. There are no conditional assumptions on the properties of solutions at later times, nor are the global solutions close to some 2D manifold. The approach is based on fast singular oscillating limits, nonlinear averaging and cancellation of oscillations in the nonlinear interactions for the vorticity field. With nonlinear averaging methods in the context of almost periodic functions, resonance conditions and a nonstandard small divisor problem, we obtain fully 3D limit resonant Euler equations. We establish the global regularity of the latter without any restriction on the size of 3D initial data and bootstrap this into the regularity on arbitrary large time intervals of the solutions of 3D Euler equations with weakly aligned uniformly large vorticity at $t = 0$.

Mathematics Subject Classification (2000). Primary 35Q35; Secondary 76B03.

Keywords. Three-dimensional Euler equations, vorticity, fast singular oscillating limits, conservation laws.

1. Introduction and main results

The long-time solvability of the 3D Cauchy problem for the Euler equations is an outstanding problem of applied analysis. At issue is the possible blow-up of vorticity in finite times [7]. Whereas local regularity and long-time regularity for small 3D initial data are well known ([32], [33], [16], [9]), there is a dearth of results for large 3D initial data without conditional assumptions on the properties of solutions at later times. Solutions in a 2D axisymmetric geometry have been constructed in [19].

We study an initial value problem for the three-dimensional Euler equations with initial data characterized by uniformly large vorticity:

$$\partial_t \mathbf{V} + (\mathbf{V}\cdot\nabla)\mathbf{V} = -\nabla p,\ \nabla\cdot\mathbf{V} = 0, \tag{1.1}$$

$$\mathbf{V}(t,y)|_{t=0} = \mathbf{V}(0) = \tilde{\mathbf{V}}_0(y) + \frac{\Omega}{2} e_3 \times y \tag{1.2}$$

where $y = (y_1, y_2, y_3)$, $\mathbf{V}(t,y) = (V_1, V_2, V_3)$ is the velocity field and p is the pressure. In Eqs. (1.1) e_3 denotes the vertical unit vector and Ω is a constant parameter. The field $\tilde{\mathbf{V}}_0(y)$ depends on three variables y_1, y_2 and y_3. Since $\mathrm{curl}(\frac{\Omega}{2} e_3 \times y) = \Omega e_3$, the vorticity vector at initial time $t = 0$ is

$$\mathrm{curl}\,\mathbf{V}(0,y) = \mathrm{curl}\,\tilde{\mathbf{V}}_0(y) + \Omega e_3, \tag{1.3}$$

and the initial vorticity has a large component weakly aligned along e_3, when $\Omega \gg 1$. These are fully three-dimensional large initial data with large initial 3D vortex stretching.

Eqs. (1.1) are studied in cylindrical domains

$$\mathbf{C} = \{(y_1, y_2, y_3) \in \mathbf{R}^3 : 0 < y_3 < 2\pi/\alpha,\ y_1^2 + y_2^2 < R^2\} \tag{1.4}$$

where α and R are positive real numbers. If h is the height of the cylinder, $\alpha = 2\pi/h$. Let

$$\Gamma = \{(y_1, y_2, y_3) \in \mathbf{R}^3 : 0 < y_3 < 2\pi/\alpha,\ y_1^2 + y_2^2 = R^2\}. \tag{1.5}$$

Without loss of generality, we can assume that $R = 1$. Eqs. (1.1) are considered with periodic boundary conditions in y_3

$$\mathbf{V}(y_1, y_2, y_3) = \mathbf{V}(y_1, y_2, y_3 + 2\pi/\alpha) \tag{1.6}$$

and vanishing normal component of velocity on Γ

$$\mathbf{V}\cdot\mathbf{N} = \tilde{\mathbf{V}}\cdot\mathbf{N} = 0 \text{ on } \Gamma; \tag{1.7}$$

where $\mathbf{N}$ is the normal vector to Γ. From the invariance of 3D Euler equations under the symmetry $y_3 \to -y_3$, $V_1 \to V_1$, $V_2 \to V_2$, $V_3 \to -V_3$, all results in this paper extend to cylindrical domains bounded by two horizontal plates. Then the boundary conditions in the vertical direction are zero flux on the vertical boundaries (zero vertical velocity on the plates). One only needs to restrict vector fields to be even in y_3 for V_1, V_2 and odd in y_3 for V_3.

We choose $\tilde{\mathbf{V}}_0(y)$ in $\mathbf{L}_2(\mathbf{C})$. We introduce $\tilde{\mathbf{V}}(t,y)$ such that

$$\mathbf{V}(t,y) = \tilde{\mathbf{V}}(t,y) + \frac{\Omega}{2} e_3 \times y,\ \ \mathrm{curl}\,\mathbf{V}(t,y) = \mathrm{curl}\,\tilde{\mathbf{V}}(t,y) + \Omega e_3. \tag{1.8}$$

For the vorticity field $\omega = \mathrm{curl}\,\mathbf{V}$ Eqs. (1.1) become

$$\frac{\partial}{\partial t}\omega + \mathbf{V}\cdot\nabla\omega = \omega\cdot\nabla\mathbf{V}, \tag{1.9}$$

$$\omega(0,y) = \mathrm{curl}\,\tilde{\mathbf{V}}_0(y) + \Omega e_3, \tag{1.10}$$

and the initial condition induces large initial vortex stretching.

We present a simple case of results obtained in our joint work with C. Bardos and F. Golse [6], where the initial value problem is solved in more general functional spaces. We establish regularity for arbitrarily large finite times for the 3D Euler solutions for Ω large, but *finite*. Our solutions are not close in any sense to those of the 2D or "quasi 2D" Euler and they are characterized by fast oscillations in the e_3 direction, together with a large vortex stretching term $\omega(t,y) \cdot \nabla \mathbf{V}(t,y) = \omega_1 \frac{\partial V_1}{\partial y_1} + \omega_2 \frac{\partial V_2}{\partial y_2} + \omega_3 \frac{\partial V_3}{\partial y_3}$, $t \geq 0$ with leading component $\Omega \frac{\partial}{\partial y_3} V_3(t,y) \gg 1$. There are no assumptions on oscillations in y_1, y_2 for our solutions (nor for the initial condition $\tilde{\mathbf{V}}_0(y)$).

Our approach is entirely based on fast singular oscillating limits of Eqs. (1.9)–(1.10), nonlinear averaging and cancellation of oscillations in the nonlinear interactions for the vorticity field for large Ω. This has been developed in [3], [4], [5] and [24] for the cases of periodic lattice domains and the infinite space $\mathbf{R}^3$. Through the canonical transformation (1.17)–(1.18) in both the field $\mathbf{V}(t,y)$ and the space coordinate $y = (y_1, y_2, y_3)$ for every Ω (not necessary large) we map every solution $\mathbf{V}(t,y)$ of Eqs. (1.1) one-to-one to a solution $\mathbf{U}(t,x)$, $x = (x_1, x_2, x_3)$ of

$$\partial_t \mathbf{U} + (\mathbf{U} \cdot \nabla_x)\mathbf{U} + \Omega \mathbf{e}_3 \times \mathbf{U} = -\nabla_x (p - \frac{\Omega^2}{4}|x_h|^2), \ \nabla_x \cdot \mathbf{U} = 0, \quad (1.11)$$

$$\mathbf{U}(t,x)|_{t=0} = \mathbf{U}(0,x) = \tilde{\mathbf{V}}_0(x), \quad (1.12)$$

where $x = y$ at $t = 0$ and $x_h = (x_1, x_2)$. For $\Omega \gg 1$ the nearly singular initial value problem (1.1)–(1.2) (that is with large initial vorticity and vortex stretching) is mapped into the problem (1.11)–(1.12) with the nearly singular Coriolis operator term restricted to solenoidal fields:

$$\frac{1}{\epsilon} e_3 \times \mathbf{U}, \ \nabla \cdot \mathbf{U} = 0, \ \epsilon = 1/\Omega << 1. \quad (1.13)$$

As detailed in Section 2, the linear part of Eq. (1.11) is the Poincaré-Sobolev nonlocal wave equations ([2], [12], [25], [29]):

$$\partial_t \mathbf{\Phi} + \Omega \mathbf{e}_3 \times \mathbf{\Phi} = -\nabla \pi, \ \nabla \cdot \mathbf{\Phi} = 0. \quad (1.14)$$

Interactions between the Poincaré waves generated by the quadratic nonlinearity in Eq. (1.11) are ruled by resonance conditions and a small divisor problem in the limit $\Omega \to \infty$. With nonlinear averaging methods in the context of Banach space valued almost periodic functions we obtain fully 3D limit resonant Euler equations. We establish the global regularity of the latter without any restriction on the size of 3D initial data and bootstrap this into the global regularity of Eqs. (1.11)–(1.12) for Ω large but finite. Then by the canonical transformation (1.17)–(1.18) of the field $\mathbf{V}$ (which is an isometry on curl-based generalizations of Sobolev spaces) we establish the long-time regularity of Eqs. (1.1)–(1.2) for large finite Ω, on arbitrarily finite large time intervals.

Our results crucially use the algebra of the curl operator with boundary conditions, for the fast singular oscillating limits of $\tilde{\omega} = \operatorname{curl} \mathbf{U}(t,x)$:

$$\partial_t \tilde{\omega} + \mathbf{U} \cdot \nabla \tilde{\omega} = \tilde{\omega} \cdot \nabla \mathbf{U} + \Omega \frac{\partial}{\partial x_3} \mathbf{U}. \tag{1.15}$$

For this we rely on deep properties of curl^{-1}, extending the early pioneering results of O.A. Ladyzhenskaya, V.A. Solonnikov and co-workers which were obtained in the context of Maxwell's equations and magneto-hydrodynamics ([18], [21], [10], [11], [30]). There are three foremost issues with the analysis of (1.1)–(1.2), (1.11)–(1.15) for large parameter Ω. First, the nature of their fast singular oscillating limit equations as $\Omega \to +\infty$ and the global regularity of their solutions (3D resonant limit Euler equations). Second, the strong convergence of solutions of (1.11)–(1.12) to those of the limit equations; and, finally, bootstrapping from analysis of the first two questions the long-time regularity of solutions of (1.1)–(1.2) for Ω large but finite.

We now detail the canonical transformation between the original vector field $\mathbf{V}(t,y)$ and the vector field $\mathbf{U}(t,x)$. Let $\mathbf{J}$ be the matrix such that $\mathbf{J}\mathbf{a} = e_3 \times \mathbf{a}$ for any vector field $\mathbf{a}$. Then

$$\mathbf{J} = \begin{pmatrix} 0 & -1 & 0 \\ 1 & 0 & 0 \\ 0 & 0 & 0 \end{pmatrix}, \quad \boldsymbol{\Upsilon}(t) \equiv e^{\Omega \mathbf{J} t/2} = \begin{pmatrix} \cos(\frac{\Omega t}{2}) & -\sin(\frac{\Omega t}{2}) & 0 \\ \sin(\frac{\Omega t}{2}) & \cos(\frac{\Omega t}{2}) & 0 \\ 0 & 0 & 1 \end{pmatrix}. \tag{1.16}$$

For any fixed parameter Ω (not necessary large) we introduce the following fundamental transformation:

$$\mathbf{V}(t,y) = e^{+\Omega \mathbf{J} t/2} \mathbf{U}(t, e^{-\Omega \mathbf{J} t/2} y) + \frac{\Omega}{2} \mathbf{J} y, \ x = e^{-\Omega \mathbf{J} t/2} y. \tag{1.17}$$

The transformation (1.17) is invertible:

$$\mathbf{U}(t,x) = e^{-\Omega \mathbf{J} t/2} \mathbf{V}(t, e^{+\Omega \mathbf{J} t/2} x) - \frac{\Omega}{2} \mathbf{J} x, \ y = e^{+\Omega \mathbf{J} t/2} x. \tag{1.18}$$

The transformations (1.17)–(1.18) establish a one-to-one correspondence between solenoidal vector fields $\mathbf{V}(t,y)$ and $\mathbf{U}(t,x)$. We note that for $t = 0$ $x = y$ and therefore $\tilde{\mathbf{V}}_0(y) = \tilde{\mathbf{V}}_0(x)$. Let $x = (x_h, x_3)$ where $x_h = (x_1, x_2)$, $|x_h|^2 = x_1^2 + x_2^2$ and similarly for y. We have:

Lemma 1.1. *The following identities hold for the vector fields $\mathbf{V}(t,y)$ and $\mathbf{U}(t,x)$ and pressure p:*

1. $\nabla_y \cdot \mathbf{V}(t,y) = \nabla_x \cdot \mathbf{U}(t,x)$.
2. $\nabla_y p = \boldsymbol{\Upsilon}(t) \nabla_x p$.
3. $\operatorname{curl}_y \mathbf{V}(t,y) = \boldsymbol{\Upsilon}(t) \operatorname{curl}_x \mathbf{U}(t,x) + \Omega e_3, \quad \operatorname{curl}_y^2 \mathbf{V}(t,y) = \boldsymbol{\Upsilon}(t) \operatorname{curl}_x^2 \mathbf{U}(t,x)$.
4. $\frac{D}{Dt} \mathbf{V}(t,y) = \boldsymbol{\Upsilon}(t) \left(\frac{D}{Dt} \mathbf{U}(t,x) + \Omega \mathbf{J} \mathbf{U} - \frac{\Omega^2}{2} x_h \right)$ *where* $\frac{D}{Dt}$ *are the corresponding Lagrangian derivatives,* $\mathbf{J}\mathbf{U} = e_3 \times \mathbf{U}$.

Lemma 1.1 establishes that the transformation (1.17)–(1.18) is canonical for (1.1)–(1.2). From the property 1 of Lemma 1.1 it follows that $\nabla_x \cdot \mathbf{U}(t,x) = 0$ since $\nabla_y \cdot \mathbf{V}(t,y) = 0$. Now using 2–4 in the above Lemma 1.1 and the fact that $\boldsymbol{\Upsilon}(t)$ is unitary we can express each term in (1.1) in x and t variables to obtain the equations for $\mathbf{U}(t,x)$ (1.11)–(1.15). Under the transformation (1.17)–(1.18) Eqs. (1.1)–(1.2) turn into Euler system (1.11)–(1.12) with an additional Coriolis term $\Omega e_3 \times \mathbf{U}$ and modified initial data and pressure. The systems Eqs. (1.1)–(1.2) and (1.11)–(1.12) are equivalent for every Ω (not necessary large) and the pair of transformations (1.17)–(1.18) establishes one-to-one correspondence between their fully three-dimensional solutions.

Remark 1.2. The canonical transformation (1.17)–(1.18) preserves the boundary conditions (1.7) which are transformed into

$$\mathbf{U} \cdot \mathbf{N} = 0, \text{ on } \Gamma. \tag{1.19}$$

Using elementary identities $(\mathbf{U} \cdot \nabla)\mathbf{U} = \operatorname{curl} \mathbf{U} \times \mathbf{U} + \nabla(\frac{|\mathbf{U}|^2}{2})$ on divergence free vector fields, Eqs. (1.11) can be rewritten in the form

$$\partial_t \mathbf{U} + (\operatorname{curl} \mathbf{U} + \Omega e_3) \times \mathbf{U} = -\nabla(p - \frac{\Omega^2}{4}|x_h|^2 + \frac{|\mathbf{U}|^2}{2}), \tag{1.20}$$

$$\nabla \cdot \mathbf{U} = 0, \ \mathbf{U}(t,x)|_{t=0} = \mathbf{U}(0) = \tilde{\mathbf{V}}_0(x). \tag{1.21}$$

Remark 1.3. For large Ω the initial value condition (1.2) can be interpreted as weak alignment of the initial vorticity at $t = 0$; in the distributional sense, for every test function $\phi(y) \in C_0^\infty(\mathbf{R}^3)$ we have:

$$\begin{aligned} |\langle \operatorname{curl} \mathbf{V}(0,y)/\Omega - e_3, \phi(y)\rangle| &= |\langle \mathbf{V}(0,y)/\Omega - \frac{1}{2}e_3 \times y, \operatorname{curl}\phi(y)\rangle| \\ &= \frac{1}{\Omega}|\langle \mathbf{U}(0,x), \operatorname{curl}\phi(x)\rangle|, \end{aligned} \tag{1.22}$$

with $\Omega \geq \Omega_1$ (Ω_1 is defined in Theorem 1.4, $\Omega_1 \gg 1$).

The fast singular oscillating limits of Eqs. (1.20)–(1.21) are investigated as $\Omega \to +\infty$, after further transformation of Eqs. (1.20)–(1.21) with the Poincaré propagator. The latter is the unitary group solution $\mathbf{E}(-\Omega t)\boldsymbol{\Phi}(0) = \boldsymbol{\Phi}(t)$ in $L_2(\mathbf{C})$ ($\mathbf{E}(0) = \mathbf{Id}$ is the identity) to the linear Poincaré wave problem ([25], [12], [29]):

$$\partial_t \boldsymbol{\Phi} + \Omega \mathbf{J} \boldsymbol{\Phi} = -\nabla \pi, \quad \nabla \cdot \boldsymbol{\Phi} = 0, \tag{1.23}$$

or, equivalently,

$$\partial_t \boldsymbol{\Phi} + \Omega \mathbf{PJP} \boldsymbol{\Phi} = 0 \tag{1.24}$$

where $\mathbf{P}$ is the Leray projection on divergence free vector fields; the solutions $\mathbf{E}(-\Omega t)\boldsymbol{\Phi}(0)$ are called Poincaré waves. Eqs. (1.23) give rise to a unitary group of transformations $\mathbf{E}(-\Omega t)$ on a space of square-integrable divergence-free vector fields $L_2(\mathbf{C})$. The spectrum of the generator of the group of motions, that is the spectrum of the skew-hermitian zero order pseudo-differential operator $\mathbf{PJP}$ is $[-i, i]$. In the case of cylindrical domains considered in this paper the eigenvalues (point spectrum) of the operator **PJP** are dense in $[-i, i]$. The operator **PJP** has norm one. Since **PJP** is bounded on $L_2(\mathbf{C})$, the solutions to (1.24) with initial condition $\mathbf{\Phi}(0)$ is given by

$$\mathbf{\Phi}(t) = \mathbf{E}(-\Omega t)\mathbf{\Phi}(0) = \sum_{j=0}^{+\infty} \frac{(-\Omega t)^j}{j!}(-\mathbf{PJP})^j\ \mathbf{\Phi}(0). \tag{1.25}$$

Applying to Eqs. (1.11)–(1.12) (equivalently, Eqs. (1.20)–(1.21)) the Leray projection **P** onto divergence free vector fields, we obtain for $\mathbf{U} = \mathbf{P}\ \mathbf{U}$

$$\begin{aligned} \partial_t \mathbf{U} + \Omega\mathbf{PJPU} &= \mathbf{B}(\mathbf{U}, \mathbf{U}), \\ \mathbf{U}|_{t=0} &= \mathbf{U}(0) = \tilde{\mathbf{V}}_0 \end{aligned} \tag{1.26}$$

where

$$\mathbf{B}(\mathbf{U}, \mathbf{U}) = -\mathbf{P}(\mathbf{U}\cdot\nabla\mathbf{U}) = \mathbf{P}(\mathbf{U}\times \operatorname{curl}\mathbf{U}). \tag{1.27}$$

The proofs of regularity rely on the analysis of the dispersion relations for Poincaré waves [25], [12] (solutions to Eqs. (1.23)–(1.24)). The resonance condition for the interactions generated by the Euler quadratic nonlinearity in the limit $\Omega \to +\infty$ takes the form (see [3] and Sections 2, 3 below):

$$\pm\frac{k_3}{\sqrt{\frac{\beta(k_1,k_2,k_3)^2}{\alpha^2} + k_3^2}} \pm \frac{m_3}{\sqrt{\frac{\beta(m_1,m_2,m_3)^2}{\alpha^2} + m_3^2}} \pm \frac{n_3}{\sqrt{\frac{\beta(n_1,n_2,n_3)^2}{\alpha^2} + n_3^2}} = 0 \tag{1.28}$$

with the convolution conditions $n_3 = k_3 + m_3$, $n_2 = k_2 + m_2$. Here $m = (m_1, m_2, m_3)$ are three-dimensional wave vectors. The integers m_1, m_2 and m_3 are for the radial, azimuthal and axial directions, respectively. Similarly, for k and n. Eqs. (1.28) are trivially satisfied for $k_3 = m_3 = n_3 = 0$ which correspond to pure two-dimensional interactions (dependence on x_1, x_2 and no dependence on x_3 in physical space). The nonlinear interactions with $k_3 m_3 n_3 = 0$, $k_3^2 + m_3^2 + n_3^2 \neq 0$ correspond to two-wave resonances and the interactions with $k_3 m_3 n_3 \neq 0$ correspond to strict three-wave resonances. The quantities β are related to zeros of certain expressions involving Bessel functions (see Eq. (2.27)).

We outline the structure of the fast oscillating limit equations obtained from Eqs. (1.26) in the limit $\Omega \to +\infty$:

$$\begin{aligned} \partial_t \mathbf{w} &= \tilde{\mathbf{B}}(\mathbf{w}, \mathbf{w}), \\ \mathbf{w}|_{t=0} &= \mathbf{w}(0) = \mathbf{U}(0) = \tilde{\mathbf{V}}_0. \end{aligned} \tag{1.29}$$

Details are given in Sections 2 and 3.

We denote the orthogonal decomposition $\mathbf{w} = \overline{\mathbf{w}} + \mathbf{w}^\perp$ where $\overline{\mathbf{w}}(t, x_1, x_2)$ is the barotropic projection (vertical averaging),

$$\overline{\mathbf{w}}(t, x_1, x_2) = \frac{1}{2\pi\alpha}\int_0^{2\pi\alpha} \mathbf{w}(t, x_1, x_2, x_3)dx_3 \tag{1.30}$$

and the orthogonal field $\mathbf{w}^\perp(t, x_1, x_2, x_3)$ verifies $\overline{\mathbf{w}}^\perp = 0$. Then

$$\mathbf{w} = \overline{\mathbf{w}} + \mathbf{w}^\perp. \tag{1.31}$$

Eqs. (1.29) conserve both energy and helicity. These equations are genuinely three-dimensional since they include all 3D modes but with wave-number interactions restricted in $\tilde{\mathbf{B}}(\mathbf{w},\mathbf{w})$. We have (see below in Sections 2 and 3)

$$\overline{\tilde{\mathbf{B}}(\mathbf{w},\mathbf{w})} = \tilde{\mathbf{B}}(\overline{\mathbf{w}},\overline{\mathbf{w}}) \tag{1.32}$$

implying

$$\tilde{\mathbf{B}}(\mathbf{w},\mathbf{w}) = \mathbf{B}_{\mathrm{III}}(\mathbf{w}^{\perp},\mathbf{w}^{\perp}) + \mathbf{B}_{\mathrm{II}}(\overline{\mathbf{w}},\mathbf{w}^{\perp}) + \mathbf{B}_{2D}(\overline{\mathbf{w}},\overline{\mathbf{w}}) \tag{1.33}$$

where $\mathbf{B}_{2D}$ corresponds to pure 2D interactions ($k_3 = m_3 = n_3 = 0$), $\mathbf{B}_{\mathrm{II}}$ is the 'catalytic' operator ($k_3 = 0$, $m_3 n_3 \neq 0$ or $m_3 = 0$, $k_3 n_3 \neq 0$). The above implies $k_3 m_3 n_3 \neq 0$ for interactions given by $\mathbf{B}_{\mathrm{III}}$. Such interactions are called strict 3-wave interactions.

Since $\overline{\tilde{\mathbf{B}}(\mathbf{w},\mathbf{w})} = \tilde{\mathbf{B}}(\overline{\mathbf{w}},\overline{\mathbf{w}})$ the solutions $\mathbf{w}(t,x_1,x_2,x_3) = (w_1,w_2,w_3)$ of the limit equations (1.29) split into an equation for $\overline{\mathbf{w}}(t,x_1,x_2)$ which decouples and an equation for $\mathbf{w}^{\perp}(t,x_1,x_2,x_3)$ with its coefficients depending on $\overline{\mathbf{w}}$. The field $\overline{\mathbf{w}}(t,x_1,x_2)$ satisfies the 2D-3C Euler equations (three components and dependence on two variables x_1, x_2). Specifically,

$$\begin{aligned} \partial_t \overline{\mathbf{w}} + (\overline{\mathbf{w}}\cdot\nabla)\overline{\mathbf{w}} &= -\nabla_h \overline{q},\ \nabla_h\cdot\overline{\mathbf{w}} = 0 \\ \overline{\mathbf{w}}|_{t=0} &= \overline{\mathbf{w}}(0) = \overline{\mathbf{U}}(0) \end{aligned} \tag{1.34}$$

where ∇_h denotes the gradient in horizontal variables x_1, x_2. The component $\mathbf{w}^{\perp}(t,x_1,x_2,x_3)$ (orthogonal to $\overline{\mathbf{w}}$, that is with zero vertical average) satisfies limit equations

$$\begin{aligned} \partial_t \mathbf{w}^{\perp} &= \mathbf{B}_{\mathrm{II}}(\overline{\mathbf{w}},\mathbf{w}^{\perp}) + \mathbf{B}_{\mathrm{III}}(\mathbf{w}^{\perp},\mathbf{w}^{\perp}) \\ \mathbf{w}^{\perp}|_{t=0} &= \mathbf{w}^{\perp}(0) = \mathbf{U}^{\perp}(0) = \mathbf{U}(0) - \overline{\mathbf{U}}(0). \end{aligned} \tag{1.35}$$

For $\overline{\mathbf{w}}(t,x_1,x_2)$ we have the usual conservation laws and global existence theorems for 2D Euler ([32], [33]).

For the generic case of no strict 3 wave resonances $\mathbf{B}_{\mathrm{III}} = 0$. In this case we have global regularity of the limit resonant equations and long time regularity of the 3D Euler equations (1.11)–(1.12) for Ω large. The set of parameters α where $\mathbf{B}_{\mathrm{III}} = 0$ has full Lebesgue measure. In such cases, global regularity of the limit resonant equations and long time regularity of the 3D Euler equations (1.11)–(1.12) is proven using the new 3D conservation laws for $\mathbf{B}_{\mathrm{II}}$ (see Section 3) and the convergence Theorem 4.6 in Section 4. More precisely, $\mathbf{B}_{\mathrm{III}} \neq 0$ for a countable discrete set of parameters α. We now state our main existence theorem. For the cylinder, denote by h the height and R the radius. We denote by $\mathbf{H}^s_\sigma(\mathbf{C})$ the usual Sobolev spaces of solenoidal vector fields in the cylinder $\mathbf{C}$, $s \geq 0$.

Theorem 1.4. *Consider the initial value problem for the 3D Euler equations* (1.1)–(1.2) *with* $\operatorname{curl}\mathbf{V}(0,y) = \Omega e_3 + \operatorname{curl}\tilde{\mathbf{V}}_0(y)$ *and* $\tilde{\mathbf{V}}_0(y) \in \mathbf{H}^s_\sigma(\mathbf{C})$, $s \geq 4$. *Let* $\operatorname{curl}^j \tilde{\mathbf{V}}_0(y)\cdot\mathbf{N} = 0$ *on* Γ, $0 \leq j \leq s$. *Let* $\|\tilde{\mathbf{V}}_0\|_{\mathbf{H}^s_\sigma(\mathbf{C})} \leq M_s$. *Let* $h/R \notin \mathcal{K}^*$ *where* $\mathcal{K}^*$ *is a countable discrete set. Let* $T_m > 0$ *fixed, arbitrary large. Then there* *exists* $\Omega_1(h/R, M_s, T_m)$ *such that for every fixed* $\Omega \geq \Omega_1$ *there exists a unique regular solution of Eqs.* (1.1)–(1.2) *for* $0 \leq t < T_m$*:*

$$||\mathbf{V}(t,y)||_{\mathbf{H}^s_\sigma(\mathbf{C})} \leq \tilde{M}_s(h/R, M_s, T_m). \tag{1.36}$$

Moreover, $\operatorname{curl}^j \mathbf{V}(t,y) \cdot \mathbf{N} = 0$ *on* Γ, $0 \leq j \leq s$. *For* M_s *fixed,* $T_m \to +\infty$ *as* $1/\Omega_1 \to 0$. *Alternatively, we can take arbitrary large but bounded sets of initial data:* $M_s \to +\infty$ *if* $1/\Omega_1 \to 0$, T_m *fixed.*

The above theorem establishes a class of genuinely 3D solutions of Euler equations which are regular on long time intervals even though the initial vorticity and the vortex stretching term are large.

2. Poincaré-Sobolev equations in cylindrical domains

In this section we consider the eigenvalue problem for the Poincaré-Sobolev equations in the cylinder $\mathbf{C}$ ([2], [25], [12], [29]):

$$\partial_t \mathbf{\Phi} + \Omega \mathbf{PJP\Phi} = 0, \quad \nabla \cdot \mathbf{\Phi} = 0. \tag{2.1}$$

The operator $\mathbf{PJP}$ is skew-symmetric with respect to the $\mathbf{L}_2$ inner product. From the fundamental identity

$$\operatorname{curl} \mathbf{PJP} = -\frac{\partial}{\partial x_3}\mathbf{P}, \tag{2.2}$$

the Poincaré problem is equivalent to the nonlocal wave operator (the Poincaré-Sobolev equation), [29] and [2]:

$$\frac{\partial^2}{\partial t^2}\operatorname{curl}^2 \Phi - \Omega^2 \frac{\partial^2}{\partial x_3^2}\mathbf{P}\Phi = 0; \tag{2.3}$$

its properties have been extensively investigated by the school of Sobolev (see references in [2], [17], [26]) for various domain geometries.

Theorem 2.1. *([17]).* **PJP** *is a bounded skew-adjoint zero-order nonlocal operator with a dense spectrum on* $[-i, +i]$.

This spectrum can be purely continuous on $[-i, +i]$ in the case of resonant domains which are ergodically filled by the characteristics of Eqs. (2.3) ([2]). The situation is simpler in periodic domains $\mathbf{T}^3$, where $\mathbf{PJP}$ does commute with the curl operator, hence with $\operatorname{curl}^{2\alpha} = (-\Delta)^\alpha$ on solenoidal fields, and $\mathbf{E}(-\Omega t) = \exp(-\Omega \mathbf{PJP}t)$ preserves all Sobolev norms ([3], [4]).

In the cylindrical domains with boundary conditions ((1.7), (1.19)), the structure of $\mathbf{PJP}$ and $\mathbf{E}(-\Omega t)$ is much more complex, as curl does not commute with the operator $\mathbf{PJP}$ (but does commute with $\mathbf{P}$), whereas the operators ∇ and $-\Delta$ do not commute with $\mathbf{P}$. The Helmholtz projection $\mathbf{U} \to \mathbf{PU}$ is such that (i) $\operatorname{div} \mathbf{PU} = 0$ and (ii) $\mathbf{PU} \cdot \mathbf{N} = 0$ on $\mathbf{\Gamma}$. The Weyl-Helmholtz decomposition theorem for $\mathbf{L}_2(D)$ where D is a bounded domain, now involves harmonic distributions:

Theorem 2.2. *Every vector field* $\mathbf{U} \in \mathbf{L}_2(D)$ *admits a unique decomposition:*

$$\mathbf{U} = \mathbf{PU} + \nabla\pi_H + \nabla\pi_0, \tag{2.4}$$

where $\nabla\pi_H$ *and* $\nabla\pi_0 \in \mathbf{L}_2(D)$ *and*

$$\Delta\pi_0 = \operatorname{div}\mathbf{U}, \ \pi_0 = 0 \text{ on } \partial D \tag{2.5}$$

$$\Delta\pi_H = 0, \frac{\partial\pi_H}{\partial N} = (\mathbf{U} - \nabla\pi_0)\cdot\mathbf{N} \text{ on } \partial D, \text{ in } \mathbf{H}^{-1/2}(\partial D). \tag{2.6}$$

Note that the set of harmonic distributions such that $\frac{\partial\pi_H}{\partial N} = 0$ on ∂D, reduces to $\{0\}$, hence the uniqueness of **PU**.

To construct the eigenfunctions and eigenvalues of **PJP** and $\mathbf{E}(-\Omega t)$, we need to invert curl in the Poincaré-Sobolev equation (2.3) subject to the boundary condition (1.19). This is where the potential theoretical result on curl inversion by O.A. Ladyzhenskaya and V.A. Solonnikov are needed in an essential way ([18], [10], [11], [21]). V.A. Solonnikov in [30] has further demonstrated that in bounded geometries the curl operator is an overdetermined elliptic system on solenoidal fields and does not admit any simple maximal self-adjoint extension. Nevertheless, two different potential theoretic inverses can be constructed for curl with different domains and ranges.

Recall the lemma for integration by parts for the curl operator

Lemma 2.3. *For* $\mathbf{U},\ \mathbf{V} \in \mathbf{H}^1(D)$

$$\int_D \operatorname{curl}\mathbf{U}\cdot\mathbf{V}dx = \int_D \mathbf{U}\cdot\operatorname{curl}\mathbf{V}dx + \int_{\partial D}(\mathbf{N}\times\mathbf{U}\cdot\mathbf{V})dS, \tag{2.7}$$

where the determinant in the boundary integral is taken in the sense $\mathbf{N}\times\mathbf{U}$ *and* $\mathbf{N}\times\mathbf{V} \in \mathbf{H}^{1/2}(\partial D)$, *and* $\mathbf{U},\ \mathbf{V} \in \mathbf{H}^{1/2}(\partial D)$ *([10], [14]).*

Lemma 2.4. *([21]). For* $\mathbf{U},\mathbf{V} \in \mathbf{J}^0 \cap \mathbf{J}_1$ *and such that* $\operatorname{curl}\mathbf{U}\cdot\mathbf{N} = \operatorname{curl}\mathbf{V}\cdot\mathbf{N} = 0$ *on* ∂D, *the operator* curl *is symmetric:*

$$\int_D \operatorname{curl}\mathbf{U}\cdot\mathbf{V}\,dx = \int_D \mathbf{U}\cdot\operatorname{curl}\mathbf{V}\,dx. \tag{2.8}$$

To briefly review the results in [18], [21], [10], [11] we introduce the notations of Ladyzhenskaya, where D is a bounded domain with boundary ∂D:

$$\mathbf{J} = \operatorname{Clos}\{\mathbf{U} \in C^\infty(\bar{D}), \operatorname{div}\mathbf{U} = 0\} \text{ in } ||\cdot||_{\mathbf{L}_2}, \mathbf{J} \equiv \mathbf{H}^0_\sigma; \tag{2.9}$$

$$\mathbf{J}_1 = \{\mathbf{U} \in \mathbf{H}^1(D), \operatorname{div}\mathbf{U} = 0\} \equiv \mathbf{H}^1_\sigma(D); \tag{2.10}$$

$$\mathbf{J}^0 = \operatorname{Clos}\{\mathbf{U} \in C^\infty(\bar{D}), \operatorname{div}\mathbf{U} = 0, \mathbf{U}\cdot\mathbf{N} = 0 \text{ on } \partial D\} \text{ in } ||\cdot||_{\mathbf{L}_2} \tag{2.11}$$

$$\mathbf{J}^0_{1,\tau} = \operatorname{Clos}\{\mathbf{U} \in C^\infty(\bar{D}), \operatorname{div}\mathbf{U} = 0, \mathbf{U}\times\mathbf{N} = 0 \text{ on } \partial D\} \text{ in } ||\cdot||_{\mathbf{H}^1(D)}. \tag{2.12}$$

Theorem 2.5. $\mathbf{J}^0_{1,\tau}(D)$ *is dense in* $\mathbf{J}(D)$.

Theorem 2.6. *For every* $\mathbf{V} \in \mathbf{J}^0(D)$, *there exists a unique* $\Psi \in \mathbf{J}^0_{1,\tau}(D)$ *such that* $\mathbf{V} = \operatorname{curl}\Psi$ *and for some* $C_1, C_2 > 0$:

$$C_1||\mathbf{V}||_{\mathbf{L}_2} \le ||\Psi||_{\mathbf{H}^1} \le C_2||\mathbf{V}||_{\mathbf{L}_2}. \tag{2.13}$$

Theorem 2.7. *For every* $\mathbf{W} \in \mathbf{J}(D)$ *there exists a unique* $\Phi \in \mathbf{J}^0(D) \cap \mathbf{H}^1(D)$ *such that* $\mathbf{W} = \operatorname{curl} \Phi$ *and for some* $C_3, C_4 > 0$*:*

$$C_3 ||\mathbf{W}||_{\mathbf{L}_2} \leq ||\Phi||_{\mathbf{H}^1} \leq C_4 ||\mathbf{W}||_{\mathbf{L}_2}. \tag{2.14}$$

Theorem 2.6 implies the existence of a bounded operator curl^{-1} with domain $\mathbf{J}^0$, range $\mathbf{J}^0_{1,\tau}$; and similarly Theorem 2.7 defines curl^{-1} with domain $\mathbf{J}$, range $\mathbf{J}^0 \cap \mathbf{H}^1$. Note that Theorem 2.7 implies the Poincaré inequality $||\Phi||_{\mathbf{L}_2} \leq C_4 ||\mathbf{W}||_{\mathbf{L}_2}$. Theorem 2.7 has been rederived by C. Bardos (see discussion in [14]) using non-potential theoretic methods of K. Friedrichs [15].

Theorem 2.8. *([21], [2]). When restricted to the domain:*

$$\mathcal{D}(\operatorname{curl}) = \operatorname{curl}^{-1} \mathbf{J}^0, \tag{2.15}$$

where the vector potential curl^{-1} *is taken as in Theorem* 2.7*, the operator* curl *is self-adjoint, invertible and with a compact inverse in* $\mathbf{J}^0$*.*

Theorem 2.8 is a straightforward corollary of Theorem 2.7, with the remark that $\mathbf{J}^0$ is a closed subspace of $\mathbf{J}$.

Remark 2.9. $\mathcal{D}(\operatorname{curl})$ is a closed proper subspace of $\mathbf{J}_1 \cap \mathbf{J}^0$. In fact,

$$\mathbf{J}^0 \cap \mathbf{J}_1 = \operatorname{curl}^{-1} \mathbf{J}^0 \bigoplus \operatorname{curl}^{-1}(\nabla \pi_H), \tag{2.16}$$

where curl^{-1} is again taken in the sense of Theorem 2.7. Theorem 2.8 has been rediscovered in many publications from the mid-seventies on. Note that whereas the eigenfunctions of curl are complete in $\mathbf{J}^0$, they are **not** complete in $\mathbf{J}^0 \cap \mathbf{J}_1$, only in $\mathcal{D}(\operatorname{curl}) \subset \mathbf{H}^1(D)$.

We now explicit the common eigenfunctions to **PJP** and curl in the cylinder. In cylindrical coordinates (r, ϕ, z) we have $\mathbf{\Phi} = (\Phi_r, \Phi_\phi, \Phi_z)$ and Eqs. (2.1) take the form

$$\partial_t \Phi_r - \Omega \Phi_\phi = -\frac{\partial p}{\partial r}, \ \partial_t \Phi_\phi + \Omega \Phi_r = -\frac{1}{r}\frac{\partial p}{\partial \phi}, \ \partial_t \Phi_z = -\frac{\partial p}{\partial z}, \tag{2.17}$$

$$\frac{1}{r}\frac{\partial}{\partial r}(r\Phi_r) + \frac{1}{r}\frac{\partial \Phi_\phi}{\partial \phi} + \frac{\partial \Phi_z}{\partial z} = 0. \tag{2.18}$$

The vector field $\mathbf{\Phi}$ is $2\pi/\alpha$ periodic in z and it satisfies $\Phi_r|_{r=R} = 0$ on Γ.

Applying curl operator to Eqs. (1.23) and using divergence free condition, we obtain

$$\partial_t \operatorname{curl} \mathbf{\Phi} = \Omega \partial_z \mathbf{\Phi}. \tag{2.19}$$

From Eqs. (2.19) we obtain Poincaré-Sobolev equations

$$\frac{\partial^2}{\partial t^2}(\operatorname{curl}^2 \mathbf{\Phi}) - \Omega^2 \frac{\partial^2}{\partial z^2}(\mathbf{\Phi}) = 0. \tag{2.20}$$

Eqs. (2.20) is a system of equations for three components of $\mathbf{\Phi}$. For the vertical component Φ_z we have the following scalar equation

$$\frac{\partial^2}{\partial t^2}\Delta \Phi_z + \Omega^2 \frac{\partial^2}{\partial z^2}\Phi_z = 0. \tag{2.21}$$

We look for normal modes in the form

$$e^{i(\Omega\sigma t+m_2\phi+m_3\alpha z)}\hat{\mathbf{\Phi}}(r). \tag{2.22}$$

Recall that without loss of generality $R=1$. Eqs. (2.21) imply

$$\frac{d^2}{dr^2}\hat{\Phi}_z+\frac{1}{r}\frac{d}{dr}\hat{\Phi}_z+(\beta^2-\frac{m_2^2}{r^2})\hat{\Phi}_z=0 \tag{2.23}$$

where

$$\beta^2=m_3^2\alpha^2(\frac{1}{\sigma^2}-1) \text{ or equivalently } \sigma^2=\frac{m_3^2\alpha^2}{\beta^2+m_3^2\alpha^2}. \tag{2.24}$$

From the boundary condition $\Phi_r|_{r=1}=0$ and Eqs. (2.17) we obtain

$$\frac{d}{dr}\hat{\Phi}_z+\frac{m_2}{\sigma r}\hat{\Phi}_z=0 \text{ at r= } 1. \tag{2.25}$$

Eqs. (2.23), (2.25) is a Sturm-Liouville eigenvalue problem.

From Eqs. (2.23) we have

$$\hat{\Phi}_z(r)=J_{m_2}(\beta r), \tag{2.26}$$

where $\mathbf{J}_{m_2}(\cdot), m_2=0,1,2,\ldots,$ are Bessel functions of the first kind; therefore, Eqs. (2.25) imply

$$\beta J'_{m_2}(\beta)\pm m_2 J_{m_2}(\beta)\sqrt{\frac{\beta^2}{m_3^2\alpha^2}+1}=0. \tag{2.27}$$

For fixed integers m_2 (azimuthal wave number) and m_3 (vertical direction) we solve Eqs. (2.27) to obtain $\beta_{m_1}(m_2,m_3)$; $m_1=1,2,3,\ldots$. Eqs. (2.27) have infinitely many solutions. Then Eqs. (2.24) imply

$$\sigma(m_1,m_2,m_3)=\pm\frac{m_3}{\sqrt{\frac{\beta(m_1,m_2,m_3)^2}{\alpha^2}+m_3^2}} \tag{2.28}$$

Clearly, $i\sigma(m_1,m_2,m_3)$ are eigenvalues of the skew-hermitian operator **PJP**. The corresponding eigenvector functions $\mathbf{\Phi}_{m_1m_2m_3}=(\Phi_r,\Phi_\phi,\Phi_z)$ form a complete set in $\mathbf{J}^0$. They are independent of Ω and are explicitly expressed in terms of Bessel functions (for example, see Eq. (2.26) for the radial component Φ_r). From Eqs. (2.19) and (2.24) it follows that the eigenfunctions $\Phi_{m_1m_2m_3}$ with $m_3\neq 0$ satisfy

$$\begin{aligned}\operatorname{curl}\Phi_{m_1m_2m_3}&=\pm\sqrt{\beta(m_1,m_2,m_3)^2+\alpha^2m_3^2}\;\Phi_{m_1m_2m_3} \\ &=\pm\lambda_{m_1m_2m_3}\mathbf{\Phi}_{m_1m_2m_3} \\ &=\pm\frac{m_3\alpha}{\sigma_{m_1m_2m_3}}\Phi_{m_1m_2m_3} \quad \text{where } \sigma^2_{m_1m_2m_3}=\frac{m_3^2\alpha^2}{\beta^2_{m_1m_2m_3}+m_3^2\alpha^2}.\end{aligned} \tag{2.29}$$

and

$$(\operatorname{curl}\Phi)\cdot\mathbf{N}|_\Gamma=0. \tag{2.30}$$

The divergence free eigenvector functions $\Phi_{m_1m_2m_3} = (\Phi_r, \Phi_\phi, \Phi_z)$ are

$$\begin{aligned}
\Phi_{r,m_1m_2m_3} &= e^{i(m_2\phi+m_3\alpha z)}\frac{i\sigma}{m_3\alpha(1-\sigma^2)}\,(\sigma\beta J'_{m_2}(\beta r)+\frac{m_2}{r}J_{m_2}(\beta r)),\\
\Phi_{\phi,m_1m_2m_3} &= e^{i(m_2\phi+m_3\alpha z)}\frac{-\sigma}{m_3\alpha(1-\sigma^2)}\,(\beta J'_{m_2}(\beta r)+\frac{\sigma m_2}{r}J_{m_2}(\beta r)),\\
\Phi_{z,m_1m_2m_3} &= e^{i(m_2\phi+m_3\alpha z)}J_{m_2}(\beta r).
\end{aligned}$$

The eigenspace corresponding to the zero eigenvalue consists of all divergence free vector fields independent of the vertical coordinate z ($\sigma = 0$ if $m_3 = 0$ in Eqs. (2.28)).

We can easily obtain asymptotic expressions of eigenvalues for large β. We recall that we have for Bessel functions

$$J'_l(\xi) = \frac{l}{\xi}J_l(\xi) - J_{l+1}(\xi), \tag{2.31}$$

$$J_{l+1}(\xi) = \frac{2l}{\xi}J_l(\xi) - J_{l-1}(\xi), \tag{2.32}$$

$$J_{l-1}(\xi) - J_{l+1}(\xi) = 2J'_l(\xi) \tag{2.33}$$

$$J_l(\xi) \sim \sqrt{\frac{2}{\pi\xi}}\cos(\xi - \frac{\pi}{4} - \frac{l\pi}{2}) \text{ as } \xi \to +\infty. \tag{2.34}$$

From Eqs. (2.27), (2.31)–(2.33) we obtain

$$\frac{J_{m_2+1}(\beta)}{J_{m_2}(\beta)} = \frac{m_2}{\beta}(1 \pm \sqrt{\frac{\beta^2}{m_3^2\alpha^2}+1}). \tag{2.35}$$

Then from Eqs. (2.35) using asymptotic expression for Bessel functions for large β we have

$$\tan(\beta - \frac{\pi}{4} - \frac{m_2\pi}{2}) \approx \pm\frac{m_2}{m_3\alpha}. \tag{2.36}$$

For fixed m_2, m_3 and α Eqs. (2.36) has infinitely many solutions $\beta_{m_1}(m_2, m_3, \alpha)$, $m_1 = 1, 2, \ldots$.

In summary, with $\check{n}_3 = \alpha n_3 = 2\pi n_3/h$ denoting the vertical Fourier wave number along x_3, we have established that:

Proposition 2.10.

(i) *On* $[-i, 0)\cup(0, +i]$ *the spectrum of* $\mathbf{PJP}$ *consists of a dense, but countable set of eigenvalues* $\pm i\sigma_n$, *with finite-dimensional eigenspaces for each eigenvalue.*

(ii) *Every eigenvector of* $\mathbf{PJP}$ *is an eigenvector of* curl *and vice versa, with eigenvalues* $\pm i\sigma_n$ *and* λ_n *mapped into each other by* $\sigma_n^2 = \frac{\check{n}_3^2}{\lambda_n^2}$.

(iii) $\ker \mathbf{PJP} =$
$$\{\mathbf{U} \in \mathbf{J}(\mathbf{C}) : \mathbf{U} \equiv \overline{\mathbf{U}}(x_1, x_2) = (U_1(x_1, x_2), U_2(x_1, x_2), U_3(x_1, x_2))\}.$$

(iv) *On* $(\ker \mathbf{PJP})^\perp$, $\mathbf{E}(-\Omega t) = \exp(-\mathbf{PJP}\Omega t)$ *is diagonalized in the* curl-*eigenvector functions basis, with eigenvalues* $\exp(\pm i\Omega\sigma_n t) = \exp(\pm i\Omega\frac{\check{n}_3}{\lambda_n}t)$

3. The structure and regularity of fast singular oscillating limit equations

3.1. Fast singular oscillating limit equations

We introduce van der Pol transformation by setting in Eqs. (1.11)–(1.20)

$$\mathbf{U}(t) = \mathbf{E}(-\Omega t)\mathbf{u}(t) \tag{3.1}$$

where $\mathbf{u}(t)$ is the "slow envelope" variable also denoted in this paper by *Poincaré variable.* We note that $\mathbf{E}(-\Omega t) = \exp(-\Omega \mathbf{PJP}t)$ reduces to the identity operator on any barotropic (vertically averaged) field implying

$$\overline{\mathbf{U}} = \overline{\mathbf{E}(-\Omega t)\mathbf{u}} = \overline{\mathbf{u}}. \tag{3.2}$$

Since $\mathbf{E}(\Omega t)|_{t=0} = \mathbf{Id}$

$$\mathbf{U}|_{t=0} = \mathbf{u}|_{t=0}. \tag{3.3}$$

Eqs. (1.26) written in $\mathbf{u}$ variables have the form ([3])

$$\begin{aligned}\partial_t \mathbf{u} &= \mathbf{B}(\Omega t, \mathbf{u}, \mathbf{u}),\\ \mathbf{B}(\Omega t, \mathbf{u}, \mathbf{u}) &= \mathbf{E}(\Omega t)\mathbf{B}(\mathbf{E}(-\Omega t)\mathbf{u}, \mathbf{E}(-\Omega t)\mathbf{u})\end{aligned} \tag{3.4}$$

where $\mathbf{B}$ is given by Eqs. (1.27). We decompose

$$\mathbf{B}(\Omega t, \mathbf{u}, \mathbf{u}) = \tilde{\mathbf{B}}(\mathbf{u}, \mathbf{u}) + \mathbf{B}^{\text{osc}}(\Omega t, \mathbf{u}, \mathbf{u}). \tag{3.5}$$

Here $\mathbf{B}^{\text{osc}}(\Omega t, \mathbf{u}, \mathbf{u})$ contains all Ωt-dependent terms (that is non-resonant) and $\tilde{\mathbf{B}}(\mathbf{u}, \mathbf{u})$ contains all resonant (that is Ωt-independent) terms.

The fast singular oscillating limit equations ([3], [4]) are obtained from (3.4) for 'slow' Poincaré variables $\mathbf{w}$ by dropping $\mathbf{B}^{\text{osc}}(\Omega t, \mathbf{u}, \mathbf{u})$ in (3.5):

$$\partial_t \mathbf{w} = \tilde{\mathbf{B}}(\mathbf{w}, \mathbf{w}), \tag{3.6}$$

$$\mathbf{w}|_{t=0} = \mathbf{w}(0) = \mathbf{U}(0) = \tilde{\mathbf{V}}_0. \tag{3.7}$$

Here the operator $\tilde{\mathbf{B}}$ is defined by (see Lemma 3.1 and Section 4 for a rigorous statement)

$$\tilde{\mathbf{B}}(\mathbf{v}, \mathbf{v}) = \lim_{\Omega\to+\infty} \frac{1}{T}\int_0^T \mathbf{B}(\Omega s, \mathbf{v}, \mathbf{v})ds = \lim_{T\to+\infty} \frac{1}{T}\int_0^T \mathbf{B}(\Omega s, \mathbf{v}, \mathbf{v})ds$$

where arguments $\mathbf{v}$ are *s-independent functions*; limits are taken in the sense of almost periodic functions in s with values in Banach spaces [8], [13], see Section 4.

The limit resonant operator $\tilde{\mathbf{B}}$ inherits properties of the operator $\mathbf{B}$:

Lemma 3.1. *([4]): Let* $(\mathbf{u}, \mathbf{v}, \mathbf{w}) \in H_1 \times H_1 \times H_1$*. Then*

$$(\tilde{\mathbf{B}}(\mathbf{u}, \mathbf{v}), \mathbf{w}) = \lim_{\Omega\to\infty} \frac{1}{T}\int_0^T (\mathbf{B}(\Omega s, \mathbf{u}, \mathbf{v}), \mathbf{w})ds \tag{3.8}$$

$$= \lim_{T\to\infty} \frac{1}{T}\int_0^T (\mathbf{B}(\Omega s, \mathbf{u}, \mathbf{v}), \mathbf{w})ds. \tag{3.9}$$

From now on we shall restrict the initial data (1.12), (1.21), (1.29), (3.3), (3.7) to the closed (proper) subspace of $\mathbf{J}^0 \cap \mathbf{H}^s_\sigma$, $s \geq 3$, s integer, defined by (with the vector potential curl^{-1} as in Theorem 2.7):

$$\mathbf{H}^s_\nu(\mathbf{C}) = \mathbf{J}^0 \cap \mathbf{H}^s_\sigma \cap \mathrm{curl}^{-s}(\mathbf{J}^0). \tag{3.10}$$

We will similarly restrict solutions of Eqs. (1.1)–(1.2) to the space $\mathbf{H}^s_\nu(\mathbf{C})$ (see Theorem 4.5).

Remark 3.2. $\mathbf{v} \in \mathbf{H}^s_\nu$ is equivalent to $\mathbf{v} \in \mathbf{J}^0 \cap \mathbf{H}^s_\sigma$ and $\mathrm{curl}^j\ \mathbf{v} \in \mathbf{J}^0, 0 \leq j \leq s$. The complement of $\mathbf{H}^s_\nu$ in $\mathbf{H}^s_\sigma$ includes functions such as $\mathrm{curl}^{-j}(\nabla \pi_H), 1 \leq j \leq s$ and is not dense even in $\mathbf{H}^1_\sigma$. The case of more general initial conditions and functional spaces for Eqs. (1.1)–(1.2) will be treated in [6].

We can explicit the limit resonant operator $\tilde{\mathbf{B}}$ with the help of the eigenfunctions $\Phi_n = \Phi_{n_1 n_2 n_3}$ of curl and **PJP**, which form a basis in the space $\mathbf{H}^s_\nu$; expand

$$\mathbf{u} = \sum_n u_n \Phi_n. \tag{3.11}$$

From the diagonalization of $\mathbf{E}(-\Omega t)$ and curl operator:

$$\partial_t u_n = \sum_{k,m,k_3+m_3=n_3} \mathbf{B}_n(\Omega t, u_k, u_m), \tag{3.12}$$

$$\mathbf{B}_n(\Omega t, u_k, u_m) = \pm\lambda_m \exp\left(i\left(\pm\frac{\check{k}_3}{\lambda_k} \pm \frac{\check{m}_3}{\lambda_m} \pm \frac{\check{n}_3}{\lambda_n}\right)\Omega t\right) (\Phi_k \times \Phi_m, \Phi_n)_{\mathbf{L}^2}\, u_k u_m, \tag{3.13}$$

where k, m, n now index the eigenvalues and eigenfunctions of curl and **PJP**.

The resonant nonlinear interactions of Poincaré waves with $\mathbf{B}(\mathbf{U}, \mathbf{U})$ are present when the Poincaré frequencies satisfy the relation $\pm\sigma_k \pm \sigma_m \pm \sigma_n = 0$, $k_3 + m_3 = n_3$, with the resonant set K now defined in terms of vertical wave numbers k_3, m_3, n_3 and eigenvalues $\pm\lambda_k$, $\pm\lambda_m$, $\pm\lambda_n$ of curl:

$$K = \{\pm\frac{k_3}{\lambda_k} \pm \frac{m_3}{\lambda_m} \pm \frac{n_3}{\lambda_n} = 0, n_3 = k_3 + m_3\}. \tag{3.14}$$

Since the eigenvalues of curl are *countable*, so are the $D_l(k, m, n) = \pm\sigma_k \pm\sigma_m \pm\sigma_n$, where k, m, n are now indexing the eigenfunctions and eigenvalues of **PJP** (and curl).

Now we prove that the nonlinear operator $\tilde{\mathbf{B}}$ commutes with vertical averaging. We have

Theorem 3.3. *The operator* $\tilde{\mathbf{B}}$ *commutes with vertical averaging. More precisely,*

$$\overline{\tilde{\mathbf{B}}(\mathbf{w}, \mathbf{w})} = \tilde{\mathbf{B}}(\overline{\mathbf{w}}, \overline{\mathbf{w}}) = \mathbf{B}_{2D}(\overline{\mathbf{w}}, \overline{\mathbf{w}}) = -\mathbf{P}(\overline{\mathbf{w}} \cdot \nabla \overline{\mathbf{w}}). \tag{3.15}$$

Proof. Let $\mathbf{w} = \overline{\mathbf{w}} + \mathbf{w}^\perp$ where the orthogonal field $\mathbf{w}^\perp$ verifies $\overline{\mathbf{w}^\perp} = 0$. Clearly,

$$\overline{\tilde{\mathbf{B}}(\mathbf{w},\mathbf{w})} = \tilde{\mathbf{B}}(\overline{\mathbf{w}},\overline{\mathbf{w}}) + \overline{\tilde{\mathbf{B}}(\mathbf{w}^\perp,\mathbf{w}^\perp)} \tag{3.16}$$

since $\overline{\mathbf{w}^\perp} = 0$. Thus, the theorem will be proven if we show that

$$\overline{\tilde{\mathbf{B}}(\mathbf{w}^\perp,\mathbf{w}^\perp)} = 0, \ \overline{\mathbf{w}^\perp} = 0. \tag{3.17}$$

In the limit $\Omega \to +\infty$ interactions in the bilinear term $\tilde{\mathbf{B}}$ are restricted to the resonant manifold

$$\pm\frac{k_3\alpha}{\sqrt{\beta(k_1,k_2,k_3)^2+k_3^2\alpha^2}} \pm \frac{m_3\alpha}{\sqrt{\beta(m_1,m_2,m_3)^2+m_3^2\alpha^2}} \pm \frac{n_3\alpha}{\sqrt{\beta(n_1,n_2,k_3)^2+n_3^2\alpha^2}} = 0 \tag{3.18}$$

and we have $n_3 = k_3 + m_3$, $n_2 = k_2 + m_2$. Now vertical averaging in Eqs. (3.17) implies $n_3 = 0$, $k_3 + m_3 = 0$. Then from Eqs. (3.18) we obtain $\beta(k_1,k_2,k_3)^2 = \beta(m_1,m_2,m_3)^2$. We use eigenvector functions $\Phi_{m_1m_2m_3}(r,\phi,z)$ to represent physical fields $\mathbf{w}^\perp(r,\phi,z)$, $\overline{\mathbf{w}^\perp} = 0$:

$$\mathbf{w}^\perp = \sum_{m_1m_2m_3} w^\perp_{m_1m_2m_3}\Phi_{m_1m_2m_3}(r,\phi,z). \tag{3.19}$$

We recall from Eqs. (2.29) that $\Phi_{m_1m_2m_3}(r,\phi,z)$, $m_3 \neq 0$ are eigenvector functions of curl. Then we obtain

$$\begin{aligned}
&\overline{\tilde{\mathbf{B}}(\mathbf{w}^\perp,\mathbf{w}^\perp)}_n \\
&= \mathbf{P}_n \sum_{k_3+m_3=0,\beta_k^2=\beta_m^2} w^\perp_{k_1k_2k_3}\Phi_{k_1k_2k_3}(r,\phi,z) \times \text{curl } w^\perp_{m_1m_2m_3}\Phi_{m_1m_2m_3}(r,\phi,z) \\
&= \pm\mathbf{P}_n \sum_{k_3+m_3=0,\beta_k^2=\beta_m^2} w^\perp_{k_1k_2k_3}\Phi_{k_1k_2k_3}(r,\phi,z)\times \\
&\qquad \sqrt{\beta(m_1,m_2,m_3)^2+\alpha^2m_3^2}\; w^\perp_{m_1m_2m_3}\Phi_{m_1m_2m_3}(r,\phi,z) \\
&= \pm\mathbf{P}_n \sum_{k_3+m_3=0,\beta_k^2=\beta_m^2} (\beta(k_1,k_2,k_3)^2+\alpha^2k_3^2)^{1/4}(\beta(m_1,m_2,m_3)^2+\alpha^2m_3^2)^{1/4} \\
&\qquad w^\perp_{k_1k_2k_3}\Phi_{k_1k_2k_3}(r,\phi,z) \times \ w^\perp_{m_1m_2m_3}\Phi_{m_1m_2m_3}(r,\phi,z) \\
&= 0.
\end{aligned}$$

Clearly, the last sum is identically zero (it changes sign if we interchange indices k and m in the summation). Then we obtain Eqs. (3.17). Theorem 3.3 is proven. □

3.2. Strict 3-wave resonances

In this section we show that for all values of α, except a countable set, the resonant sets lie in $\{k_3m_3n_3 = 0\}$. This is generic case of no strict 3-wave resonances.

In the case of strict 3-wave resonances we have $k_3m_3n_3 \neq 0$ and Eqs. (3.18) become

$$\pm\frac{1}{\sqrt{\frac{\beta^2_{k_1}(k_2,k_3\alpha)}{k_3^2\alpha^2}+1}} \pm \frac{1}{\sqrt{\frac{\beta^2_{m_1}(m_2,m_3\alpha)}{m_3^2\alpha^2}+1}} \pm \frac{1}{\sqrt{\frac{\beta^2_{n_1}(n_2,n_3\alpha)}{n_3^2\alpha^2}+1}} = 0. \tag{3.20}$$

We also have convolutions in the azimuthal ϕ and the axial z directions implying $n_3 = k_3 + m_3$, $n_2 = k_2 + m_2$. We recall that for every pair of integers k_2 and k_3 the quantities $\beta_{k_1}(k_2, k_3\alpha)$ are found from the equation

$$\beta J_{k_2}{}'(\beta) \pm k_2 J_{k_2}(\beta)\sqrt{\frac{\beta_{k_1}^2(k_2, k_3\alpha)}{k_3^2\alpha^2} + 1} = 0. \tag{3.21}$$

For every pair of integers k_2 and k_3, Eqs. (3.21) have a countable number of solutions denoted by $\beta(k_1, k_2, k_3\alpha)$; $k_1 = 1, 2, 3, \ldots$. Similarly, for $\beta(m_1, m_2, m_3\alpha)$ and $\beta(n_1, n_2, n_3\alpha)$.

Eqs. (3.20) can be written in the form

$$\pm\frac{1}{\sqrt{X}} \pm \frac{1}{\sqrt{Y}} \pm \frac{1}{\sqrt{Z}} = 0 \tag{3.22}$$

where

$$\frac{\beta^2(k_1, k_2, k_3\alpha)}{k_3^2\alpha^2} + 1 = X \leftrightarrow \left(\frac{\beta J_{k_2}{}'(\beta)}{k_2 J_{k_2}(\beta)}\right)^2 = X, \tag{3.23}$$

with similar expressions for Y and Z.

Substituting Eqs. (3.21) in Eqs. (3.20) we obtain

$$\pm\frac{k_2 J_{k_2}(\beta(k_1, k_2, k_3\alpha))}{\beta(k_1, k_2, k_3\alpha) J_{k_2}{}'(\beta(k_1, k_2, k_3\alpha))} \pm \frac{m_2 J_{m_2}(\beta(m_1, m_2, m_3\alpha))}{\beta(m_1, m_2, m_3\alpha) J_{m_2}{}'(\beta(m_1, m_2, m_3\alpha))}$$
$$\pm \frac{n_2 J_{n_2}(\beta(n_1, n_2, n_3\alpha))}{\beta(n_1, n_2, n_3\alpha) J_{n_2}{}'(\beta(n_1, n_2, n_3\alpha))} = 0. \tag{3.24}$$

In Eqs. (3.24) k_2, m_2, n_2, k_3, m_3, $n_3 \in \mathbf{Z}$ and $k_1, m_1, n_1 = 1, 2, 3, \ldots$. Also, $n_2 = k_2 + m_2$ and $n_3 = k_3 + m_3$. In fact, we can think of Eqs. (3.24) as a countable set of nonlinear equations for α. Clearly, for every fixed k_j, m_j, n_j Eq. (3.24) has at most a countable number of solutions α. Thus, we have a countable number of equations and each equation has at most a countable number of solutions α. Therefore, the set of parameters α's for which strict 3-wave resonances can occur is countable.

Proposition 3.4. *The set $\mathcal{K}^*$ of parameters α's for which strict 3-wave resonances can occur is countable and discrete.*

3.3. Regularity of fast singular oscillating limit equations

In the generic case of no strict 3-wave resonances $\mathbf{B}_{\text{III}} = 0$ and the limit Euler equations (1.35) become

$$\partial_t \mathbf{w}^\perp = \mathbf{B}_{\text{II}}(\overline{\mathbf{w}}(t), \mathbf{w}^\perp), \ \mathbf{w}^\perp|_{t=0} = \mathbf{w}^\perp(0) = \mathbf{U}^\perp(0) = \mathbf{U}(0) - \overline{\mathbf{U}}(0). \tag{3.25}$$

where $\overline{\mathbf{w}}(t)$ satisfies 2D Euler equations with vertically averaged initial data $\overline{\mathbf{w}}|_{t=0} = \overline{\mathbf{w}}(0) = \overline{\mathbf{U}}(0)$. Eqs. (3.25) for $\mathbf{w}^\perp(t)$ are solved with periodic boundary conditions in the third coordinate and $\mathbf{w}^\perp \cdot \mathbf{N}|_\Gamma = 0$. We also have $\operatorname{curl} \mathbf{w}^\perp \cdot \mathbf{N}|_\Gamma = 0$.

Eqs. (3.25) possess new 3D conservation laws:

Theorem 3.5. *Let* $\overline{\mathbf{w}}(t)$ *be a solution of 2D-3C Euler Eqs.* (1.34). *Then for every* $\mathbf{w}^{\perp}(t)$ *solution of Eqs.* (3.25) *with initial data* $\mathbf{w}^{\perp}(0)$ *we have:*

$$||\partial_3 \mathbf{w}^{\perp}(t)||^2 = ||\partial_3 \mathbf{w}^{\perp}(0)||^2, \tag{3.26}$$

where ∂_3 *denotes the partial derivative with respect to* x_3.

Proof. Applying ∂_3 to Eqs. (3.25) and using skew-symmetry property

$$(\mathbf{B}_{\mathrm{II}}(\overline{\mathbf{w}}, \partial_3 \mathbf{w}^{\perp}), \partial_3 \mathbf{w}^{\perp}) = 0$$

we obtain

$$\frac{d}{dt}||\partial_3 \mathbf{w}^{\perp}||^2 = 0. \qquad \square \tag{3.27}$$

Moreover, for the initial data and solutions in the function space $\mathbf{H}^s_{\nu}$, we have further conservation laws:

Theorem 3.6. *Let* $\mathbf{w}^{\perp}(t)$ *be solutions of the limit equations* (3.25) *in* $\mathbf{H}^s_{\nu}$. *We have, for* $0 \leq j \leq s$, j *even:*

$$||\operatorname{curl}^j \mathbf{w}^{\perp}(t)||^2 = ||\operatorname{curl}^j \mathbf{w}^{\perp}(0)||^2. \tag{3.28}$$

Proof. Proceed as in the proof of Theorem 3.3, but with $k_3 = n_3$ (and $m_3 = 0$) or with $m_3 = n_3$ (and $k_3 = 0$), together with $\lambda_k^2 = \lambda_n^2$, $\beta_k^2 = \beta_n^2$ (resp. $\lambda_m^2 = \lambda_n^2, \beta_m^2 = \beta_n^2$). Note that expansion along the eigenfunctions of curl and **PJP** requires their completeness at least in $\mathbf{H}^1_{\nu}$, cf. Remark 3.2. $\square$

Remark 3.7. Note that 2D-3C Euler equations only admit conservation of energy and enstrophy. The above conservation laws (3.26)–(3.28) ensure global regularity of the limit Euler equations (3.25).

Theorem 3.8. *Let* $h/R \notin \mathcal{K}^*$. *Let* $||\mathbf{w}(0)||_{\mathbf{H}^s_{\nu}} \leq M_s, s \geq 1$. *Let* $T_1 > 0$ *fixed, arbitrary large. Then there exists a unique regular solution* $\mathbf{w}(t)$ *of the limit resonant 3D Euler equations* (3.6)–(3.7), *for* $0 \leq t \leq T_1$:

$$||\mathbf{w}(t)||_{\mathbf{H}^s_{\nu}} \leq \ddot{M}_s(h/R, M_s, T_1). \tag{3.29}$$

4. Long time regularity for finite large Ω

Two major obstacles in extending the fast singular oscillating limit methods developed in [3]–[5] from the periodic lattice case to the cylinder (as well as other axisymmetric domains) are that: (i) **PJP** is not skew-symmetric with respect to the inner product of classical Sobolev spaces $\mathbf{H}^s_{\sigma}(\mathbf{C})$, $s \geq 1$; (ii) $\mathbf{E}(\Omega t)$ is not an isometry in these spaces (∇ does not commute with **PJP** and $\mathbf{E}(\Omega t)$). Item (i) implies that a priori estimates of Eqs. (1.11)–(1.12) in Sobolev spaces are $1/\epsilon = \Omega$ dependent; and (ii) that estimates for $\mathbf{u}(t, x)$, the Poincaré slow variable (van der Pol transformation of $\mathbf{U}(t, x)$, Eq. (3.1)) are not invariant for the physical variable $\mathbf{U}(t, x)$. That invariance was used in an essential way in the convergence proofs of [3]–[5] (periodic case). The resolution of the above requires the introduction of Hilbert spaces with the metric based on the operator curl-norms, with norms equivalent to that of $\mathbf{H}^s_\sigma(\mathbf{C})$, s integer, $s \geq 1$.

As before (cf. Eq. 3.10), we restrict ourselves to initial data and solutions in the spaces ($s \geq 3$):

$$\mathbf{H}^s_\nu(\mathbf{C}) = \mathbf{J}^0 \cap \mathbf{H}^s_\sigma \cap \operatorname{curl}^{-s}(\mathbf{J}^0), \tag{4.1}$$

such that $\mathbf{v} \in \mathbf{H}^s_\nu$ implies $\operatorname{curl}^j \mathbf{v} \cdot \mathbf{N} = 0$ on Γ, $0 \leq j \leq s$ (cf., Remark 3.2). More general functional spaces dense in $\mathbf{H}^1_\sigma$ will be treated in [6].

Lemma 4.1. *Let* $\mathbf{v} \in \mathbf{H}^s_\nu, s \geq 1$. *Then there exist constants* $C_1, C_2 > 0$ *such that:*

$$C_1 ||\mathbf{v}||_{\mathbf{H}^s_\sigma} \leq ||\mathbf{v}||_{\mathbf{H}^s_\nu} \leq C_2 ||\mathbf{v}||_{\mathbf{H}^s_\sigma}, \tag{4.2}$$

where

$$||\mathbf{v}||^2_{\mathbf{H}^s_\nu} = ||\mathbf{v}||^2_{L_2} + ||\operatorname{curl}^s \mathbf{v}||^2_{L_2}. \tag{4.3}$$

Proof. Iterated applications of Theorem 2.7 and 2.8, we have equivalence of the "curl-norms" with the usual Sobolev space norms. □

From now on, we designate by $||\mathbf{v}||_s$ the curl-norm of $\mathbf{v}$ defined in Eq. 4.3, and by $\langle \mathbf{u}, \mathbf{v} \rangle_s$ the corresponding inner product of $\mathbf{u}, \mathbf{v}$ in $\mathbf{H}^s_\nu$. We have the:

Lemma 4.2. *Let* $\mathbf{u}, \mathbf{v} \in \mathbf{H}^s_\nu, s \geq 0$; *then we have skew-symmetry in the curl-norms:*

$$\langle \mathbf{PJPu}, \mathbf{v} \rangle_s = -\langle \mathbf{u}, \mathbf{PJPv} \rangle_s \tag{4.4}$$

and

$$\langle \mathbf{PJPu}, \mathbf{u} \rangle_s = 0. \tag{4.5}$$

Proof. Obvious for $s = 0$; we outline the case $s = 1$:

$$\begin{aligned} (\operatorname{curl} \mathbf{PJPu}, \operatorname{curl} \mathbf{v})_{L_2} &= \left(-\frac{\partial \mathbf{u}}{\partial z}, \operatorname{curl} \mathbf{v} \right)_{L_2} = \left(\mathbf{u}, \operatorname{curl} \frac{\partial \mathbf{v}}{\partial z} \right)_{L_2} \\ &= \left(\operatorname{curl} \mathbf{u}, \frac{\partial \mathbf{v}}{\partial z} \right)_{L_2} = -(\operatorname{curl} \mathbf{u}, \operatorname{curl} \mathbf{PJPv})_{L_2}, \end{aligned}$$

since both $\mathbf{u}, \mathbf{v}, \frac{\partial \mathbf{u}}{\partial z}, \frac{\partial \mathbf{v}}{\partial z}$ satisfies the conditions of Lemma 2.4. The cases $s > 1$ follows a similar proof. □

Remark 4.3. The all important Lemmas 4.1, 4.2 allow for local estimates of solutions to the 3D Euler equations (1.11)–(1.12) which are $1/\epsilon = \Omega$ independent.

Corollary 4.4. *The Poincaré-Sobolev unitary operator* $E(\Omega t)$ *is an isometry on the curl spaces* $\mathbf{H}^s_\nu, s \geq 0$; *in particular,*

$$||\mathbf{U}||_s = ||\mathbf{u}||_s, \tag{4.6}$$

where $\mathbf{U}(t) = \mathbf{E}(-\Omega t)\mathbf{u}(t)$.

To establish long time regularity of the 3D Euler equations Eqs. (1.20)–(1.21) on $0 \leq t \leq T_M$, T_M fixed, arbitrary large, we first establish convergence in $\mathbf{H}^s_\nu$ (as $\Omega \to \infty$) of the solution to that of the limit resonant equations (1.34)–(3.25) on the interval $[0, T_s]$, where T_s is some local time of existence of (1.20)–(1.21). We only consider the case of "catalytic resonances", $h/R \notin \mathcal{K}^*$. With the help of the long time existence of solutions to the limit resonant equations on $[0, T_M]$, cf. Theorem 3.8, we extend local regularity on $[0, T_s]$ to long-time regularity on $[0, T_M]$ by partitioning $[0, T_M]$ into subintervals of length T_s and bootstrapping estimates.

Theorem 4.5. *Let* $\mathbf{U}(0) \in \mathbf{H}_\nu^\beta, \beta \geq 3$ *and* $||\mathbf{U}(0)||_\beta \leq M_{0\beta}$, *where the* β*-norm is the curl-norm* (4.3). *Then:*

(i) *there exists* $T_\beta > 0$ *such that there exists a unique regular solution of the* 3*D Euler equations on* $0 \leq t \leq T_\beta$ *which satisfies*

$$||\mathbf{U}(t)||_\beta^2 \leq M_\beta^2, \quad 0 \leq t \leq T_\beta; \tag{4.7}$$

moreover M_β, T_β *do not depend on* Ω, *but only on* $M_{0\beta}, h/R$.

(ii) *For every* α *such that* $\beta \geq \alpha \geq 3$, *there exists a constant* $C(\beta)$ *such that*

$$||\mathbf{U}(t)||_\beta^2 \leq (||\mathbf{U}(0)||_\beta^2) \exp\left(\left(C(\beta)\int_0^{T_\beta} ||\mathbf{V}(\tau)||_\alpha \, d\tau\right) + T_\beta\right), \tag{4.8}$$

where α *can be fixed independently of* β.

Proof. (i) is proven by a straightforward adaptation of the proof of Kato [16] in $\mathbf{R}^3$ to the cylinder $\mathbf{C}$, replacing the usual Sobolev spaces by the spaces $\mathbf{H}_\nu^\beta$. Kato's method is a vanishing viscosity limit via local existence for the Navier-Stokes equations (the latter via fixed-point construction, not a Galerkin approximation). That estimates are uniform in Ω stems from Lemma 4.1 and 4.2. (ii) can then be derived exactly as in Theorem 4.1 of [3], replacing Fourier methods by curl eigenvector function expansions. □

We now proceed to estimate the error between solutions $\mathbf{u}(t)$ of Eqs. (3.4)–(3.5) with finite large Ω

$$\partial_t \mathbf{u} = \mathbf{B}(\Omega t, \mathbf{u}, \mathbf{u}), \tag{4.9}$$

$$\mathbf{B}(\Omega t, \mathbf{u}, \mathbf{u}) = \mathbf{E}(\Omega t)\mathbf{B}(\mathbf{E}(-\Omega t)\mathbf{u}, \mathbf{E}(-\Omega t)\mathbf{u}) = \tilde{\mathbf{B}}(\mathbf{u}, \mathbf{u}) + \mathbf{B}^{\text{osc}}(\Omega t, \mathbf{u}, \mathbf{u}) \tag{4.10}$$

and solutions $\mathbf{w}(t)$ of the limit resonant 3D Navier-Stokes equations

$$\partial_t \mathbf{w} = \tilde{\mathbf{B}}(\mathbf{w}, \mathbf{w}), \tag{4.11}$$

$$\mathbf{w}|_{t=0} = \mathbf{w}(0) = \mathbf{U}(0) = \mathbf{u}(0). \tag{4.12}$$

Let

$$\mathbf{r}(t) = \mathbf{u}(t) - \mathbf{w}(t), \ \mathbf{r}(0) = 0. \tag{4.13}$$

Recall that $\mathbf{E}(\Omega t)$ is an isometry on the curl spaces $\mathbf{H}_\nu^s$ spaces. Therefore, estimates for norms of $\mathbf{r}(t)$ yield estimates for the norms of the error

$$\mathbf{R}(t) = \mathbf{E}(-\Omega t)(\mathbf{u}(t) - \mathbf{w}(t)) = \mathbf{U}(t) - \mathbf{E}(-\Omega t)\mathbf{w}(t). \tag{4.14}$$

The equation for the error $\mathbf{r}(t)$ is

$$\mathbf{r}(t) = \int_0^t \left(\tilde{\mathbf{B}}(\mathbf{u}, \mathbf{r}) + \tilde{\mathbf{B}}(\mathbf{r}, \mathbf{w}) + \mathbf{B}^{\text{osc}}(\Omega s, \mathbf{u}(s), \mathbf{u}(s))\right) ds, \tag{4.15}$$

where

$$\mathbf{B}_n^{\mathrm{osc}}(\Omega t, \mathbf{u}(t), \mathbf{u}(t)) = \sum_{\substack{l, k_3+m_3=n_3, non-resonant \\ k_2+m_2=n_2}} \exp(i\Omega t D_l(k,m,n))\mathbf{u}_k(t)\mathbf{u}_m(t) \quad (\mathrm{curl}\, \Phi_k \times \Phi_m, \Phi_n). \tag{4.16}$$

We have

Theorem 4.6. *Assume that the regular solution* $\mathbf{U}(t)$ *of Eq.* (1.11), (1.20) *with initial condition* $||\mathbf{U}(0)||_s \le M_{s0}$ *exists on* $0 \le t \le T_s$, *for some* T_s *(not necessary small), with* $||\mathbf{U}(t)||_s \le \tilde{\mathbf{M}}_s(\mathbf{M}_{s0}, T_s, h/R)$. *Then under conditions* $\alpha \ge 3$, $s-\alpha \ge 1$ *we have*

$$||\mathbf{r}(t)||_\alpha \le \delta(\Omega), \ \forall t \in [0, T_s], \tag{4.17}$$

where $\delta(\Omega) \to 0$ *as* $\Omega \to +\infty$; T_s *is independent from* Ω; $\delta(\Omega)$ *depends on* M_{s0}, T_s, α, s, *and* h/R.

Recall that **PJP** is skew-symmetric under inner product in the curl-norm spaces, therefore, local small time existence for (3.4)–(3.5) is independent of Ω.

To prove this convergence result we notice that the first two terms inside the integral on the right-hand side of (4.15) are linear in $\mathbf{r}$ and we only need to show that the contribution of $\mathbf{B}^{\mathrm{osc}}$ can be made arbitrary small as $\epsilon = 1/\Omega \to 0$. From (4.16) note that $\mathbf{B}_n^{\mathrm{osc}}(\tau/\epsilon, \mathbf{u}(t), \mathbf{u}(t))$ is an almost periodic function of τ with values in $\mathbf{L}^\infty(t; \mathbf{H}_\nu^\alpha)$ since the set $D_l(k,m,n) = \pm\sigma_k \pm \sigma_m \pm \sigma_n$ is countable. We need the following lemma for almost periodic functions with values in Banach spaces ([6], [8], [13], [27]).

Lemma 4.7. *([6], [27]). Let* $f(t,\tau) \in AP(\mathbf{R}; C_0(t; \mathbf{E}))$ *be almost periodic functions of the variable* τ, *with values in* $C_0(t; \mathbf{E})$, $t \in [0, T_0]$, $\mathbf{E}$ *is a Hilbert space. Let*

$$f(t,\tau) \sim \sum_{j \in J} f_j(t) \exp(i\omega_j \tau), \tag{4.18}$$

in the sense of Banach space-valued almost periodic functions in τ, $f_j(t) \in C_0(t; \mathbf{E})$, *over the (countable) set* J *of frequencies. If*

$$sup_{0 \le t \le T_0} ||f(t,\tau)||_{\mathbf{E}} \le M_f, \ \text{and if } |\omega_j| \ge \eta > 0 \text{ on } J \tag{4.19}$$

then:

$$\mathbf{E} - lim_{\epsilon \to 0} \int_0^T f(t, t/\epsilon) dt = 0; \tag{4.20}$$

the limit also converges uniformly on $0 \le T \le T_0$ *provided* $||f(t_2,\tau) - f(t_1,\tau)||_{\mathbf{E}} \le c|t_2 - t_1|$, *uniformly on* $[0, T_0]$.

To apply Lemma 4.7 to $\mathbf{B}^{\mathrm{osc}}(\Omega t, \mathbf{u}(t), \mathbf{u}(t))$ one needs $|\omega_j| = |\sigma_k \pm \sigma_m \pm \sigma_n| > \eta$ uniformly in k, m, n. For this, define $\pi_R \mathbf{u}$ (similarly $\pi_R \mathbf{w}$) the projection of $\mathbf{u}$ onto the curl eigenvector functions with $|\lambda k|, |\lambda m|, |\lambda n| \le R$, with:

$$||\mathbf{u} - \pi_R \mathbf{u}||_{\mathbf{H}_\nu^\alpha} \le M_s R^{\alpha - s}, \ s > \alpha. \tag{4.21}$$

Then $| \pm \sigma_k \pm \sigma_m \pm \sigma_n| > \eta(R)$ for $\pi_R \mathbf{u}$ in Eq. (4.16), which controls the small divisor estimate ([1]). $\mathbf{B}^{\rm osc}(\Omega t, \mathbf{u}(t), \mathbf{u}(t))$ is decomposed as

$$\begin{aligned}\mathbf{B}^{\rm osc}(\Omega t, \mathbf{u}, \mathbf{u}) = \pi_R \mathbf{B}^{\rm osc}(\Omega t, \pi_R \mathbf{u}, \pi_R \mathbf{u}) + \pi_R \mathbf{B}^{\rm osc}(\Omega t, \mathbf{u}, (I - \pi_R)\mathbf{u}) \\ + \pi_R \mathbf{B}^{\rm osc}(\Omega t, (I - \pi_R)\mathbf{u}, \mathbf{u}) + (I - \pi_R)\mathbf{B}^{\rm osc}(\Omega t, \mathbf{u}, \mathbf{u}).\end{aligned} \tag{4.22}$$

Finally, apply Lemma 4.7 to the integral

$$\int_0^t \pi_R \mathbf{B}^{\rm osc}(\Omega s, \pi_R \mathbf{u}(s), \pi_R \mathbf{u}(s)) ds, \text{ with } \mathbf{E} = \mathbf{H}^\alpha_\nu, \tag{4.23}$$

and use (4.21) to bound the error terms involving $(I - \pi_R)$. This completes the outline of the proof of the local convergence Theorem 4.6, with details following along the proof of Theorem 6.3, p. 139–140 in [3], albeit curl eigenvector functions replacing Fourier modes.

Remark 4.8. In Theorem 4.6 $\mathbf{u}(t)$ converges strongly to $\mathbf{w}(t)$. The original $\mathbf{U}(t)$ also converges strongly to $\mathbf{E}(-\Omega t)\mathbf{w}(t)$. But $\mathbf{U}(t)$ has no strong limit in $\mathbf{H}^s_\nu$ as $\Omega \to \infty$, hence the singular nature of the fast oscillating limit.

From the global regularity Theorem 3.8 for the 3D resonant Euler equations and Theorem 4.6 we bootstrap long-time regularity for Eqs.(1.11)–(1.12) for Ω large but finite.

Proof of Theorem 1.4. We complete in some detail the proof of Theorem 1.4, as the bootstrapping arguments are rather different from the usual classical procedure for Navier-Stokes equations [20]. Let T_m fixed, arbitrary large. Let $||\tilde{\mathbf{V}}_0(y)||_s = ||\mathbf{U}_0(x)||_s \le M_{s0}$, $s \ge 4$. From Theorem 3.8, with $||\mathbf{w}(0)||_\alpha = ||\mathbf{U}_0||_\alpha \le M_{\alpha 0} \le M_{s0}$, for some fixed $\alpha \ge 3, s - \alpha \ge 1$, there exists $\tilde{M}_\alpha$ such that:

$$||\mathbf{w}(t)||_\alpha \le \tilde{\mathbf{M}}_\alpha(M_{s0}, T_M, h/R) \text{ on } 0 \le t \le T_m, \tag{4.24}$$

for the solution of the resonant 3D Euler equations. We need the:

Corollary 4.9. *Let $||\mathbf{U}(t_1)||_\alpha \le R_\alpha$ at some $t_1 \ge 0, \alpha \ge 3, R_\alpha > 0$ given. Then there exists $T_\alpha(R_\alpha, h/R)$ such that*

$$||\mathbf{U}(t)||_\alpha \le 4R_\alpha \quad \text{on } [t_1, t_1 + T_\alpha]; \tag{4.25}$$

moreover if $\mathbf{U}(t_1) \in \mathbf{H}^s_\nu$ for some $s > \alpha$:

$$\begin{aligned}||\mathbf{U}(t)||^2_s &\le ||\mathbf{U}(t_1)||^2_s \exp\left\{\left(C(s)\left(\sup_{t_1 \le t \le t_1 + T_\alpha} ||\mathbf{U}||_\alpha\right) + 1\right)(t - t_1)\right\} \\ &\le ||\mathbf{U}(t_1)||^2_s \exp\{(4C(s)R_\alpha + 1)(t - t_1)\} \quad \text{on } [t_1, t_1 + T_\alpha].\end{aligned} \tag{4.26}$$

Proof. First, apply Theorem 4.5 with $\alpha = \beta$ to derive Eqs. (4.25). Then apply (ii) in Theorem 4.5 with $\beta = s$. □

We now choose, with $\tilde{M}_\alpha$ given by Eqs. (4.24):

$$R_\alpha = 3\tilde{M}_\alpha, \tag{4.27}$$

hence $T_\alpha = T_\alpha(M_{s0}, T_m, \alpha, h/R)$. Let $Q_\alpha = 4C(s)R_\alpha + 1$. We define:

$$\tilde{M}_s^2 = M_{s0}^2 \exp Q_\alpha(T_m + T_\alpha), \tag{4.28}$$

and we shall demonstrate that:

$$||\mathbf{U}(t)||_s^2 \leq \tilde{M}_s^2 \quad \text{on } 0 \leq t \leq T_m, \tag{4.29}$$

by choosing Ω large enough to make the error $\delta(\Omega)$ (in Theorem 4.6) uniformly small on the sequence of intervals $[0, T_\alpha], [T_\alpha, 2T_\alpha], \ldots, [nT_\alpha, (n+1)T_\alpha]$, where $nT_\alpha \leq T_m < (n+1)T_\alpha$. We apply Theorem 4.6 on the global interval $[0, T_m]$, assuming a priori the estimate (4.29) (which will be shown self-consistent under bootstrapping); we choose such large Ω that $\delta(\Omega)$ is so small and the assertion of Theorem 4.6 holds with $T_s \equiv T_m$, and for $\Omega \geq \Omega_1$, the constant $\delta(\Omega)$ satisfies:

$$\delta(\Omega) \leq R_\alpha/2 \quad \text{for } \Omega \geq \Omega_1(\tilde{M}_s, T_m, \alpha, s, h/R); \tag{4.30}$$

equivalently, Ω_1 depends only on $M_{s0}, T_m, \alpha, s, h/R$. Hence, for $0 \leq t \leq T_m$:

$$||\mathbf{U}(t)||_\alpha \leq ||\mathbf{U}(t) - \mathbf{E}(-\Omega t)\mathbf{w}(t)||_\alpha + ||\mathbf{w}(t)||_\alpha \leq R_\alpha/2 + R_\alpha/3 \leq R_\alpha. \tag{4.31}$$

We now apply Corollary 4.9 on $[0, T_\alpha]$, with the choice of R_α in Eqs. (4.27):

$$||\mathbf{U}(T_\alpha)||_s^2 \leq M_{s0}^2 \exp(Q_\alpha T_\alpha). \tag{4.32}$$

From the uniform estimate (4.31) for $||\mathbf{U}(t)||_\alpha$, valid at $t = T_\alpha$, we apply again Corollary 4.9 on $[T_\alpha, 2T_\alpha]$:

$$||\mathbf{U}(t)||_s^2 \leq M_{s0}^2 \exp(Q_\alpha T_\alpha) \exp(Q_\alpha T_\alpha); \tag{4.33}$$

and repeating the argument on the interval $[kT_\alpha, (k+1)T_\alpha], k \leq n$:

$$||\mathbf{U}(t)||_s^2 \leq M_{s0}^2 \exp(kQ_\alpha T_\alpha) \exp(Q_\alpha T_\alpha), \tag{4.34}$$

since $||\mathbf{U}(kT_\alpha)||_\alpha \leq R_\alpha$. Completing the bootstrapping, the maximal estimate for $||\mathbf{U}(t)||_s^2$ occurs on $[nT_\alpha, T_m]$:

$$\begin{aligned} ||\mathbf{U}(t)||_s^2 &\leq M_{s0}^2 \exp(nQ_\alpha T_\alpha) \exp(Q_\alpha(T_m - nT_\alpha)) \\ &\leq \tilde{M}_s^2, \end{aligned}$$

which corroborates the self-consistency of the choice $||\mathbf{U}(t)||_s \leq \tilde{M}_s$ in the application of Theorem 4.6 for a uniform $\delta(\Omega)$. The proof of Theorem 1.4 is then completed with the canonical transformation (1.17)–(1.18) between $\mathbf{V}(t, y)$ and $\mathbf{U}(t, x)$ and its isometry properties. □

Remark 4.10. The case $h/R \in \mathcal{K}^*$ includes the quadratic resonant operator $\mathbf{B}_{III}(\mathbf{w}^\perp, \mathbf{w}^\perp)$ in Eqs. (1.35). We have not found new conservation laws for the latter besides energy and helicity. A most interesting issue is the possibility of singularity and blow-up for the full resonant Euler equations (1.35). Partial results in the periodic lattice geometry are derived in [5], where (1.35) is demonstrated to be equivalent to a countable sequence of uncoupled finite dimension dynamical systems, in the generic case.

Acknowledgment

We would like to thank Professors V.I. Arnold, C. Foias, A. Fursikov, O.A. Ladyzhenskaya, H.K. Moffatt, Ya. Sinai and V.A. Solonnikov for the very useful discussions. V.A. Solonnikov has been most helpful in bringing to our attention the earlier work of his and his colleagues on magnetohydrodynamics. We acknowledge the support of the AFOSR grant F49620-02-10026.

References

[1] V.I. Arnold (1965), Small denominators. I. Mappings of the circumference onto itself, *Amer. Math. Soc. Transl. Ser. 2*, **46**, p. 213–284.

[2] V.I. Arnold and B.A. Khesin (1997), Topological Methods in Hydrodynamics, Applied Mathematical Sciences, **125**, Springer.

[3] A. Babin, A. Mahalov, and B. Nicolaenko (1997), Global regularity and integrability of 3D Euler and Navier-Stokes equations for uniformly rotating fluids, *Asymptotic Analysis*, **15**, p. 103–150.

[4] A. Babin, A. Mahalov and B. Nicolaenko (1999), Global regularity of 3D rotating Navier-Stokes equations for resonant domains, *Indiana Univ. Math. J.*, **48**, No. 3, p. 1133–1176.

[5] A. Babin, A. Mahalov and B. Nicolaenko (2001), 3D Navier-Stokes and Euler equations with initial data characterized by uniformly large vorticity, *Indiana Univ. Math. J.*, **50**, p. 1–35.

[6] C. Bardos, F. Golse, A. Mahalov and B. Nicolaenko (2004), Long-time regularity of 3D Euler equations with initial data characterized by uniformly large vorticity in cylindrical domains, to appear.

[7] J.T. Beale, T. Kato and A. Majda (1984), Remarks on the breakdown of smooth solutions for the 3D Euler equations, *Commun. Math. Phys.*, **94**, p. 61–66.

[8] A.S. Besicovitch (1954), Almost Periodic Functions, Dover, New York.

[9] J.P. Bourguignon and H. Brezis (1974), Remark on the Euler equations, *J. Func. Anal.*, **15**, p. 341–363.

[10] E.B. Bykhovskii (1957), Solutions of problems of mixed type of the Maxwell's equations for ideally conducting boundaries, *Vestnik Leningrad Univ.*, **13**, p. 50–66.

[11] E.B. Bykhovskii and N.V. Smirnov (1960), On orthogonal expansions in spaces of square integrable vector functions and operators of vector analysis, *Trudy Math. Inst. V.A. Steklov, Special Volume on Mathematical Problems in Hydrodynamics and Magnetohydrodynamics*, **59**, p. 5–36.

[12] E. Cartan (1922), Sur les petites oscillations d'une masse fluide, *Bull. Sci. Math.*, **46**, p. 317–352 and p. 356–369.

[13] C. Corduneanu (1968), Almost periodic Functions, Wiley-Interscience, New York.

[14] G. Duvaut and J.L. Lions (1976), The Inequalities in Mechanics and Physics, Springer-Verlag.

[15] K. Friedrichs (1955), Differential forms on Riemannian manifolds, *Comm. Pure Appl. Math*, **8**, No. 2.

[16] T. Kato (1972), Nonstationary flows of viscous and ideal fluids in $\mathbf{R}^3$, *J. Func. Anal.*, **9**, p. 296–305.

[17] N.D. Kopachevsky and S.G. Krein (2001), *Operator Approach to Linear Problems of Hydrodynamics*, Operator Theory: Advances and Applications, **128**, Birkhäuser Verlag.

[18] O.A. Ladyzhenskaya ed. (1960), *Trudy Math. Inst. V.A. Steklov, Special Volume on Mathematical Problems in Hydrodynamics and Magnetohydrodynamics*, **59**.

[19] O.A. Ladyzhenskaya (1968), On the unique global solvability of the 3D Cauchy problem for the Navier-Stokes equations under conditions of axisymmetry, *Boundary Value problems of Math. Phys. and Related Problems No. 2, Zapiski Seminarov LOMI*, **7**, p. 155–177.

[20] O.A. Ladyzhenskaya (1969), *The Mathematical Theory of Viscous Incompressible Flow*, 2nd ed., Gordon and Breach, New York.

[21] O.A. Ladyzhenskaya and V.A. Solonnikov (1960), Solution of some nonstationary problems of magnetohydrodynamics for viscous incompressible fluids, *Trudy Math. Inst. V.A. Steklov, Special Volume on Mathematical Problems in Hydrodynamics and Magnetohydrodynamics*, **59**, p. 115–173.

[22] J. Leray (1934), Sur le mouvement d'un liquide visqueux emplissant l'espace, *Acta Math.*, **63**, p. 193–248.

[23] A. Mahalov, S. Leibovich and E.S. Titi (1990), Invariant helical subspaces for the Navier–Stokes Equations, *Arch. for Rational Mech. and Anal.*, **112**, No. 3, p. 193–222.

[24] A. Mahalov and B. Nicolaenko (2003), Global solvability of the three-dimensional Navier-Stokes equations with uniformly large initial vorticity, *Russian Math. Surveys*, **58:2**, p. 287–318.

[25] H. Poincaré (1910), Sur la précession des corps déformables, *Bull. Astronomique*, **27**, p. 321–356.

[26] J.V. Ralston (1973), On stationary modes in inviscid rotating fluids, *J. Math. Anal. and Appl.*, **44**, p. 366–383.

[27] S. Schochet (1994), Resonant nonlinear geometric optics for weak solutions of conservation laws, *J. Diff. Eq.*, **113**, p. 473–503.

[28] S. Schochet (1994), Fast singular limits of hyperbolic PDE's, *J. Diff. Eq.*, **114**, p. 476–512.

[29] S.L. Sobolev (1954), Ob odnoi novoi zadache matematicheskoi fiziki, *Izvestiia Akademii Nauk SSSR, Ser. Matematicheskaia*, **18**, No. 1, p. 3–50.

[30] V.A. Solonnikov (1972), Overdetermined elliptic boundary value problems, *Boundary Value problems of Math. Phys. and Related Problems No. 5, Zapiski Seminarov LOMI*, **21**, p. 112–158.

[31] E.M. Stein (1970), *Singular integrals and differentiability properties of functions*, Princeton University Press.

[32] W. Wolibner (1933), Un théorème sur l'existence du mouvement plan d'un fluide parfait, homogène et incompressible, pendant un temps infiniment long, *Mat. Z.*, **37**, p. 698–726.

[33] V.I. Yudovich (1963), Non stationary flow of an ideal incompressible liquid, *Zb. Vych. Mat.*, **3**, p. 1032–1066.

A. Mahalov and B. Nicolaenko
Department of Mathematics and Statistics
Arizona State University
Tempe, AZ 85287-1804, USA
e-mail: mahalov@asu.edu, byn@stokes.la.asu.edu

C. Bardos and F. Golse
U.M.R. of Mathematics
University of Paris-7
F-75251 Paris Cedex-05
and
Laboratoire J.L. Lions
University of Paris-6
France
e-mail: bardos@math.jussieu.fr, Francois.Golse@ens.fr

Progress in Nonlinear Differential Equations
and Their Applications, Vol. 61, 187–195

A Model of a Two-dimensional Pump

Piotr Bogusław Mucha

Dedicated to Professor V.A. Solonnikov

The aim of this note is to study the system of equations of motion of viscous incompressible fluid in a time-dependent domain in the two-space-dimensional case governed by the classical Navier-Stokes equations with the slip conditions at the boundary. We look for periodic in time solutions. We are interested in data with inhomogeneous boundary conditions. Also the dependence of the viscosity coefficient will be important, since we want to study the inviscid limit for the solutions of our model.

The equations read

$$\begin{array}{lll} v_t + v \cdot \nabla v - \nu \Delta v + \nabla p = 0 & \text{in} & P, \\ \text{div } v = 0 & \text{in} & P, \\ n \cdot \mathbf{T}(v,p) \cdot \tau = 0 & \text{on} & \partial P, \\ n \cdot v = V & \text{on} & \partial P, \\ v(\cdot, t) = v(\cdot, t+1) & \text{on} & p(t), \end{array} \tag{1}$$

where $v = (v^1, v^2)$ describes the velocity vector, p the pressure. Time-dependent domain is denoted by $p(t) \subset \mathbf{R}^2$ and

$$P = \bigcup_{0 \le t < 1} p(t) \times \{t\}, \quad \partial P = \bigcup_{0 \le t < 1} \partial p(t) \times \{t\}. \tag{2}$$

By the periodicity with respect to time

$$p(t) = p(t+1). \tag{3}$$

Vectors n and τ are the normal and tangent unit vectors to boundary $\partial p(t)$. $\mathbf{T}$ is the stress tensor and

$$\mathbf{T}(v,p) = \{\nu(v^i_{,j} + v^j_{,i}) - p\delta_{ij}\}_{i,j=1,2}, \tag{4}$$

where ν is the viscous coefficient.

We may divide the boundary into two parts

$$\partial p(t) = \Gamma_r \cup \Gamma_m(t), \tag{5}$$

where Γ_r is the rigid part and Γ_m is the moving part of the boundary. We assume that the boundary is sufficiently smooth at least $C^{2,1}$-piecewise and interior angles are $\pi/2$ – see the picture below. For simplicity we require the domain to be simply connected.

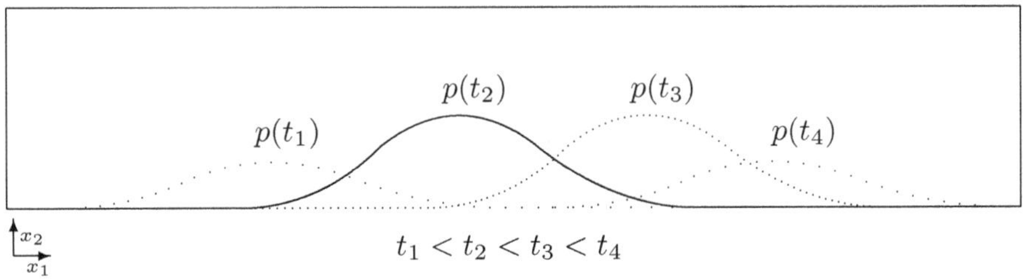

This decomposition divides also the boundary datum V. On the rigid part $(1)_4$ decibels the inflow/outflow condition

$$V|_{\Gamma_r} = V_0 \tag{6}$$

and on Γ_m we assume that

$$V|_{\Gamma_m} = V_1 = \frac{\partial}{\partial t}\varphi(x,t)\cdot n|_{\Gamma_m}, \tag{7}$$

where φ describes the surface of the boundary. Relation (7) says that there is no inflow through Γ_m.

Moreover, to keep compatibility conditions with $(7)_2$ we require that

$$\int_{\partial p(t)} V(\cdot,t)d\sigma = 0. \tag{8}$$

The system described above may be treated as a model of two-dimensional pump. Question concerns existence of solutions for the large data is connected with the effectiveness of the pump.

The first result of the paper is the following.

Theorem A. *Let* $V \in C^{1,1/2}(\partial P)$ *and*

$$||A_0(p(t))||_{C(0,1)}||2\chi(x,t)||_{C(\partial P)} < 1, \tag{9}$$

where χ *is the curvature of* $\partial p(t)$ *and*

$$A_0(p(t)) = ||W||_{C(\partial p(t))}, \tag{10}$$

where W *is the solution to the following problem*

$$\begin{array}{lll} rot\, W = 1 & in & p(t), \\ div\, W = 0 & in & p(t), \\ n\cdot W = 0 & on & \partial p(t). \end{array}$$

Then system (1) *admits at least one weak solution such that*

$$rot\, v \in L_\infty(P), \quad v \in C^{a,a/2}(P)$$

for any $0 < a < 1$. *Moreover*

$$||rot\, v||_{L_\infty(P)} \le B(1 - ||A_0||_C||2\chi||_C)^{-1}||V||_{C^{1,0}}, \tag{11}$$

where B *is independent of the viscous coefficient* ν.

Under a geometrical constraint, given by assumption (9), we are able to show existence of solutions for any large inflow conditions. Of course, (9) restricts V_m, however in a mild sense, since the effectiveness of the pump modelled by the equations can be any large. To measure this quantity we may write the following definition

$$\mathcal{E} = \int_0^1 \int_{\Gamma_m(t)} V_b(t)n \cdot e_1 d\sigma dt.$$

Adding to V_b term $g(t)e_1 \cdot n$ with suitable large smooth function $g(t)$ we may increase $\mathcal{E}$ as we wish. Quantity V_0 is not restricted.

A key element of our system is the slip boundary condition $(1)_{3,4}$. Only for this type of boundary relations it is possible to show estimate (11).

For the different boundary conditions, as for the most popular Dirichlet ones, the theory gives a similar result [2, 3, 4, 6, 9]. The technique for this type of problems are based on the energy approach and the Hopf extension. This method delivers us main estimate which is highly dependent of the viscosity coefficient.

The initial-boundary value problem for the system with the external force has been considered in [8].

Our model admits estimate (11) and properties of this bound (independence of the viscosity) allows to examine the inviscid limit for the system (1).

The next result concerns this subject.

Theorem B. *Let assumptions from Theorem A be fulfilled. Consider system* (1) *for viscous coefficients* $\nu \to 0$.

Then for a subsequence

$$\nu_k \to 0 \quad with \quad k \to \infty;$$

we have

$$v^{\nu_k} \to v_E \quad strongly\ in\ C^{a,a/2}(P)$$

and

$$rot\, v^{\nu_k} \to rot\, v_E \quad weakly\text{-}*\ in\ L_\infty(P),$$

where v^{ν_k} *is the solution to problem* (1) *with viscous coefficient equals* ν_k *and* v_E *fulfills the Euler system*

$$\begin{array}{lll} v_{E,t} + v_E \cdot \nabla v_E + \nabla p_E = 0 & in & P, \\ div\, v_E = 0 & in & P, \\ n \cdot v_E = V & on & P. \end{array} \tag{12}$$

Moreover

$$||rot\, v_E||_{L_\infty} \le B(1 - ||A_0||_C||2\chi||_C)^{-1}||V||_{C^{1,0}(\partial P)}. \tag{13}$$

The inviscid limit is not unique, since even for the Navier-Stokes equations we have not got this feature. Also it is impossible, because the inflow condition is non zero (on the rigid part of the boundary), thus to obtain the uniqueness of the Euler system we need an extra condition. The result is so strong as for the torus [5], although here our domain is time-dependent. It is a consequence of estimate (11), which can be proved, provided condition (9).

Question concerns the Eulerian limit for the Navier-Stokes equations with the slip boundary condition shave been investigated in [1, 7].

Throughout the paper we try to use standard notations as in [4, 10].

A priori bound

We want to show estimate (11) assuming that the solutions exist. The first step is to reformulate the problem. By the two-dimensional properties the vorticity of the velocity is a scalar function

$$\alpha = \mathrm{rot}\; v = v^2_{,1} - v^1_{,2}.$$

Also the vorticity equation obtained form $(1)_{1,2}$ has a special form

$$\alpha_t + v \cdot \nabla\alpha - \nu\Delta\alpha = 0. \tag{14}$$

To obtain a boundary condition to equations (14) we use properties of the slip boundary condition $(1)_{3,4}$, which enables to calculate the value of the vorticity at the very boundary as follows

$$\alpha = 2\chi v \cdot \tau - 2V_{,s}, \tag{15}$$

where s is the unit length parameter of $\partial p(t)$ (it is enough to differentiate $(1)_4$ with respect to s and use $(1)_3$) and recall χ is the curvature of $\partial p(t)$.

Examining system (14)–(15), we can apply the maximum principle, since the form of the nonlinear term is proper for this technique. Then we get

$$||\alpha||_{L_\infty(P)} \leq ||2\chi||_{C(\partial P)}||v||_{C(\partial P)} + 2||V_{,s}||_{C(P)}. \tag{16}$$

To find estimate on the velocity we consider the elliptic problem,

$$\begin{array}{lll} \mathrm{rot}\; v = \alpha & \text{in} & p(t), \\ \mathrm{div}\; v = 0 & \text{in} & p(t), \\ n \cdot v = V(\cdot, t) & \text{on} & \partial p(t) \end{array} \tag{17}$$

for $t \in [0, 1)$.

The standard theory delivers us the following bound

$$||v||_{C(p(t))} \leq A_0(p(t))||\alpha||_{L_\infty(p(t))} + c||V||_{C^1(p(t))}, \tag{18}$$

where A_0 is the constant given by (10).

Combining (16) and (18) we obtain

$$||\alpha||_{L_\infty(P)} \leq ||A_0(p(t))||_{C(0,1)}||2\chi||_{C(\partial P)}||\alpha||_{L_\infty(P)} + B||V||_{C^{1,0}(\partial P)}, \tag{19}$$

but we assumed that

$$||A_0(p(t))||_{C(0,1)}||2\chi||_{C(\partial P)} < 1.$$

Thus (19) gives

$$||\alpha||_{L_\infty(P)} \leq B(1 - ||A_0(p(t))||_{C(0,1)}||2\chi||_{C(\partial P)})^{-1}||V||_{V^{1,0}}. \tag{20}$$

The above considerations lead us to the following result.

Lemma 1. *For sufficiently smooth solutions to problem* (1) *estimate* (11) *is valid, provided constraint* (9).

To clarify existence to the system we need regularity of solutions with respect to time. It is possible to prove using the weak-$*$ formulation to the coupled system (14), (15) and (17).

Definition 1. *We say that $v \in C^{a,0}(P)$ such that $\mathrm{rot}\, v \in L_\infty(P)$, is the weak-$*$ solution to problem* (1), *iff the following identity holds*

$$\int_P \alpha\phi_t dxdt + \int_P \alpha v \cdot \nabla\phi dxdt$$

$$+\nu \int_P \alpha\Delta\phi dxdt - 2\nu \int_{\partial P} (\chi v \cdot \tau - V_{,s}) d\sigma dt = 0 \tag{21}$$

for any

$$\phi \in W_1^2(P) \cap \{\phi|_{\partial P} = 0\}$$

and v fulfilling the following system

$$\begin{array}{lll} rot\ v = \alpha & in & p(t), \\ div\ v = 0 & in & p(t), \\ n \cdot v = V(\cdot, t) & on & \partial p(t) \end{array} \tag{22}$$

for $t \in [0, 1)$.

By the weak-$*$ formulation we prove the following Lemma.

Lemma 2. *Weak-$*$ solutions fulfilling Definition* 1 *belong to $C^{a,a/2}(P)$.*

Proof. We need to show regularity with respect to time. By (21) we have that

$$\int_P \alpha_t \phi dxdt = -\int_P \alpha v \cdot \nabla\phi dxdt$$

$$-\nu \int_P \alpha\Delta\phi dxdt + 2\nu \int_{\partial P} (\chi v \cdot \tau - V_{,s}) \frac{\partial\phi}{\partial n} d\sigma dt \tag{23}$$

which holds for ϕ as in Definition 1.

However we may examine only the r.h.s. of (23). Assuming that v is a weak-$*$ solution, the r.h.s. of (23) is well defined for any

$$\phi \in W_p^{2,0}(P) \cap \{\phi|_{\partial P} = 0\}$$

for any $p > 1$.

Hence

$$|\text{the r.h.s. of (23)}| \leq c||V||_{C^{1,0}(\partial P)}(\nu + 1 + ||V||_{C^{1,0}(\partial P)})||\phi||_{W^{2,0}(P)}. \tag{24}$$

Hence by the definition of the weak derivative from the l.h.s. of (23) we conclude that

$$\alpha_t \in L_{p^*}(0, 1; (W_p^2(p(t)))^*)$$

with $1/p + 1/p^* = 1$.

Putting this information to the investigation of problem (22), remembering that $V \in C^{1,1/2}(\partial P)$ we obtain the following bound on the stream function of the velocity filed

$$||\psi_t||_{L_{p^*}(P)} \le c||\alpha_t||_{L_{p^*}(0,1;(W^2_p(p(t)))^*)} + DATA, \tag{25}$$

where ψ defines the velocity

$$v = \nabla^\perp \psi = (-\partial_{x_2}\psi, \partial_{x_1}\psi).$$

Such a scalar function can be found since our domain is simply connected.

To obtain bound (25) we need to transform the time-dependent domain into a rigid one to keep well-posedness of the differentiation with respect to time. Since boundary $\partial P \in C^{2,1}$ and also the curvature of $\partial p(t)$ is also controlled, this procedure delivers us only technical difficulties which we omit here.

Since bound (24) depends on ν in a mild sense (the estimate is valid for $\nu \to 0$) and holds for any $p > 1$. Thus by estimate (20) and (24), and properties of solvability of the elliptic system (22) we have

$$\psi \in W^{2,1}_q(P) \ \text{ for any } \ q < \infty$$

which follows that

$$v = \nabla^\perp \psi \in W^{1,1/2}_q(P);$$

and the embedding theorem gives that;

$$v \in C^{a,a/2}(P) \ \text{ for } \ 0 < a < 1.$$

By (24) we deduce that

$$||v||_{C^{a,a/2}(P)} \le c||V||_{C^{1,1/2}(\partial P)}(||V||_{C^{1,1/2}(P)} + 1 + \nu). \tag{26}$$

Lemma 2 is proved.

Existence

To show existence we apply the Leray-Schauder theorem. Lemmas 1 and 2 give us a priori bound for the solutions. Now we need to consider a linearization of the system.

For a function w we consider the following coupled system

$$\begin{array}{lll} \beta_t + w \cdot \nabla \beta - \nu \Delta \beta = 0 & \text{in} & P, \\ \beta = 2\chi u \cdot \tau - 2V_{,s} & \text{on} & \partial P, \end{array} \tag{27}$$

$$\begin{array}{lll} \text{rot } u = \beta & \text{in} & p(t), \\ \text{div } u = 0 & \text{in} & p(t), \\ n \cdot u = V(\cdot, t) & \text{on} & \partial p(t) \end{array} \tag{28}$$

for $t \in [0, 1)$.

Lemma 3. *Let $w \in C^{a,a/2}(P)$, then if assumptions of Theorem A are fulfilled then*

$$||u||_{C^{a,a/2}(P)} \le c||V||_{C^{1,1/2}(\partial P)}(||V||_{C^{1,1/2}(P)} + 1 + \nu). \tag{29}$$

System (27)–(28) describes us the following map

$$\Xi : C^{a,a/2}(P) \to C^{a,a/2}(P),$$

such that

$$\Xi(w) = u.$$

It is easy to check that the fixed point of the map will be the solution to problem (1) in the sense of Definition 1. By considerations in Lemma 1 and 2, we can prove the same bound (with the same constants) for system (27)–(28) which shows well-posedness of map ξ.

One point, which will not be shown, is the continuity of the map. We should show that if

$$w \to \bar{w} \text{ in } C^{a,a/2}(P),$$

then

$$\Xi(w) \to \Xi(\bar{w}) \text{ in } C^{a,a/2}(P).$$

We omit the proof of this fact, since it follows from the standard energy approach.

Since P is bounded by the Ascoli-Arzeli theorem bounded set in $C^{a,a/2}(P)$ is compact. Moreover the set of solutions $\Xi(u) = \lambda u$ for $\lambda \in [0,1]$ is bounded in $C^{a,a/2}(P)$ which is a consequence of a priori bounds obtained above. Thus by the Leray-Schuader theorem we find at least one fixed point of map ξ.

It follows that we show existence of weak-$*$ solution to problem (1). Theorem A is proved.

The inviscid limit

Restate problem (1) underlining dependence of the viscosity

$$\begin{array}{lll} v^\nu_t + v^\nu \cdot \nabla v^\nu - \nu\Delta v^\nu + \nabla p^\nu = 0 & \text{in} & P, \\ \text{div } v^\nu = 0 & \text{in} & P, \\ n \cdot \mathbf{T}(v^\nu, p^\nu) \cdot \tau = 0 & \text{on} & \partial P, \\ n \cdot v^\nu = V & \text{on} & \partial P, \\ v^\nu(\cdot, t) = v^\nu(\cdot, t+1) & \text{on} & p(t). \end{array} \tag{30}$$

By Theorem A the following bounds hold

$$\|\text{ rot } v^\nu\|_{L_\infty(P)} \le S, \quad \|v^\nu\|_{C^{a,a/2}(P)} \le S \tag{31}$$

for $\nu < 1$. The restriction on the viscosity coefficient comes from (26). It is not essential since we investigate the limit

$$\nu \to 0.$$

As we mentioned at the very beginning we will not obtain any uniqueness result, hence we concentrate our attention to a subsequence

$$\nu_k \to 0 \quad \text{for} \quad k \to +\infty.$$

The subsequence has been chosen in such a way that

$$\text{rot } v^{\nu_k} \rightharpoonup \text{rot } v_E \quad \text{weakly-}* \text{ in } L_\infty(P)$$

and

$$v^{\nu_k} \to v_E \quad \text{strongly in } C^{a,a/2}(P).$$

Introduce a definition of weak solutions to the Euler system (12).

Definition 2. *Let $p > 2$. We say that*

$$v \in C(P) \cap W_p^{1,0}(P) \quad \textit{such that} \quad v|_{\partial P} = V$$

is the weak solution to problem (12), *iff the following identity is valid*

$$\int_P v\Phi_t dxdt - \int_P v\nabla v\Phi dxdt = 0 \tag{32}$$

for any

$$\Phi \in C^\infty(P) \cap \{div\, \Phi = 0\} \cap \{n \cdot \Phi|_{\partial P} = 0\}.$$

Since domains $p(t)$ are simply connected we characterize vector functions Φ by scalar functions φ as follows

$$\Phi = \nabla^\perp \varphi = (-\partial_{x_2}\varphi, \partial_{x_1}\varphi)$$

and $\varphi \in C^\infty(P) \cap \{\varphi|_{\partial P} = 0\}$.

Putting form of Φ into formula (32) we obtain

$$\int_P (\text{rot } v)\varphi_t dxdt + \int_P (\text{rot } v)v \cdot \nabla\varphi dxdt = 0. \tag{33}$$

Next, we recall Definition 1. The solution of (30) fulfills

$$\int_P (\text{rot } v^{\nu_k})\phi_t dxdt + \int_P (\text{rot } v^{\nu_k})v^{\nu_k}\nabla\phi dxdt$$

$$+\nu_k \int_P (\text{rot } v^{\nu_k})\Delta\phi dxdt - 2\nu_k \int_P (\chi v^{\nu_k} \cdot \tau - V_{,s})d\sigma dt = 0 \tag{34}$$

for ϕ as in Definition 1.

Then by (31) we see that if $k \to \infty$, then

$$\int_P (\text{rot } v^{\nu_k})\phi_t dxdt \to \int_P (\text{rot } v_E)\phi_t dxdt,$$

$$\int_P (\text{rot } v^{\nu_k})v^{\nu_k}\nabla\phi dxdt \to \int_P (\text{rot } v_E)v_E \cdot \nabla\phi dxdt,$$

$$|\nu_k \int_P (\text{rot } v^{\nu_k})\Delta\phi dxdt - 2\nu_k \int_P (\chi v^{\nu_k} \cdot \tau + V_{,s})d\sigma dt| \to 0.$$

Thus v_E fulfills Definition 2.

Next we increase the regularity of the obtained solution. By considerations in the proof of Lemma 2 we deduce that

$$v_E \in W_p^{1,1/2}(P)$$

for any $p < \infty$. Thus by the embedding theorem

$$v_E \cdot \nabla v_E \in L_p(P).$$

By the definition of the distributional derivative and (32) we conclude that

$$v_{E,t} \in L_p(P),$$

too. Construction of Φ together with Definition 2 guarantee existence of a scalar function p_E such that

$$v_{E,t} + v_E \cdot \nabla v_E + \nabla p_E = 0$$

a.e. in P and $\nabla p \in L_p(\Omega)$. Theorem B is proved.

Acknowledgment

The work has been supported by Polish KBN grant No 2 PO3A 002 23.

References

[1] Clopeau T.; Mikelić A.; Robert R.: On the vanishing viscosity limit for the 2D incompressible Navier-Stokes equations with the friction type boundary conditions, Nonlinearity 11 (1998), 1625–1636.

[2] Fursikov, A.V.; Gunzburger, M.D.; Hou, L.S.: Inhomogeneous boundary value problems for three-dimensional evolutionary Navier-Stokes equations. J. Math. Fluid Mech. 4, (2002) 45–75.

[3] Galdi, G.P.: An introduction to the mathematical theory of the Navier-Stokes equations, Springer-Verlag, 1994.

[4] Ladyzhenskaya, O.A.; Solonnikov V.A.; Uraltseva N.N.: Linear and Quasilinear Equations of Parabolic Type, AMS, 1968.

[5] Lions P.L.: Mathematical topics in fluid mechanics, Vol. 1 Incompressible Models, Oxford University Press, 1996.

[6] Łukaszewicz, G.: On the Navier-Stokes equations in time-dependent domains and with boundary conditions involving the pressure. J. Math. Sci. Univ. Tokyo 4, No. 3, (1997) 529–550.

[7] Mucha, P.B.: On the inviscid limit of the Navier-Stokes equations for flows with large flux, Nonlinearity 16, No 5, (2003) 1715–1732.

[8] Mucha, P.B.: On a pump, to appear.

[9] Resendiz, B.A.J.; Rojas-Medar, M.A.: Existence of weak solutions of micropolar fluid equations in a time dependent domain. Rev. Mat. Appl. 23 (2002), 27–46.

[10] Temam, R.: Navier-Stokes equations. North-Holland, Amsterdam 1977.

Piotr Bogusław Mucha
Institute of Applied Mathematics and Mechanics
Warsaw University
ul. Banacha 2
02-097 Warszawa, Poland
e-mail: p.mucha@mimuw.edu.pl

Progress in Nonlinear Differential Equations
and Their Applications, Vol. 61, 197–212

Regularity of a Weak Solution to the Navier-Stokes Equation in Dependence on Eigenvalues and Eigenvectors of the Rate of Deformation Tensor

Jiří Neustupa and Patrick Penel

Dedicated to V.A. Solonnikov on occasion of his seventies.

Abstract. We formulate sufficient conditions for regularity of a so-called suitable weak solution $(v; p)$ in a sub-domain D of the time-space cylinder Q_T by means of requirements on one of the eigenvalues or on the eigenvectors of the rate of deformation tensor.

Mathematics Subject Classification (2000). 35Q30, 76D05.

Keywords. Navier-Stokes equations.

1. Introduction

This paper develops the announced results of our note in C. R. Acad. Sci. Paris [11] and provides all detailed proofs.

Let Ω be a domain in $\mathbb{R}^3$, T be a positive number and $Q_T = \Omega\times\,]0,T[$. We deal with the Navier-Stokes initial-boundary value problem

$$\frac{\partial \boldsymbol{v}}{\partial t} + (\boldsymbol{v}\cdot\nabla)\boldsymbol{v} = -\nabla p + \nu\,\Delta\boldsymbol{v} \qquad \text{in } Q_T, \tag{1.1}$$

$$\operatorname{div}\boldsymbol{v} = 0 \qquad \text{in } Q_T, \tag{1.2}$$

$$\boldsymbol{v} = 0 \qquad \text{on } \partial\Omega\times\,]0,T[\,, \tag{1.3}$$

$$\boldsymbol{v}|_{t=0} = \boldsymbol{v}_0 \tag{1.4}$$

where $\boldsymbol{v} = (v_1,\, v_2,\, v_3)$ denotes the velocity, p denotes the pressure and $\nu > 0$ is the viscosity coefficient.

The notions of a weak solution and a suitable weak solution to the problem (1.1)–(1.4) are well known. The readers can find the definitions and surveys of their important properties, e.g., in G.P. Galdi [6] and in J. Neustupa and P. Penel [10].

A point $(x,t) \in Q_T$ is called a *regular point* of a weak solution $\boldsymbol{v}$ of (1.1)–(1.4) if there exists a neighborhood U of (x,t) in Q_T such that $\boldsymbol{v}$ is essentially bounded in U. Points of Q_T which are not regular are called *singular.* We shall denote by $S(\boldsymbol{v})$ the set of all singular points of $\boldsymbol{v}$. $S(\boldsymbol{v})$ is a closed set in Q_T.

We suppose that $(\boldsymbol{v};p)$ is a suitable weak solution of the problem (1.1)–(1.4). Then the singular set $S(\boldsymbol{v})$ has the 1-dimensional parabolic measure, and consequently also the 1-dimensional Hausdorff measure, equal to zero. (See L. Caffarelli, R. Kohn and L. Nirenberg [3].)

We prove the local regularity of the solution $(\boldsymbol{v};p)$ under a certain assumption about one of the eigenvalues of the rate of deformation tensor in Section 2. The result is an improvement of Theorem 2 from [10]. In Section 3, we study the regularity of the solution $(\boldsymbol{v};p)$ under the assumption about a certain smoothness of the eigenvectors of the rate of deformation tensor.

We assume for simplicity that the external force in the Navier-Stokes equation (1.1) is zero. However, the results can be extended to the case of a nonzero force $\mathbf{f} \in L^q(Q_T)^3$ (for some $q > \frac{5}{2}$).

Both the eigenvalues and the eigenvectors of the rate of deformation tensor are the notions which are naturally connected with the flow field and are independent of a system of coordinates. The results give the information on "what types of deformations" of infinitely small volumes of the flow either contribute to regularity or support a hypothetical singularity.

Suppose that D is a sub-domain of Q_T and D' is a domain in D such that $D' \subset \overline{D'} \subset D$. Since $S(\boldsymbol{v})$ is closed in Q_T, $S(\boldsymbol{v}) \cap \overline{D'}$ is a closed set. Denote by $T(D')$, respectively G', the projection of D', respectively the projection of $S(\boldsymbol{v}) \cap \overline{D'}$, onto the time axis. Then the 1-dimensional Hausdorff measure of G' is zero and

$$T(D') = \bigcup_{\gamma \in \Gamma'}]a'_\gamma, b'_\gamma[\cup G' \tag{1.5}$$

where the sets on the right-hand side are mutually disjoint. In accordance with the terminology from [7] and [6], we will call time instants b'_γ *D'-epochs of irregularity.* Existence of a singular point of solution $(v;p)$ in D' implies the existence of at least one D'-epoch of irregularity and vice versa. Thus, let us further assume that t_0 is a D'-epoch of irregularity of solution $(\boldsymbol{v};p)$ and (x_0,t_0) (where $x_0 = (x_{01}, x_{02}, x_{03})$) is a singular point of $(\boldsymbol{v};p)$ in D'. Then t_0 is equal to b'_γ for some $\gamma \in \Gamma'$. We will later show that these assumptions are in a contradiction with the conditions of Theorem 1 in Section 2 or Theorem 2 in Section 3. However, we need several auxiliary lemmas at first.

Lemma 1.1. *There exist positive numbers τ, r_1, r_2 such that $r_1 < r_2$ and if we denote $C_1 =]x_{01} - r_1, x_{01} + r_1[\times]x_{02} - r_1, x_{02} + r_1[\times]x_{03} - r_1, x_{03} + r_1[$ and $C_2 =]x_{01} - r_2, x_{01} + r_2[\times]x_{02} - r_2, x_{02} + r_2[\times]x_{03} - r_2, x_{03} + r_2[$, then*

1. *τ is so small that $a'_\gamma < b'_\gamma - \tau = t_0 - \tau$,*
2. *$\overline{C_2} \times [t_0 - \tau, t_0] \subset D'$,*
3. *$\{(\overline{C_2} - C_1) \times [t_0 - \tau, t_0]\} \cap S(\boldsymbol{v}) = \emptyset$,*
4. *$\boldsymbol{v}$ and all its space derivatives are bounded on $(\overline{C_2} - C_1) \times [t_0 - \tau, t_0]$.*
5. *p, respectively $\partial \boldsymbol{v}/\partial t$, has all space derivatives in $L^\alpha(t_0 - \tau, t_0;\, L^\infty(C_2 - \overline{C_1}))$, respectively in $L^\alpha(t_0 - \tau, t_0;\, L^\infty(C_2 - \overline{C_1})^3)$, for each α such that $1 \le \alpha < 2$.*

Items 1–4 of Lemma 1.1 were proved in [8] with the unimportant difference that we worked with the balls $B_1 = B_{r_1}(x_0)$ and $B_2 = B_{r_2}(x_0)$ instead the cubes C_1 and C_2 in [8]. The proof was commented in [10], too. Item 5 was proved in [10] with B_1 and B_2 instead of C_1 and C_2. It is worth of mentioning that in the case when $\Omega = \mathbb{R}^3$, the proof of statement 5 can be modified so that it is valid for all $\alpha = \infty$. However, the same approach cannot be applied if $\partial\Omega$ is not empty and the improvement of statement 5 remains an open problem. This is remarkable, because statement 5 concerns a local behavior of the solution and in spite of it we are not able to exclude the influence of the boundary, no matter how far it is from the considered domain $C_2 - \overline{C_1}$.

Numbers r_1, r_2 and τ given by Lemma 1.1 are not unique. On the other hand, there exist decreasing sequences $\{r_1^n\}$, $\{r_2^n\}$, $\{\tau^n\}$ of numbers with the properties of r_1, r_2 and τ stated in Lemma 1.1 which tend to zero. This follows from the fact that the 1-dimensional Hausdorff measure of the set $S(\boldsymbol{v})$ is zero. We shall use the possibility of choosing r_1 as small as we need in Section 3.

Put $r_3 = (2r_1 + r_2)/3$, $r_4 = (r_1 + 2r_2)/3$ and
$C_3 =]x_{01} - r_3, x_{01} + r_3[\times]x_{02} - r_3, x_{02} + r_3[\times]x_{03} - r_3, x_{03} + r_3[$,
$C_4 =]x_{01} - r_4, x_{01} + r_4[\times]x_{02} - r_4, x_{02} + r_4[\times]x_{03} - r_4, x_{03} + r_4[$.

We shall use the spaces $L^{a,b}(C_2 \times]t_0 - \tau, t_0[) \equiv L^a(t_0 - \tau, t_0;\, L^b(C_2))$. We shall abbreviate their denotation to $L^{a,b}$. We shall also denote by $|||\,.\,|||_{a,b}$ the norm in $L^{a,b}$. $|||\,.\,|||_{(\infty,2)\cap(2,6)}$ will mean the sum $|||\,.\,|||_{\infty,2} + |||\,.\,|||_{2,6}$. Analogously, $\|\,.\,\|_k$ will denote the norm in $L^k(C_2)$.

The restrictions of functions defined a.e. in Q_T to subsets of Q_T will be denoted by the same letters. Thus, for example, $\boldsymbol{v} \in L^\infty(t_0 - \tau, t_0;\, W^{1,2}(C_2)^3)$ is the statement about the restriction of $\boldsymbol{v}$ to $C_2 \times]t_0 - \tau, t_0[$.

The next lemma can be easily proved by means of the Hölder inequality.

Lemma 1.2. *If $g \in L^{\infty,2} \cap L^{2,6}$, $2 \le a \le +\infty$, $2 \le b \le 6$ and $\frac{3}{2} \le 2/a + 3/b \le \frac{5}{2}$ then*

$$|||g|||_{a,b} \le |||g|||_{2,2}^{2/a+3/b-3/2}\ |||g|||_{(\infty,2)\cap(2,6)}^{5/2-(2/a+3/b)}. \tag{1.6}$$

Analogously as in our previous paper [10], we can localize the Navier-Stokes initial-boundary value problem to C_2 in the spatial variables. We shall therefore use an infinitely differentiable cut-off function η such that $\eta = 0$ on $\mathbb{R}^3 - C_4$, $\eta = 1$ on $\overline{C_3}$ and $0 \le \eta \le 1$ on $\overline{C_4} - C_3$. Since the product $\eta\boldsymbol{v}$ does not satisfy the equation of continuity, we put $\boldsymbol{u} = \eta\boldsymbol{v} - \boldsymbol{V}$ where $\boldsymbol{V}$ is an appropriate function such that $\operatorname{div} \boldsymbol{V} = \operatorname{div}(\eta\boldsymbol{v}) = \nabla\eta\cdot\boldsymbol{v}$. The existence of function $\boldsymbol{V}$ follows, e.g., from G.P. Galdi [5] (Theorem 3.2, Chap. III.3) and from W. Borchers and H. Sohr [2]

(Theorem 2.4). One can observe (see, e.g., W. Borchers and H. Sohr [2], pp. 73–76), that since $\int_{C_2} \nabla\eta \cdot \boldsymbol{v}\; dx = \int_{\partial C_2} \eta\boldsymbol{v} \cdot \boldsymbol{n}\; dS = 0$ (where $\boldsymbol{n}$ is the outer normal vector to ∂C_2) and $\nabla\eta$ has a compact support in $C_2 - \overline{C_1}$, $\boldsymbol{V}(.\,,t)$ also has a compact support in $C_2 - \overline{C_1}$. Moreover, it follows from the smoothness of the function $\nabla\eta\cdot\boldsymbol{v}$ (which is a consequence of item 2 of Lemma 1.1 and the smoothness of $\boldsymbol{v}$ on $\mathrm{supp}\,(\nabla\eta) \times [t_0 - \tau, t_0]$ – see item 4 of Lemma 1.1) that all space derivatives of $\boldsymbol{V}$ are bounded on $\overline{C_2} \times [t_0 - \tau, t_0]$.

It can be verified that $\boldsymbol{u}$ $(= \eta\boldsymbol{v} - \boldsymbol{V})$ satisfies in a strong sense the equations

$$\frac{\partial \boldsymbol{u}}{\partial t} + (\boldsymbol{u}\cdot\nabla)\boldsymbol{u} = \boldsymbol{h} - \nabla(\eta\, p) + \nu\,\Delta\boldsymbol{u} \tag{1.7}$$

$$\operatorname{div}\boldsymbol{u} = 0 \tag{1.8}$$

where

$$\begin{aligned}\boldsymbol{h} = & -\frac{\partial \boldsymbol{V}}{\partial t} - \boldsymbol{V}\cdot\nabla(\eta\,\boldsymbol{v}) - (\eta\,\boldsymbol{v})\cdot\nabla\boldsymbol{V} + \boldsymbol{V}\cdot\nabla\boldsymbol{V} + (\eta\,\boldsymbol{v}\cdot\nabla\eta)\,\boldsymbol{v} \\ & - \eta\,(1-\eta)\,\boldsymbol{v}\cdot\nabla\boldsymbol{v} - 2\nu\,\nabla\eta\cdot\nabla\boldsymbol{v} - \nu\,\boldsymbol{v}\,\Delta\eta + \nu\,\Delta\boldsymbol{V} + p\,\nabla\eta.\end{aligned}$$

$\boldsymbol{u}$ satisfies the initial and boundary conditions

$$\boldsymbol{u}(.\,,t_0-\tau) = \eta\,\boldsymbol{v}(.\,,t_0-\tau) - \boldsymbol{V}(.\,,t_0-\tau), \tag{1.9}$$

$$\boldsymbol{u} = \boldsymbol{0} \qquad \text{on } \partial C_2 \times\,]t_0-\tau, t_0[\tag{1.10}$$

Moreover, since $\eta\,\boldsymbol{v}(.\,,t)$ and $\boldsymbol{V}(.\,,t)$ have a compact support in C_2 for all $t \in\,]t_0 - \tau, t_0[$, $\boldsymbol{u}$ has all derivatives equal to zero on $\partial C_2\times\,]t_0 - \tau, t_0[$.

If we also use item 5 of Lemma 1.1 and take into account that $\mathrm{supp}\,\nabla\eta \subset C_2 - \overline{C_1}$, we can observe that $\nabla\eta \cdot \partial\boldsymbol{v}/\partial t$ has all space derivatives in $L^\alpha(t_0 - \tau, t_0;\, L^\infty(C_2))$ for each α between 1 and 2. Thus, $\partial\boldsymbol{V}/\partial t$ has all space derivatives in $L^\alpha(t_0 - \tau, t_0;\, L^\infty(C_2)^3)$, too. The same holds about $p\,\nabla\eta$. All other terms in function $\boldsymbol{h}$ have all space derivatives bounded in $C_2 \times [t_0 - \tau, t_0]$. We can therefore conclude that function $\boldsymbol{h}$ has all its space derivatives in $L^\alpha(t_0 - \tau, t_0;\, L^\infty(C_2)^3)$ for every $\alpha \in\,]1, 2[$. Moreover, $\boldsymbol{h}$ has a compact support in $\left(C_2 - \overline{C_1}\right) \times [t_0 - \tau, t_0]$.

The components of $\boldsymbol{u}$ will be denoted by u_1, u_2 and u_3. Partial derivatives of u_i with respect to x_j will be denoted by $u_{i,j}$. All these partial derivatives (for $i, j = 1, 2, 3$) belong to $L^{2,2}$. $\boldsymbol{\omega} = (\omega_1, \omega_2, \omega_3)$ will denote $\operatorname{curl}\boldsymbol{u}$.

2. Regularity in dependence on eigenvalues of the rate of deformation tensor

Suppose that $t \in\,]t_0 - \tau, t_0[$. Multiplying equation (1.7) by $\Delta\boldsymbol{u}$ and integrating on C_2, we obtain

$$\frac{d}{dt}\frac{1}{2}\int_{C_2} |\nabla\boldsymbol{u}|^2\, dx + \nu\int_{C_2} |\Delta\boldsymbol{u}|^2\, dx = -\int_{C_2} h\cdot\Delta\boldsymbol{u}\, dx \tag{2.1}$$

$$+\int_{C_2} (\boldsymbol{u}\cdot\nabla)\boldsymbol{u}\cdot\Delta\boldsymbol{u}\, dx = -\int_{C_2} \boldsymbol{h}\cdot\Delta\boldsymbol{u}\, dx - \int_{C_2} u_{i,j}\, u_{i,k}\, u_{j,k}\, dx.$$

Let us denote $\sigma_{ij} = \frac{1}{2}(u_{i,j} + u_{j,i})$ and $\tau_{ij} = \frac{1}{2}(u_{i,j} - u_{j,i})$ $(i,j = 1,2,3)$. The last integral on the right-hand side of (2.1) can be written in the form

$$-\int_{C_2} u_{i,j}\, u_{i,k}\, u_{j,k}\, dx = -\int_{C_2} \sigma_{ij}\, u_{i,k}\, u_{j,k}\, dx$$

$$= -\int_{C_2} \sigma_{ij}\,(\sigma_{ik} + \tau_{ik})\,(\sigma_{jk} + \tau_{jk})\, dx = -\int_{C_2} \sigma_{ij}\,\sigma_{ik}\,\sigma_{jk}\, dx$$

$$-\int_{C_2} \sigma_{ij}\,\tau_{ik}\,\tau_{jk}\, dx = -\int_{C_2} \sigma_{ij}\,\sigma_{ik}\,\sigma_{jk}\, dx + \frac{1}{4}\int_{C_2} \sigma_{ij}\,\omega_i\,\omega_j\, dx.$$

(We have used the symmetry of (σ_{ij}) and the skew-symmetry of (τ_{ij}) which implies that $\sigma_{ij}\,\sigma_{ik}\,\tau_{jk} = -\sigma_{ij}\,\sigma_{ik}\,\tau_{kj} = -\sigma_{ik}\,\sigma_{ij}\,\tau_{jk} = -\sigma_{ij}\,\sigma_{ik}\,\tau_{jk}$. Hence $\sigma_{ij}\,\sigma_{ik}\,\tau_{jk} = 0$. Similarly, $\sigma_{ij}\,\tau_{ik}\,\sigma_{jk} = 0$.) Thus

$$\frac{d}{dt}\frac{1}{2}\int_{C_2} |\nabla \boldsymbol{u}|^2\, dx + \nu \int_{C_2} |\Delta \boldsymbol{u}|^2\, dx \tag{2.2}$$

$$= -\int_{C_2} \boldsymbol{h}\cdot\Delta\boldsymbol{u}\, dx - \int_{C_2} \sigma_{ij}\,\sigma_{ik}\,\sigma_{jk}\, dx + \frac{1}{4}\int_{C_2} \sigma_{ij}\,\omega_i\,\omega_j\, dx.$$

Applying operator curl to equation (1.7), multiplying the equation by $\boldsymbol{\omega}$ and integrating on C_2, we obtain

$$\frac{d}{dt}\frac{1}{2}\int_{C_2} |\boldsymbol{\omega}|^2\, dx + \nu \int_{C_2} |\nabla\boldsymbol{\omega}|^2\, dx = \int_{C_2} \sigma_{ij}\,\omega_i\,\omega_j\, dx + \int_{C_2} \operatorname{curl}\boldsymbol{h}\cdot\boldsymbol{\omega}\, dx. \tag{2.3}$$

Since

$$\int_{C_2} |\boldsymbol{\omega}|^2\, dx = \int_{C_2} |\nabla\boldsymbol{u}|^2\, dx \qquad \text{and} \qquad \int_{C_2} |\nabla\boldsymbol{\omega}|^2\, dx = \int_{C_2} |\Delta\boldsymbol{u}|^2\, dx,$$

(2.3) implies

$$\frac{d}{dt}\frac{1}{2}\int_{C_2} |\nabla\boldsymbol{u}|^2\, dx + \nu \int_{C_2} |\Delta\boldsymbol{u}|^2\, dx = \int_{C_2} \sigma_{ij}\,\omega_i\,\omega_j\, dx + \int_{C_2} \operatorname{curl}\boldsymbol{h}\cdot\boldsymbol{\omega}\, dx. \tag{2.4}$$

Multiplying equation (2.4) by $-\frac{1}{4}$ and summing with equation (2.2), we obtain

$$\frac{d}{dt}\frac{3}{8}\int_{C_2} |\nabla\boldsymbol{u}|^2\, dx + \frac{3\nu}{4}\int_{C_2} |\Delta\boldsymbol{u}|^2\, dx \tag{2.5}$$

$$= -\int_{C_2} \sigma_{ij}\,\sigma_{ik}\,\sigma_{jk}\, dx - \int_{C_2} h\cdot\Delta\boldsymbol{u}\, dx - \frac{1}{4}\int_{C_2} \operatorname{curl}\boldsymbol{h}\cdot\boldsymbol{\omega}\, dx$$

$$= -\int_{C_2} \sigma_{ij}\,\sigma_{ik}\,\sigma_{jk}\, dx - \frac{5}{4}\int_{C_2} \boldsymbol{h}\cdot\Delta\boldsymbol{u}\, dx\,.$$

Assume that x is a given point in C_2 for a while. Then the system of coordinates can be chosen so that the tensor (σ_{ij}) has a diagonal representation at point x:

$$(\sigma_{ij})(x,t) = \begin{pmatrix} \lambda_1 & 0 & 0 \\ 0 & \lambda_2 & 0 \\ 0 & 0 & \lambda_3 \end{pmatrix}$$

where λ_1, λ_2, λ_3 are the eigenvalues of $(\sigma_{ij})(x,t)$. We can suppose without loss of generality that

$$\lambda_1 \le \lambda_2 \le \lambda_3 . \tag{2.6}$$

Let us denote by $\mathbf{e}^1$, $\mathbf{e}^2$, $\mathbf{e}^3$ the corresponding eigenvectors. The eigenvectors are (or can be chosen to be) orthogonal because (σ_{ij}) is symmetric. The equation of continuity (1.8) implies that

$$\lambda_1 + \lambda_2 + \lambda_3 = 0. \tag{2.7}$$

Then

$$(\sigma_{ij}\,\sigma_{ik}\,\sigma_{jk})(x,t) = \lambda_1^3 + \lambda_2^3 + \lambda_3^3 = 3\,\lambda_1\,\lambda_2\,\lambda_3 . \tag{2.8}$$

The product $\lambda_1\,\lambda_2\,\lambda_3$ is an invariant of the tensor $(\sigma_{ij})(x,t)$ and so it is independent of the choice of the system of coordinates. Hence (2.8) holds in all points $x \in C_2$ and we have

$$\begin{aligned} \frac{d}{dt}\frac{3}{8}\int_{C_2} |\nabla \boldsymbol{u}|^2\,dx + \frac{3\nu}{4}\int_{C_2} |\Delta \boldsymbol{u}|^2\,dx &= -3\int_{C_2} \lambda_1\,\lambda_2\,\lambda_3\,dx \\ -\frac{5}{4}\int_{C_2} \boldsymbol{h}\cdot\Delta\boldsymbol{u}\,dx &\le -3\int_{C_2} \lambda_1\,\lambda_2\,\lambda_3\,dx + c_1\,\|\boldsymbol{h}(\,.\,,t)\|_1 . \end{aligned} \tag{2.9}$$

($\boldsymbol{h}$ has the support in $(\overline{C}_2 - C_1)\times\,]t_0-\tau, t_0[$ where $\boldsymbol{v}$ and $\boldsymbol{V}$ have all their space derivatives bounded. Since $\boldsymbol{u} = \eta\boldsymbol{v} - \boldsymbol{V}$, $\boldsymbol{u}$ has all its space derivatives bounded in $(\overline{C}_2 - C_1)\times\,]t_0-\tau, t_0[$, too. Hence the integral of $\boldsymbol{h}\,\Delta\boldsymbol{u}$ can be estimated by $c_1\,\|\boldsymbol{h}(\,.\,,t)\|_1$.) Note that an analogous equality as the first part of (2.9) was already obtained by R. Betchow in [1], p. 502. The eigenvalues λ_1, λ_2, λ_3 are functions of x and t and the ordering (2.6) is supposed to be valid a.e. in $C_2\times\,]t_0-\tau, t_0[$. Thus, naturally, the corresponding eigenvectors can have jumps in those points (x,t) where at least two of the eigenvalues coincide and this can happen even in the situation when $\boldsymbol{u}$ is smooth.

We can now formulate the theorem. However, we wish to formulate it in terms of solution $\boldsymbol{v}$ and not $\boldsymbol{u}$ $(= \eta\boldsymbol{v} - \boldsymbol{V})$. So we will speak about eigenvalues ζ_1, ζ_2 and ζ_3 of the symmetric tensor $\frac{1}{2}\,(v_{i,j} + v_{j,i})$ (so-called "rate of deformation tensor") and not about eigenvalues λ_1, λ_2, λ_3 of the tensor $(\sigma_{ij}) = \frac{1}{2}\,(u_{i,j} + u_{j,i})$.

Theorem 2.1. *Suppose that D is an open sub-domain of Q_T, $(\boldsymbol{v};p)$ is a suitable weak solution of the problem (1.1)–(1.4), $\zeta_1 \le \zeta_2 \le \zeta_3$ are the eigenvalues of the tensor $\frac{1}{2}\,(v_{i,j} + v_{j,i})$ in D and*

(i) *one of the functions ζ_1, $(\zeta_2)_+$, ζ_3 belongs to $L^{r,s}_{loc}(D)$ for some real numbers r, s such that $1 \le r \le +\infty$, $\frac{3}{2} < s \le +\infty$ and $2/r + 3/s \le 2$.*

($(\zeta_2)_+$ denotes the positive part of ζ_2.) Then the solution $(\boldsymbol{v};p)$ is regular in D.

Proof. Suppose that $(\zeta_2)_+ \in L^{r,s}(D)$ and $r > 1$. The cases $\zeta_1 \in L^{r,s}(D)$ or $\zeta_3 \in L^{r,s}(D)$ or $r = 1$ could be treated analogously.

Eigenvalues ζ_1, ζ_2 and ζ_3 satisfy, due to the equation of continuity and analogously to (2.7), the equation $\zeta_1 + \zeta_2 + \zeta_3 = 0$. Thus, since $\zeta_1 \leq \zeta_2 \leq \zeta_3$, ζ_1 is non-positive and ζ_3 is non-negative. The eigenvalues λ_1, λ_2, λ_3 of (σ_{ij}) coincide with the eigenvalues ζ_1, ζ_2, ζ_3 of $\left(\frac{1}{2}\,(v_{i,j}+v_{j,i})\right)$ on C_1 and λ_1, λ_2, λ_3 are bounded (due to the smoothness of $\boldsymbol{v}$) on $(C_2 - C_1)\times\,]t_0-\tau, t_0[$. It follows from (2.9) that

$$\frac{d}{dt}\frac{3}{8}\int_{C_2}|\nabla \boldsymbol{u}|^2\,dx + \frac{3\nu}{4}\int_{C_2}|\Delta \boldsymbol{u}|^2\,dx \leq 3\int_{C_1}(-\zeta_1)\zeta_2\zeta_3\,dx + c_2 \tag{2.10}$$

$$+\,c_1\,\|\boldsymbol{h}(.\,,t)\|_1 \leq 3\int_{C_1}(-\zeta_1)(\zeta_2)_+\zeta_3\,dx + c_2 + c_1\,\|\boldsymbol{h}(.\,,t)\|_1\,.$$

Integrating now with respect to time on the interval $]t_0, t_0-\tau[$, we obtain:

$$\begin{aligned}\|\!|\nabla \boldsymbol{u}|\!\|^2_{(\infty,2)\cap(2,6)} &\leq c_3\int_{t_0-\tau}^{t_0}\int_{C_1}(-\zeta_1)(\zeta_2)_+\zeta_3\,dx\,dt + c_4\\ &\leq c_3\,\|\!|(\zeta_2)_+|\!\|_{r,s}\,\|\!|(-\zeta_1)\zeta_3|\!\|_{r/(r-1),s/(s-1)} + c_4\,.\end{aligned}$$

The eigenvalues ζ_1 and ζ_3 obviously satisfy the inequality $(-\zeta_1)\zeta_3 \leq c_5\,|\nabla \boldsymbol{v}|^2 = c_5\,|\nabla \boldsymbol{u}|^2$ with some positive constant c_5 a.e. in $C_1\times\,]t_0-\tau, t_0[$. Then

$$\|\!|\nabla \boldsymbol{u}|\!\|^2_{(\infty,2)\cap(2,6)} \leq c_3\,c_5\,\|\!|(\zeta_2)_+|\!\|_{r,s}\,\|\!|\nabla \boldsymbol{u}|\!\|^2_{2r/(r-1),2s/(s-1)} + c_4\,.$$

Since

$$2\,\frac{r-1}{2r} + 3\,\frac{s-1}{2s} = \frac{5}{2} - \frac{1}{2}\Big(\frac{2}{r}+\frac{3}{s}\Big),$$

Lemma 1.2 gives:

$$\begin{aligned}\|\!|\nabla \boldsymbol{u}|\!\|^2_{(\infty,2)\cap(2,6)} &\leq c_3\,c_5\,\|\!|(\zeta_2)_+|\!\|_{r,s}\,\|\!|\nabla \boldsymbol{u}|\!\|^{2-(2/r+3/s)}_{2,2}\,\|\!|\nabla \boldsymbol{u}|\!\|^{2/r+3/s}_{(\infty,2)\cap(2,6)} + c_4\\ &\leq c_6\,\|\!|(\zeta_2)_+|\!\|_{r,s}\,\|\!|\nabla \boldsymbol{u}|\!\|^{2/r+3/s}_{(\infty,2)\cap(2,6)} + c_4\,.\end{aligned} \tag{2.11}$$

If we choose τ sufficiently small (i.e., the interval $]t_0-\tau, t_0[$ sufficiently short), we can achieve $c_6\,\|\!|(\zeta_2)_+|\!\|_{r,s}$ to be less than 1. Then (2.11) implies that $\|\!|\nabla \boldsymbol{u}|\!\|_{(\infty,2)\cap(2,6)} < +\infty$. Now it can be easily verified that

$$\lim_{\delta\to 0+}\ \sup\ \frac{1}{\delta}\int_{t_0-\delta^2}^{t_0}\int_{|x-x_0|<\delta}|\nabla \boldsymbol{u}|^2\,dx\,dt = 0. \tag{2.12}$$

Moreover, there exists a locally in time strong solution to the problem (1.7), (1.8), (1.10) which coincides with $\boldsymbol{u}$ at a certain instant of time $t' \leq t_0$ and which can be also identified with $\boldsymbol{u}$ (due to the known results about uniqueness) on a time interval $]t', t_0+\tau'[$ (for some $\tau' > 0$). Hence $\boldsymbol{u}$ satisfies the condition

$$\lim_{\delta\to 0+}\ \sup\ \frac{1}{\delta}\int_{t_0}^{t_0+\delta^2/8}\int_{|x-x_0|<\delta}|\nabla \boldsymbol{u}|^2\,dx\,dt = 0, \tag{2.13}$$

too. Thus, $\boldsymbol{v}$ also satisfies both conditions. Conditions (2.12) and (2.13) are known to be sufficient for the regularity of the solution $(\boldsymbol{v}, p)$ at the point (x_0, t_0) (see, e.g., L. Caffarelli, R. Kohn and L. Nirenberg [3], p. 776). Consequently, this means that $(\boldsymbol{v}; p)$ can have no singular point in D. □

It follows from the proof of Theorem 1 that the case $\zeta_2 \leq 0$ supports regularity, while the case $\zeta_2 > 0$ supports a hypothetical singularity of the solution $(\boldsymbol{v}; p)$. Since the sign of ζ_2 is connected with the types of deformation of "infinitely small" volumes of the fluid, we can also interpret our result in this way: *Deformations where the "infinitely small" volumes of the fluid are compressed in two dimensions and stretched in one dimension support regularity, while the cases when the "infinitely small" volumes of the fluid are compressed in one dimension and stretched in two dimensions support the hypothetical "blow up".*

The proof of Theorem 2.1 is based on the control of the first integral on the right-hand side of (2.10) which obviously equals negative integral of the determinant of the matrix $\left(\frac{1}{2}\,(v_{i,j} + v_{j,i})\right)$. Although various numerical experiments indicate a possibility to estimate this integral without additional assumptions, this observation has not been rigorously confirmed yet. This is connected with a deep geometric question whether the local deformations of the two types mentioned above are always in a balance so that their total contribution over some part of the flow field enables to derive a sufficient estimate of the discussed integral.

It is also seen from (2.3) that the integral $\int_{C_2} \sigma_{ij}\, \omega_i\, \omega_j\, dx$ has the same importance. D. Chae and H.J. Choe [4] have shown that an appropriate additional information on only two components of vorticity enables to control this integral. There arises a challenging question whether an information on only one component of vorticity (analogously to the one-velocity-regularity-criterion, see J. Neustupa, A. Novotny and P. Penel [9] or J. Neustupa and P. Penel [10]) can play the same role or whether we should look for another description of the vorticity dynamics, possibly in a different coordinate system or in terms of quantities which is more naturally connected with the fluid dynamics.

3. Regularity in dependence on eigenvectors of the rate of deformation tensor

Suppose that $\boldsymbol{e}^i = (e_1^i, e_2^i, e_3^i)$ $(i = 1, 2, 3)$ are the eigenvectors of the rate of deformation tensor $\frac{1}{2}(v_{i,j} + v_{j,i})$ in D such that $\boldsymbol{e}^i \cdot \boldsymbol{e}^j = \delta_{ij}$ $(i, j = 1, 2, 3)$ for a.a. $(x, t) \in D$. The denotation of the eigenvectors is independent of the ordering of the associated eigenvalues in this section. Suppose further that

(ii) $\boldsymbol{e}^i$ $(i = 1, 2, 3)$ *are continuous in* D *and their* 1st*-order derivatives with respect to the space variables* x_1, x_2, x_3 *are essentially bounded in* D.

Then

$$(v_{k,l} + v_{l,k})\, e_k^i\, e_l^j = 0 \qquad (i, j = 1, 2, 3;\ i \neq j) \tag{3.1}$$

in a.a. points $(x, t) \in D$. As in Section 1, we assume that D' is a sub-domain of D such that $D' \subset \overline{D'} \subset D \subset Q_T$, t_0 is a D'-epoch of irregularity and (x_0, t_0) is a singular point of $(\boldsymbol{v}; p)$ in D'. We can assume without loss of generality that

$$\boldsymbol{e}^1(x_0, t_0) = (1, 0, 0), \quad \boldsymbol{e}^2(x_0, t_0) = (0, 1, 0), \quad \boldsymbol{e}^3(x_0, t_0) = (0, 0, 1). \tag{3.2}$$

Function $\boldsymbol{u}$ $(= \eta \boldsymbol{v} - \boldsymbol{V})$ coincides with $\boldsymbol{v}$ in C_1, hence

$$(u_{k,l} + u_{l,k})\, e_k^i\, e_l^j \;=\; 0 \qquad (i,j = 1,2,3;\; i \neq j) \tag{3.3}$$

in $C_1 \times\,]t_0 - \tau, t_0[$. System (3.3) can be written in the form

$$A \cdot \begin{pmatrix} u_{1,2} + u_{2,1} \\ u_{1,3} + u_{3,1} \\ u_{2,3} + u_{3,2} \end{pmatrix} = -2u_{1,1} \begin{pmatrix} e_1^1\, e_1^2 \\ e_1^1\, e_1^3 \\ e_1^2\, e_1^3 \end{pmatrix} - 2u_{2,2} \begin{pmatrix} e_2^1\, e_2^2 \\ e_2^1\, e_2^3 \\ e_2^2\, e_2^3 \end{pmatrix} - 2u_{3,3} \begin{pmatrix} e_3^1\, e_3^2 \\ e_3^1\, e_3^3 \\ e_3^2\, e_3^3 \end{pmatrix}$$

where

$$A = \begin{pmatrix} e_1^1\, e_2^2 + e_2^1\, e_1^2, & e_1^1\, e_3^2 + e_3^1\, e_1^2, & e_2^1\, e_3^2 + e_3^1\, e_2^2 \\ e_1^1\, e_2^3 + e_2^1\, e_1^3, & e_1^1\, e_3^3 + e_3^1\, e_1^3, & e_2^1\, e_3^3 + e_3^1\, e_2^3 \\ e_1^2\, e_2^3 + e_2^2\, e_1^3, & e_1^2\, e_3^3 + e_3^2\, e_1^3, & e_2^2\, e_3^3 + e_3^2\, e_2^3 \end{pmatrix}.$$

Obviously, $A(x_0, t_0)$ is the 3×3 unit matrix and due to the continuity of the eigenvectors $\mathbf{e}^i$ $(i = 1,2,3)$, A is regular in $C_1 \times\,]t_0 - \tau, t_0[$ if r_1 and τ are sufficiently small. Then

$$u_{1,2} + u_{2,1} \;=\; a_k^{1,2}\, u_{k,k} \;=\; a_1^{1,2}\, u_{1,1} + a_2^{1,2}\, u_{2,2} + a_3^{1,2}\, u_{3,3}, \tag{3.4}$$

$$u_{1,3} + u_{3,1} \;=\; a_k^{1,3}\, u_{k,k} \;=\; a_1^{1,3}\, u_{1,1} + a_2^{1,3}\, u_{2,2} + a_3^{1,3}\, u_{3,3}, \tag{3.5}$$

$$u_{2,3} + u_{3,2} \;=\; a_k^{2,3}\, u_{k,k} \;=\; a_1^{2,3}\, u_{1,1} + a_2^{2,3}\, u_{2,2} + a_3^{2,3}\, u_{3,3} \tag{3.6}$$

where

$$\begin{pmatrix} a_i^{1,2} \\ a_i^{1,3} \\ a_i^{2,3} \end{pmatrix} = -2A^{-1} \cdot \begin{pmatrix} e_i^1\, e_i^2 \\ e_i^1\, e_i^3 \\ e_i^2\, e_i^3 \end{pmatrix} \qquad (i = 1,2,3).$$

(We do not sum over i on the right-hand side.)

By saying that r_1 is sufficiently small we mean that r_1 takes the value of r_1^n for n sufficiently large. (The sequence $\{r_1^n\}$ was discussed in Section 1.)

The continuity of the eigenvectors $\mathbf{e}^1$, $\mathbf{e}^2$, $\mathbf{e}^3$ and (3.2) also imply that

$$\sum_{k=1}^{3} \Big(|a_k^{1,2}| + |a_k^{1,3}| + |a_k^{2,3}| \Big) \;\leq\; \epsilon(r_1^n, \tau^n) \tag{3.7}$$

in $C_1 \times\,]t_0 - \tau^n, t_0[$, where $\epsilon(r_1^n, \tau^n) \to 0$ as $n \to +\infty$ (i.e., as $r_1^n \to 0+$ and $\tau^n \to 0+$).

Theorem 3.1. *Suppose that D is an open sub-domain of Q_T, $(\boldsymbol{v}; p)$ is a suitable weak solution of the problem (1.1)–(1.4), $\mathbf{e}^i = (e_1^i, e_2^i, e_3^i)$ $(i = 1,2,3)$ are the eigenvectors of the rate of deformation tensor $\frac{1}{2}(v_{i,j} + v_{j,i})$ in D such that $\mathbf{e}^i \cdot \mathbf{e}^j = \delta_{ij}$ $(i,j = 1,2,3)$ for a.a. $(x,t) \in D$ and $\mathbf{e}^i$ $(i = 1,2,3)$ satisfy condition (ii). Then the solution $(\boldsymbol{v}; p)$ is regular in D.*

Proof. Differentiating the equation of continuity (1.8) with respect to x_1 and substituting for $u_{2,21}$ and $u_{3,31}$ from (3.4) and (3.5), we obtain the spatial wave equation

$$u_{1,11} - u_{1,22} - u_{1,33} \;=\; (a_k^{1,2}\, u_{k,k})_{,2} + (a_k^{1,3}\, u_{k,k})_{,3}\,. \tag{3.8}$$

This equation, as well as analogous equations which can be obtained by differentiating equation (1.8) with respect to x_2 and x_3, plays a fundamental role in this proof. Standard estimates from the theory of the wave equation, together with delicate and laborious manipulation with the terms on the right-hand side of equation (3.8), will lead to estimate (3.16) which will further imply the validity of the regularity criteria (2.12) and (2.13).

Let us denote by $C_1^{2,3}$ the square $]x_{02}-r_1, x_{02}+r_1[\times]x_{03}-r_1, x_{03}+r_1[$. Multiplying equation (3.8) by $u_{1,1}$ and integrating on $C_1^{2,3}$ with respect to x_2 and x_3, we get:

$$\begin{aligned}
&\frac{\partial}{\partial x_1}\frac{1}{2}\int_{C_1^{2,3}}(u_{1,1}^2+u_{1,2}^2+u_{1,3}^2)\,dx_2\,dx_3\\
&\quad=\int_{x_{03}-r_1}^{x_{03}+r_1}\Big[u_{1,2}\,u_{1,1}\Big]_{x_2=x_{02}-r_1}^{x_2=x_{02}+r_1}dx_3+\int_{x_{02}-r_1}^{x_{02}+r_1}\Big[u_{1,3}\,u_{1,1}\Big]_{x_3=x_{03}-r_1}^{x_3=x_{03}+r_1}dx_2\\
&\quad=\int_{C_1^{2,3}}\Big[(a_k^{1,2}\,u_{k,k})_{,2}+(a_k^{1,3}\,u_{k,k})_{,3}\Big]\,u_{1,1}\,dx_2\,dx_3\,.
\end{aligned}$$

In order to simplify the notation, we shall denote all the boundary integrals together by (BI). (BI) will change its value as we shall repeatedly use the integration by parts and it will absorb more and more boundary terms. It can generally depend on r_1. However, as a consequence of Lemma 1.1, for each r_1 fixed, it will be a bounded function of x_1 and t on $]x_{01}-r_1, x_{01}+r_1[\times]t_0-\tau, t_0[$. We shall also further omit writing $dx_2\,dx_3$ behind the integrals on $C_1^{2,3}$ and we shall write only ∂_1 instead of $\partial/\partial x_1$. Thus,

$$\begin{aligned}
\partial_1\frac{1}{2}\int_{C_1^{2,3}}(u_{1,1}^2+u_{1,2}^2+u_{1,3}^2)&=\int_{C_1^{2,3}}(a_{k,2}^{1,2}+a_{k,3}^{1,3})\,u_{k,k}\,u_{1,1}\\
&+\int_{C_1^{2,3}}\Big[(a_1^{1,2}-a_3^{1,2})\,u_{1,12}\,u_{1,1}+(a_2^{1,2}-a_3^{1,2})\,u_{2,22}\,u_{1,1}\Big]\\
&+\int_{C_1^{2,3}}\Big[(a_1^{1,3}-a_2^{1,3})\,u_{1,13}\,u_{1,1}+(a_3^{1,3}-a_2^{1,3})\,u_{3,33}\,u_{1,1}\Big]+(BI),
\end{aligned}$$

$$\begin{aligned}
\partial_1\frac{1}{2}\int_{C_1^{2,3}}(u_{1,1}^2+u_{1,2}^2+u_{1,3}^2)&=\int_{C_1^{2,3}}(a_{k,2}^{1,2}+a_{k,3}^{1,3})\,u_{k,k}\,u_{1,1}\\
&-\frac{1}{2}\int_{C_1^{2,3}}(a_{1,2}^{1,2}-a_{3,2}^{1,2}+a_{1,3}^{1,3}-a_{2,3}^{1,3})\,u_{1,1}^2-\int_{C_1^{2,3}}(a_{2,2}^{1,2}-a_{3,2}^{1,2})\,u_{2,2}\,u_{1,1}\\
&-\int_{C_1^{2,3}}(a_2^{1,2}-a_3^{1,2})\,u_{2,2}\,u_{1,12}-\int_{C_1^{2,3}}(a_{3,3}^{1,3}-a_{2,3}^{1,3})\,u_{3,3}\,u_{1,1}\\
&-\int_{C_1^{2,3}}(a_3^{1,3}-a_2^{1,3})\,u_{3,3}\,u_{1,13}+(BI)
\end{aligned}$$

$$= \frac{1}{2}\int_{C_1^{2,3}} (a^{1,2}_{1,2} + a^{1,2}_{3,2} + a^{1,3}_{1,3} + a^{1,3}_{2,3})\, u_{1,1}^2$$
$$+ \int_{C_1^{2,3}} \Big[(a^{1,3}_{2,3} + a^{1,2}_{3,2})\, u_{2,2} + (a^{1,2}_{3,2} + a^{1,3}_{2,3})\, u_{3,3}\Big]\, u_{1,1}$$
$$- \int_{C_1^{2,3}} (a^{1,2}_2 - a^{1,2}_3)\, u_{2,2}\, u_{1,12} - \int_{C_1^{2,3}} (a^{1,3}_3 - a^{1,3}_2)\, u_{3,3}\, u_{1,13} + (BI).$$

Thus, we have

$$\partial_1 \frac{1}{2}\int_{C_1^{2,3}} |\nabla u_1|^2 + \int_{C_1^{2,3}} (a^{1,2}_2 - a^{1,2}_3)\, u_{2,2}\, u_{1,12} \tag{3.9}$$
$$+ \int_{C_1^{2,3}} (a^{1,3}_3 - a^{1,3}_2)\, u_{3,3}\, u_{1,13} = I_1(\nabla u_1, \nabla u_2, \nabla u_3) + (BI)$$

where

$$|I_1(\nabla u_1, \nabla u_2, \nabla u_3)| \le \Big(c_7 + \frac{c_8}{\xi}\Big)\int_{C_1^{2,3}} |\nabla u_1|^2 + c_9\,\xi \int_{C_1^{2,3}} \Big[|\nabla u_2|^2 + |\nabla u_3|^2\Big].$$

ξ is an arbitrarily small number and constants c_7–c_9 do not depend on ξ, r_1 and τ if ξ, r_1 and τ are sufficiently small. Let us now deal with the integral of $b_2\, u_{2,2}\, u_{1,12}$ in order to eliminate the second integral on the left-hand side of (3.9).

$$\int_{C_1^{2,3}} b_2\, u_{2,2}\, u_{1,12} = \partial_1 \int_{C_1^{2,3}} b_2\, u_{2,2}\, u_{1,2} - \int_{C_1^{2,3}} b_{2,1}\, u_{2,2}\, u_{1,2} - \int_{C_1^{2,3}} b_2\, u_{2,21}\, u_{1,2}$$
$$= \partial_1 \int_{C_1^{2,3}} b_2\, u_{2,2}\, u_{1,2} - \int_{C_1^{2,3}} b_{2,1}\, u_{2,2}\, u_{1,2} + \int_{C_1^{2,3}} b_2\, u_{1,22}\, u_{1,2}$$
$$- \int_{C_1^{2,3}} b_2\, (a^{1,2}_k\, u_{k,k})_{,2}\, u_{1,2}$$
$$= \partial_1 \int_{C_1^{2,3}} b_2\, u_{2,2}\, u_{1,2} - \int_{C_1^{2,3}} b_{2,1}\, u_{2,2}\, u_{1,2} - \frac{1}{2}\int_{C_1^{2,3}} b_{2,2}\, u_{1,2}^2$$
$$- \int_{C_1^{2,3}} b_2\, (a^{1,2}_{1,2} - a^{1,2}_{3,2})\, u_{1,1}\, u_{1,2} - \int_{C_1^{2,3}} b_2\, (a^{1,2}_{2,2} - a^{1,2}_{3,2})\, u_{2,2}\, u_{1,2}$$
$$- \int_{C_1^{2,3}} b_2\, (a^{1,2}_1 - a^{1,2}_3)\, u_{1,12}\, u_{1,2} - \int_{C_1^{2,3}} b_2\, (a^{1,2}_2 - a^{1,2}_3)\, u_{2,22}\, u_{1,2} + (BI)$$
$$= \partial_1 \int_{C_1^{2,3}} \Big[b_2\, u_{2,2}\, u_{1,2} - \tfrac{1}{2}\, b_2\, (a^{1,2}_1 - a^{1,2}_3)\, u_{1,2}^2\Big] + \frac{1}{2}\int_{C_1^{2,3}} (b_2\, (a^{1,2}_1 - a^{1,2}_3))_{,1}\, u_{1,2}^2$$
$$- \int_{C_1^{2,3}} b_{2,1}\, u_{2,2}\, u_{1,2} - \frac{1}{2}\int_{C_1^{2,3}} b_{2,2}\, u_{1,2}^2 - \int_{C_1^{2,3}} b_2\, (a^{1,2}_{1,2} - a^{1,2}_{3,2})\, u_{1,1}\, u_{1,2}$$

$$
\begin{aligned}
& - \int_{C_1^{2,3}} b_2\,(a^{1,2}_{2,2} - a^{1,2}_{3,2})\, u_{2,2}\, u_{1,2} + \int_{C_1^{2,3}} (b_2\,(a^{1,2}_2 - a^{1,2}_3))_{,2}\, u_{2,2}\, u_{1,2} \\
& + \int_{C_1^{2,3}} b_2\,(a^{1,2}_2 - a^{1,2}_3)\, u_{2,2}\, u_{1,22} + (BI) \\
= {} & \partial_1 \int_{C_1^{2,3}} \Big[b_2\, u_{2,2}\, u_{1,2} - \tfrac{1}{2}\, b_2\,(a^{1,2}_1 - a^{1,2}_3)\, u_{1,2} \Big] + \frac{1}{2} \int_{C_1^{2,3}} (b_2\,(a^{1,2}_1 - a^{1,2}_3))_{,1}\, u^2_{1,2} \\
& - \int_{C_1^{2,3}} b_{2,1}\, u_{2,2}\, u_{1,2} - \frac{1}{2} \int_{C_1^{2,3}} b_{2,2}\, u^2_{1,2} - \int_{C_1^{2,3}} b_2\,(a^{1,2}_{1,2} - a^{1,2}_{3,2})\, u_{1,1}\, u_{1,2} \\
& - \int_{C_1^{2,3}} b_2\,(a^{1,2}_{2,2} - a^{1,2}_{3,2})\, u_{2,2}\, u_{1,2} + \int_{C_1^{2,3}} (b_2\,(a^{1,2}_2 - a^{1,2}_3))_{,2}\, u_{2,2}\, u_{1,2} \\
& - \int_{C_1^{2,3}} b_2\,(a^{1,2}_2 - a^{1,2}_3)\, u_{2,2}\, u_{2,12} + \int_{C_1^{2,3}} b_2\,(a^{1,2}_2 - a^{1,2}_3)\, u_{2,2}\,(a^{1,2}_k\, u_{k,k})_{,2} \\
& + (BI) \\
= {} & \partial_1 \int_{C_1^{2,3}} \Big[b_2\, u_{2,2}\, u_{1,2} - \tfrac{1}{2}\, b_2\,(a^{1,2}_1 - a^{1,2}_3)\, u^2_{1,2} - \tfrac{1}{2}\, b_2\,(a^{1,2}_2 - a^{1,2}_3)\, u^2_{2,2} \Big] \\
& + \frac{1}{2} \int_{C_1^{2,3}} (b_2\,(a^{1,2}_1 - a^{1,2}_3))_{,1}\, u^2_{1,2} - \int_{C_1^{2,3}} b_{2,1}\, u_{2,2}\, u_{1,2} \\
& - \int_{C_1^{2,3}} b_2\,(a^{1,2}_{1,2} - a^{1,2}_{3,2})\, u_{1,1}\, u_{1,2} - \frac{1}{2} \int_{C_1^{2,3}} b_{2,2}\, u^2_{1,2} \\
& - \int_{C_1^{2,3}} b_2\,(a^{1,2}_{2,2} - a^{1,2}_{3,2})\, u_{2,2}\, u_{1,2} + \int_{C_1^{2,3}} (b_2\,(a^{1,2}_2 - a^{1,2}_3))_{,2}\, u_{2,2}\, u_{1,2} \\
& + \frac{1}{2} \int_{C_1^{2,3}} (b_2\,(a^{1,2}_2 - a^{1,2}_3))_{,1}\, u^2_{2,2} + \int_{C_1^{2,3}} b_2\,(a^{1,2}_2 - a^{1,2}_3)\, u_{2,2}\,(a^{1,2}_{1,2} - a^{1,2}_{3,2})\, u_{1,1} \\
& + \int_{C_1^{2,3}} b_2\,(a^{1,2}_2 - a^{1,2}_3)\, u_{2,2}\,(a^{1,2}_{2,2} - a^{1,2}_{3,2})\, u_{2,2} \\
& + \int_{C_1^{2,3}} b_2\,(a^{1,2}_2 - a^{1,2}_3)\, u_{2,2}\,(a^{1,2}_1 - a^{1,2}_3)\, u_{1,12} \\
& + \int_{C_1^{2,3}} b_2\,(a^{1,2}_2 - a^{1,2}_3)^2\, u_{2,2}\, u_{2,22} + (BI).
\end{aligned}
$$

Thus, we have:

$$
\begin{aligned}
& \int_{C_1^{2,3}} b_2\,[1 - (a^{1,2}_2 - a^{1,2}_3)\,(a^{1,2}_1 - a^{1,2}_3)]\, u_{2,2}\, u_{1,12} \qquad (3.10) \\
= {} & \partial_1 \int_{C_1^{2,3}} \Big[b_2\, u_{2,2}\, u_{1,2} - \tfrac{1}{2}\, b_2\,(a^{1,2}_1 - a^{1,2}_3)\, u^2_{1,2} - \tfrac{1}{2}\, b_2\,(a^{1,2}_2 - a^{1,2}_3)\, u^2_{2,2} \Big] \\
& + \frac{1}{2} \int_{C_1^{2,3}} (b_2\,(a^{1,2}_1 - a^{1,2}_3))_{,1}\, u^2_{1,2} - \int_{C_1^{2,3}} b_{2,1}\, u_{2,2}\, u_{1,2} - \frac{1}{2} \int_{C_1^{2,3}} b_{2,2}\, u^2_{1,2}
\end{aligned}
$$

$$- \int_{C_1^{2,3}} b_2 \, (a_{1,2}^{1,2} - a_{3,2}^{1,2}) \, u_{1,1} \, u_{1,2} + \int_{C_1^{2,3}} b_{2,2} \, (a_2^{1,2} - a_3^{1,2}) \, u_{2,2} \, u_{1,2}$$
$$+ \frac{1}{2} \int_{C_1^{2,3}} (b_2 \, (a_2^{1,2} - a_3^{1,2}))_{,1} \, u_{2,2}^2 + \int_{C_1^{2,3}} b_2 \, (a_2^{1,2} - a_3^{1,2}) \, u_{2,2} \, (a_{1,2}^{1,2} - a_{3,2}^{1,2}) \, u_{1,1}$$
$$- \frac{1}{2} \int_{C_1^{2,3}} b_{2,2} \, (a_2^{1,2} - a_3^{1,2})^2 \, u_{2,2}^2 + (BI).$$

If we now choose b_2 so that

$$b_2 \, [1 - (a_2^{1,2} - a_3^{1,2}) \, (a_1^{1,2} - a_3^{1,2})] = a_2^{1,2} - a_3^{1,2}$$

then we can use (3.10) and express the second integral on the left-hand side of (3.9):

$$\int_{C_1^{2,3}} (a_2^{1,2} - a_3^{1,2}) \, u_{2,2} \, u_{1,12} = \partial_1 \int_{C_1^{2,3}} b_2 \, u_{2,2} \, u_{1,2} \tag{3.11}$$
$$- \partial_1 \frac{1}{2} \int_{C_1^{2,3}} \Big[b_2 \, (a_1^{1,2} - a_3^{1,2}) \, u_{1,2}^2 + b_2 \, (a_2^{1,2} - a_3^{1,2}) \, u_{2,2}^2 \Big] + I_2(\nabla u_1, \nabla u_2)$$
$$+ (BI) = \partial_1 \int_{C_1^{2,3}} b_2 \, u_{2,2} \, u_{1,2} + \partial_1 \int_{C_1^{2,3}} \Big[\beta^{1,2} u_{1,2}^2 + \gamma^{1,2} u_{2,2}^2 \Big]$$
$$+ I_2(\nabla u_1, \nabla u_2) + (BI)$$

where

$$|I_2(\nabla u_1, \nabla u_2)| \leq \Big(c_{10} + \frac{c_{11}}{\xi} \Big) \int_{C_1^{2,3}} |\nabla u_1|^2 + [c_{12} \, \xi + c_{13} \, \epsilon(r_1, \tau)] \int_{C_1^{2,3}} |\nabla u_2|^2.$$

Constants c_{12}–c_{11} do not depend on ξ, r_1 and τ if ξ, r_1 and τ are sufficiently small. We can analogously derive that

$$\int_{C_1^{2,3}} (a_3^{1,3} - a_2^{1,3}) \, u_{3,3} \, u_{1,13} = \partial_1 \int_{C_1^{2,3}} b_3 \, u_{3,3} \, u_{1,3} \tag{3.12}$$
$$+ \partial_1 \int_{C_1^{2,3}} \Big[\beta^{1,3} u_{1,3}^2 + \gamma^{1,3} u_{3,3}^2 \Big] + I_3(\nabla u_1, \nabla u_3) + (BI)$$

where

$$b_3 \, [1 - (a_3^{1,3} - a_2^{1,3}) \, (a_1^{1,3} - a_2^{1,3})] = a_3^{1,3} - a_2^{1,3}, \quad \beta^{1,3} = -\frac{1}{2} \, b_3 \, (a_1^{1,3} - a_2^{1,3}),$$
$$\gamma^{1.3} = -\frac{1}{2} \, b_3 \, (a_3^{1,3} - a_2^{1,3})$$

and

$$|I_3(\nabla u_1, \nabla u_3)| \leq \Big(c_{14} + \frac{c_{15}}{\xi} \Big) \int_{C_1^{2,3}} |\nabla u_1|^2 + [c_{16} \, \xi + c_{17} \, \epsilon(r_1, \tau)] \int_{C_1^{2,3}} |\nabla u_3|^2.$$

Constants c_{14}–c_{17} are independent of ξ, r_1 and τ if ξ, r_1 and τ are sufficiently small. Substituting from (3.11) and (3.12) into (3.9), we obtain:

$$\partial_1 \int_{C_1^{2,3}} \Big[\tfrac{1}{2}|\nabla u_1|^2 + \beta^{1,2}\, u_{1,2}^2 + \beta^{1,3}\, u_{1,3}^2\Big] = -\partial_1 \int_{C_1^{2,3}} \Big[b_2\, u_{2,2}\, u_{1,2} + b_3\, u_{3,3}\, u_{1,3}\Big]$$
$$+ \partial_1 \int_{C_1^{2,3}} \Big[\gamma^{1,2} u_{2,2}^2 + \gamma^{1,3} u_{3,3}^2\Big] + I_1 - I_2 - I_3 + (BI).$$

Integrating with respect to x_1 from $x_{01} - r_1$ to x_1, we get:

$$\int_{C_1^{2,3}} \Big[\tfrac{1}{2}|\nabla u_1|^2 + \beta^{1,2}\, u_{1,2}^2 + \beta^{1,3}\, u_{1,3}^2\Big]\Big|_{x_1} = -\int_{C_1^{2,3}} \Big[b_2\, u_{2,2}\, u_{1,2} + b_3\, u_{3,3}\, u_{1,3}\Big]\Big|_{x_1}$$
$$+ \int_{C_1^{2,3}} \Big[\gamma^{1,2}\, u_{2,2}^2 + \gamma^{1,3}\, u_{3,3}^2\Big]\Big|_{x_1} + \int_{x_{01}-r_1}^{x_1} (I_1 - I_2 - I_3)\, ds_1 + (BI)$$
$$\leq \xi \int_{C_1^{2,3}} [u_{1,2}^2 + u_{1,3}^2]\Big|_{x_1} + \int_{C_1^{2,3}} \Big[\Big(\gamma^{1,2} + \frac{1}{4\xi}\, b_2^2\Big)\, u_{2,2}^2 + \Big(\gamma^{1,3} + \frac{1}{4\xi}\, b_3^2\Big)\, u_{3,3}^2\Big]\Big|_{x_1}$$
$$+ \Big(c_{18} + \frac{c_{19}}{\xi}\Big) \int_{x_{01}-r_1}^{x_1} \Big(\int_{C_1^{2,3}} |\nabla u_1|^2\Big)\, ds_1$$
$$+ [c_{20}\xi + c_{21}\epsilon(r_1, \tau)] \int_{x_{01}-r_1}^{x_1} \Big(\int_{C_1^{2,3}} \Big[|\nabla u_2|^2 + |\nabla u_3|^2\Big]\Big)\, ds_1 + (BI)$$

where $c_{18} = c_7 + c_{10} + c_{14}$, $c_{19} = c_8 + c_{11} + c_{15}$, $c_{20} = c_9 + c_{12} + c_{16}$ and $c_{21} = c_{13} + c_{17}$. If we use the positive definiteness of the quadratic form $\frac{1}{2}\,|\nabla u_1|^2 + \beta^{1,2}\, u_{1,2}^2 + \beta^{1,3}\, u_{1,3}^2 - \xi\, u_{1,2}^2 - \xi\, u_{1,3}^2$ in the range of sufficiently small $\beta^{1,2}$, $\beta^{1,3}$ and ξ (which is true if r_1 and τ are sufficiently small), we can further obtain:

$$\frac{1}{2} \int_{C_1^{2,3}} |\nabla u_1|^2\Big|_{x_1} \leq c_{22} \int_{C_1^{2,3}} \Big[\tfrac{1}{2}|\nabla u_1|^2 + \beta^{1,2}\, u_{1,2}^2 + \beta^{1,3}\, u_{1,3}^2 - \xi\, u_{1,2}^2 - \xi\, u_{1,3}^2\Big]\Big|_{x_1}$$
$$+ c_{22}\Big(c_{18} + \frac{c_{19}}{\xi}\Big) \int_{x_{01}-r_1}^{x_1} \Big(\int_{C_1^{2,3}} |\nabla u_1|^2\Big)\, ds_1 + g(x_1)$$

where
$$g(x_1) = c_{22} \int_{C_1^{2,3}} \Big[\Big(|\gamma^{1,2}| + \frac{1}{4\xi}\, b_2^2\Big)\, u_{2,2}^2 + \Big(|\gamma^{1,3}| + \frac{1}{4\xi}\, b_3^2\Big)\, u_{3,3}^2\Big]\Big|_{x_1}$$
$$+ c_{22}\, [c_{20}\xi + c_{21}\epsilon(r_1, \tau)] \int_{x_{01}-r_1}^{x_1} \Big(\int_{C_1^{2,3}} \Big[|\nabla u_2|^2 + |\nabla u_3|^2\Big]\Big)\, ds_1 + (BI).$$

Applying Gronwall's lemma, we get:

$$\frac{1}{2} \int_{C_1^{2,3}} |\nabla u_1|^2\Big|_{x_1}$$
$$\leq g(x_1) + c_{22}\Big(c_{18} + \frac{c_{19}}{\xi}\Big) \int_{x_{01}-r_1}^{x_1} g(s_1)\, \exp\Big[c_{22}\Big(c_{18} + \frac{c_{19}}{\xi}\Big) s_1\Big]\, ds_1$$
$$\leq c_{23} \int_{C_1^{2,3}} \Big[\Big(|\gamma^{1,2}| + \frac{1}{4\xi}\, b_2^2\Big)\, u_{2,2}^2 + \Big(|\gamma^{1,3}| + \frac{1}{4\xi}\, b_3^2\Big)\, u_{3,3}^2\Big]\Big|_{x_1}$$
$$+ c_{24}(\xi, r_1, \tau) \int_{x_{01}-r_1}^{x_{01}+r_1} \Big(\int_{C_1^{2,3}} [|\nabla u_2|^2 + |\nabla u_3|^2]\Big)\, ds_1 + c_{25}$$

where $c_{24}(\xi, r_1, \tau) \to 0$ if $\xi \to 0$, $r_1 \to 0$ and $\tau \to 0$. If ξ, r_1 and τ are so small that $c_{24}(\xi, r_1, \tau) \le 1$ and

$$c_{23} \int_{C_1^{2,3}} \left[\left(|\gamma^{1,2}| + \frac{1}{4\xi}\, b_2^2 \right) u_{2,2}^2 + \left(|\gamma^{1,3}| + \frac{1}{4\xi}\, b_3^2 \right) u_{3,3}^2 \right] \bigg|_{x_1} \le \frac{1}{8} \int_{C_1^{2,3}} \left[u_{2,2}^2 + u_{3,3}^2 \right] \bigg|_{x_1}$$

then

$$\frac{1}{2} \int_{C_1^{2,3}} |\nabla u_1|^2 \bigg|_{x_1}$$
$$\le \frac{1}{8} \int_{C_1^{2,3}} \left[u_{2,2}^2 + u_{3,3}^2 \right] \bigg|_{x_1} + \int_{x_{01}-r_1}^{x_{01}+r_1} \left(\int_{C_1^{2,3}} \left[|\nabla u_2|^2 + |\nabla u_3|^2 \right] \right) ds_1 + c_{25}.$$

If we integrate with respect to x_1 on the interval $]x_{01} - r_1, x_{01} + r_1[$ and take r_1 so small that $\frac{1}{8} + 2r_1 \le \frac{1}{6}$ then we obtain the estimate

$$\frac{1}{2} \int_{C_1} |\nabla u_1|^2\, dx_1\, dx_2\, dx_3 \le \left(\frac{1}{8} + 2r_1 \right) \int_{C_1} \left[|\nabla u_2|^2 + |\nabla u_3|^2 \right] dx_1\, dx_2\, dx_3 \tag{3.13}$$
$$+ c_{26} \le \frac{1}{6} \int_{C_1} \left[|\nabla u_2|^2 + |\nabla u_3|^2 \right] dx_1\, dx_2\, dx_3 + c_{26}.$$

If we differentiate the equation of continuity (1.8) with respect to x_2 and substitute for $u_{1,12}$ and $u_{3,32}$ from (3.4) and (3.6), we obtain the wave equation

$$u_{2,22} - u_{2,11} - u_{2,33} = (a_k^{1,2}\, u_{k,k})_{,1} + (a_k^{2,3}\, u_{k,k})_{,3}$$

instead of (3.8). Then, using the same approach as the one which has lead to (3.14), we can derive the estimate

$$\frac{1}{2} \int_{C_1} |\nabla u_2|^2\, dx_1\, dx_2\, dx_3 \le \frac{1}{6} \int_{C_1} \left[|\nabla u_1|^2 + |\nabla u_3|^2 \right] dx_1\, dx_2\, dx_3 + c_{26}. \tag{3.14}$$

Analogously, we can also obtain the inequality

$$\frac{1}{2} \int_{C_1} |\nabla u_3|^2\, dx_1\, dx_2\, dx_3 \le \frac{1}{6} \int_{C_1} \left[|\nabla u_1|^2 + |\nabla u_2|^2 \right] dx_1\, dx_2\, dx_3 + c_{26}. \tag{3.15}$$

Summing (3.13), (3.14) and (3.15), we get:

$$\int_{C_1} |\nabla \boldsymbol{u}|^2\, dx \le 18\, c_{26}\,. \tag{3.16}$$

This estimate is valid for all $t \in]t_0 - \tau, t_0[$. $\boldsymbol{v}$ satisfies the same estimate (possibly with a different constant on the right-hand side) because $\boldsymbol{u}$ and $\boldsymbol{v}$ coincide on $C_1 \times]t_0 - \tau, t_0[$. Now we can easily verify that conditions (2.12) and (2.13) are again satisfied and consequently, the theory of the Navier-Stokes equation ((see, e.g., L. Caffarelli, R. Kohn and L. Nirenberg [3], p. 776) says that the solution $(\boldsymbol{v}; p)$ cannot have a singularity at the point (x_0, t_0). Since this point can be chosen arbitrarily in D' on the time level t_0, t_0 cannot be a D'-epoch of irregularity. Hence the solution $(\boldsymbol{v}; p)$ has no epoch of irregularity in D' and therefore it is regular in D'. Consequently, it is also regular in D. □

Acknowledgement. The research was supported by the University of Toulon-Var and in the case of the first author, it was also supported by the Grant Agency of the Czech Republic (grant No. 201/02/0684) and by the research plan of the Ministry of Education of the Czech Republic No. MSM 98/210000010.

References

[1] R. Betchow, *An inequality concerning the production of vorticity in isotropic turbulence.* Physics of Fluids (1956), 497–504.

[2] W. Borchers and H. Sohr, *On the equations* $\operatorname{rot} v = g$ *and* $\operatorname{div} u = f$ *with zero boundary conditions.* Hokkaido Math. J. **19** (1990), 67–87.

[3] L. Caffarelli, R. Kohn and L. Nirenberg, *Partial regularity of suitable weak solutions of the Navier-Stokes equations.* Comm. on Pure and Appl. Math. **35** (1982), 771–831.

[4] D. Chae, H.J. Choe, *Regularity of solutions to the Navier-Stokes equations.* Electronic J. of Diff. Equations **5** (1999), 1–7.

[5] G.P. Galdi, *An Introduction to the Mathematical Theory of the Navier-Stokes Equations, Vol. I: Linearized Steady Problems.* Springer-Verlag, New York – Berlin – Heidelberg (1994).

[6] G.P. Galdi, *An Introduction to the Navier-Stokes initial-boundary value problem.* In *Fundamental Directions in Mathematical Fluid Mechanics*, editors G.P. Galdi, J. Heywood, R. Rannacher, series "Advances in Mathematical Fluid Mechanics", Birkhäuser Verlag, Basel (2000), 1–98.

[7] J. Leray, *Sur le mouvements d'un liquide visqueux emplissant l'espace.* Acta Mathematica **62** (1934), 193–248.

[8] J. Neustupa, *Partial regularity of weak solutions to the Navier-Stokes equations in the class* $L^\infty(0,T;\, L^3(\Omega)^3)$. J. of Mathematical Fluid Mechanics **1** (1999), 309–325

[9] J. Neustupa, A. Novotný, P. Penel, *An Interior Regularity of a Weak Solution to the Navier-Stokes Equations in Dependence on One Component of Velocity.* Quaderni di Matematica, vol. 10, Topics in Mathematical Fluid Mechanics, Eds. G.P. Galdi and R. Ranacher, 2002, 163–183.

[10] J. Neustupa and P. Penel, *Anisotropic and geometric criteria for interior regularity of weak solutions to the 3D Navier-Stokes equations.* In *Mathematical Fluid Mechanics, Recent Results and Open Problems*, editors J. Neustupa and P. Penel, series "Advances in Mathematical Fluid Mechanics", Birkhäuser Verlag, Basel (2001), 237–268.

[11] J. Neustupa and P. Penel, *The role of eigenvalues and eigenvectors of the symmetrized gradient of velocity in the theory of the Navier-Stokes equations.* Comptes Rendus Acad. Sci. Paris, Sér. I 336(10), 2003, 805–810.

Jiří Neustupa
Czech Technical University, Faculty of Mechanical Engineering
Department of Technical Mathematics, Karlovo nám. 13, 121 35 Praha 2, Czech Republic
e-mail: `neustupa@marian.fsik.cvut.cz`

Patrick Penel
Université de Toulon et du Var, Mathématique, BP 132, F-83957 La Garde, France
e-mail: `penel@univ-tln.fr`

Progress in Nonlinear Differential Equations
and Their Applications, Vol. 61, 213–223

Free Work and Control of Equilibrium Configurations

M. Padula

1. Introduction

We set $y \in C_k$ (reference configuration), $\mathbf{x}_* \in C_*$ (equilibrium configuration), and define by $\chi^t(y,t)$ the story of the particle. Let us begin with the general frame of indefinite equations governing motions of a general continuum, that is,

$$\begin{aligned} \rho\Big[\frac{\partial \mathbf{v}}{\partial t} + (\mathbf{v}\cdot\nabla)\mathbf{v}\Big] &= \nabla\cdot\mathbf{T} + \rho\mathbf{f}, \\ \frac{\partial \mathbf{x}}{\partial t} + (\mathbf{v}\cdot\nabla)\mathbf{x} &= \mathbf{v}, \end{aligned} \tag{1.1}$$

where $\mathbf{x}$, ρ, and $\mathbf{v}$ denote the position, the density, and the velocity of a particle at configuration C. Also, $\mathbf{f}$ is the external body force, and

$$\mathbf{T} = -\mathbf{T}(\chi^t(y,t), \nabla_y\chi^t(y,t), \nabla_y \otimes \nabla_y\chi^t(y,t)))$$

is the stress.

These equations represent a coupled hyperbolic-parabolic system.

Weak formulation is deduced by the equation

$$\int_C \rho\Big[\frac{\partial \mathbf{v}}{\partial t} + (\mathbf{v}\cdot\nabla)\mathbf{v}\Big]\cdot\varphi dc = \int_C \nabla\cdot\mathbf{T}\cdot\varphi dc + \int_C \rho\mathbf{f}\cdot\varphi dc. \tag{1.2}$$

If the elastic part of stress $\mathbf{T}$, say $\mathbf{T}_0$ derives from an energy, putting in (1.2) $\varphi = \mathbf{v}$ it is easy to deduce a balance equation for the kinetic plus internal energy, see [3], [13].

Indeed, (1.2) contains much more information. Employing arbitrariness of choice for the test function φ, we consider φ as a spatial vector $\mathbf{w}$, say difference between the coordinates of two points, in this way, we obtain a quantity dimensionally equivalent to a finite work, and here named *free work*:

$$\int_C \rho\frac{d\mathbf{v}}{dt}\cdot\varphi dc = \Delta_w L^e + \Delta_w L^i, \tag{1.3}$$

and $\Delta_w L^e$, $\Delta_w L^i$ denote the finite works of external and internal forces, respectively, corresponding to the given displacement $\mathbf{w}(x)$. In particular, if the forces are conservative, the work of external and internal forces corresponding to the given displacement $\mathbf{w} = \mathbf{x}_1 - \mathbf{x}_2$, $\mathbf{x}_i \in C_i$, represents the opposite of the internal plus potential energy E, and equation (1.3) reduces to

$$\int_C \rho \frac{d\mathbf{v}}{dt} \cdot \mathbf{w} dc = -[E(C_2) - E(C_1)], \tag{1.4}$$

Some remarks on free work are needed to understand its novelty. Free work is only dimensionally a work, not a real work done by the force, because the position vector $\mathbf{w}(x)$ may have no relation with the motion of x.

Usually, an equilibrium configuration C_* of a continuum system is said stable if its energy assumes a proper minimum in it. Some time this statement is not proved dynamically. Indeed, this statement is a natural consequence of a correct reading of (1.4). Here, we analyze a case where dynamical stability is not known, to our knowledge: phase transitions. In the paper we consider only regular functions.

The internal energy E is usually not convex, one general expression for such problems is given by, see [14]. Setting $F_{sl} = \frac{\partial x_r}{\partial y_l}$, we consider the following energy

$$E = \int_{C_k} \rho_k \Big[\epsilon_0(\mathbf{F}) + \epsilon_1(\nabla_y \otimes \mathbf{F}) + \epsilon_2(\mathbf{x})\Big] dc_k = \int_{C_k} \rho_k \epsilon dc_k. \tag{1.5}$$

Here E is the elastic energy stored in the body (non convex functional), $\int C_k \rho_k \epsilon_1 dc_k$ models the interfacial energy, and $\int C_k \rho_k \epsilon_2 dc_k$ is the energy of the elastic foundation.

Let the position vector $\mathbf{w}$ be zero at boundary. To make assumptions on the elastic behavior of the body we need the Lagrangian representation of the stress which is furnished by the first Piola-Kirchoff tensor $\mathbf{K}(\mathbf{x})$. The link between the Cauchy $\mathbf{T}_0$ and Piola-Kirchoff $\mathbf{K}$ tensors is given by:

$$(\mathbf{T}_0)_{rl} = \frac{1}{J} K_{rs}(\mathbf{x}) F_{sl}, \qquad F_{sl} = \frac{\partial x_s}{\partial y_l}. \tag{1.6}$$

We set $\frac{\partial}{\partial y_l} = \partial_l$, $(\int^{yl} F_{rs} = (\int^{y_l} F_{rs}(y', t) dy_l')$, with $y' = (y_i, y_l')$, $i \neq l$, and shall assume that the material is hyperelastic in the following sense Let us compute the first variation of $E(\mathbf{x} + t\mathbf{w})$ at $t = 0$, calling $\mathbf{A} = \nabla_{\mathbf{y}} \mathbf{w} = \mathbf{F}(\mathbf{w})$, we get

$$\begin{aligned} \delta E(\mathbf{w}) &= \int_{C_k} \rho_k \Big[\frac{\partial \epsilon_0}{\partial F_{rs}} A_{rs} + \frac{\partial \epsilon_1}{\partial\, \partial_l F_{rs}} \partial_l A_{rs} + \frac{\partial \epsilon_2}{\partial \int^{y_l} F_{rs}} \int^{y_l} A_{rs}\Big] dc_k \\ &= -\int_{C_k} \rho_k w^r \partial_s \Big[\frac{\partial \epsilon_0}{\partial F_{rs}} - \partial_l \frac{\partial \epsilon_1}{\partial\, \partial_l F_{rs}} - \int^{y_l} \frac{\partial \epsilon_2}{\partial \int^{y_l} F_{rs}}\Big] dc_k = -\Delta_w L^i. \end{aligned} \tag{1.7}$$

Hence, we obtain the expression for the Piola-Kirchoff tensor

$$K_{rs}(\mathbf{x}) = \frac{\partial \epsilon_0}{\partial F_{rs}} - \partial_l \frac{\partial \epsilon_1}{\partial\, \partial_l F_{rs}} - \int^{y_l} \frac{\partial \epsilon_2}{\partial \int^{y_l} F_{rs}}. \tag{1.8}$$

Substituting (1.7) into (1.3), and choosing $\varphi = \mathbf{v}$, it is straightforward to obtain

$$\frac{d}{dt}\Big\{\int_C \rho \frac{\mathbf{v}^2}{2} dc + \int_{C_k} \rho_k \epsilon dc_k\Big\} = \int_C \rho \mathbf{f} \cdot \mathbf{v} dc. \tag{1.9}$$

If external forces are conservative $\mathbf{f} = \nabla U$, U is the potential, then (1.9) reduces to the conservation of total energy

$$\frac{d}{dt}\int_C \rho\Big\{\frac{\mathbf{v}^2}{2} + \epsilon - \nabla U\Big\} dc = 0. \tag{1.10}$$

If forces are conservative, we get a conservation law for the total energy $\mathcal{E} = T+E-U$, where T is the kinetic energy. In particular, since $\mathcal{E}(C_*) = E(C_*)-U(C_*)$ is constant in time, we know that

$$\frac{d[\mathcal{E}(C) - \mathcal{E}(C_*)]}{dt} = 0.$$

Let $\xi \to \mathcal{E}$ be function of $\mathbf{x}$, $\mathbf{v}$, $\nabla_y \mathbf{x}$, $\nabla_y \mathbf{x} \otimes \nabla_y \mathbf{x}$, and denote by $\Lambda = \mathbf{x}$, $\mathbf{v}$, $\nabla_y \mathbf{x}$, $\nabla_y \mathbf{x} \otimes \nabla_y \mathbf{x}$. Assume that $\mathcal{E}$ has a local minimum at C_* then, by Taylor expansion it results

$$\mathcal{E}(C) - \mathcal{E}(C_*) = \int_{C_*} \rho_k \left[\nabla \mathcal{E}(\Lambda_*) \cdot \Lambda + \frac{1}{2}\Lambda \cdot \nabla \otimes \nabla \mathcal{E}(\bar{\Lambda}) \cdot \Lambda\right], \tag{1.11}$$

where $\bar{\Lambda}$ is a point between Λ and Λ_*. If $\mathcal{E}$ has a proper minimum at C_*, then $\mathcal{E}(C) - \mathcal{E}(C_*)$ is equivalent to a L^2 norm of motion, and it is a correct Lyapunov functional.

Let us remark that our stability theorem is local with respect to initial data, as it should be expected. We only consider small perturbations of the rest, because for our proof to work we must fix initial data $C(0)$ so close to C_* that $\mathcal{E}(\mathbf{x}(0))$ follows in the neighborhood of $\mathbf{x}_*$ where C_* is a proper minimum.

Now we change again the displacement in a more physical way. Let us write the displacement $\mathbf{x} = \mathbf{x}_* + \mathbf{u}$, where $\mathbf{u}$ is the perturbation. We choose $\mathbf{w}$ as the displacement $\mathbf{w} = \mathbf{x} - \mathbf{x}_*$ and we obtain the **pivot equation**. The name is due to its direct relation with the proof of nonlinear instability. In case there is a internal energy E, then we have

$$\Delta_w L^i = -\Delta_w E.$$

To understand the reason of our definition, it is enough to consider a single material particle, assume zero external forces, and observe the following elementary things, [5]:

(i) The displacement $\mathbf{w}$ represents the difference between the position of the same particle $y \in C_k$ in two different motions $\mathbf{x}$, $\mathbf{x}_*$.

(ii) For a single point $\mathbf{x}$, we have only kinetic and potential energy E. In order a equilibrium position $\mathbf{x}_*$ to be stable, it must occur that, if we remove the particle from $\mathbf{x}_*$ to a position $\mathbf{x}$ belonging to a small finite neighborhood of $\mathbf{x}_*$ the force must tend to restore the old situation, say, it must be

$$\mathbf{f}^i \cdot (\mathbf{x} - \mathbf{x}_*) = E(\mathbf{x}_*) - E(\mathbf{x}) < 0.$$

In equivalent way $\mathbf{x}_*$ must be a proper minimum for the potential energy E.

(iii) For a single point $\mathbf{x}$, in order an equilibrium position $\mathbf{x}_*$ to be unstable, if we remove a particle from $\mathbf{x}_*$ to a position $\mathbf{x}$ belonging to a small neighborhood of $\mathbf{x}_*$ there must be an initial data such that forces tend to send the particle far away from the initial position, say, it must be

$$\mathbf{f}^i \cdot (\mathbf{x} - \mathbf{x}_*) = E(\mathbf{x}_*) - E(\mathbf{x}) > 0.$$

This is certainly true if $\mathbf{x}_*$ is a proper maximum for the energy.

These remarks explain the name we gave to this work, which is just ruling the property of stability for some equilibrium configuration.

Concerning previous results, we remind that in [7] it is formally proved exponential stability of the rest state of a compressible isothermal fluids in the L^2 norm, while in [2], [4], the same problem is solved for thermally conducting fluids. The use of such weak L^2 norm was realized by using as test function in (1.2) the function φ such that its product with the difference between the gradient of pressure between the two motions C and C_*, $-\nabla(\rho - \rho_*)$, equals the L^2 norm of $\rho - \rho_*$, say $\int_C \varphi \cdot \nabla(\rho - \rho_*) dc = \int_C (\rho - \rho_*)^2 dc$. In this way a (dissipative-like) term for $(\rho - \rho_*)$ was deduced, useful for the decay of perturbations $(\rho - \rho_*)$ in the L^2 norm. From this research it has been realized that the methods there employed contain a deeper physical meaning, actually they are successful in several more general problems [8], [10], [12], [6], [13]. This paper is a natural continuation of this research, in particular the results of [7] can be obtained in much simpler way with the use of lounching free equation.

More general extensions to steady (not rest) motions have been considered in [8], [10].

The first part studies stability in the mean for hyperelastic materials, here $\mathbf{T}_1$ may vanish, and the resulting stability is not asymptotic. This result is not new, it was also analyzed deeply by Wang and Truesdell in [15].

The second part studies exponential stability in case of linear dissipative hyperelastic materials, that is $\mathbf{T}_1 \neq 0$, e.g. $\mathbf{T}_1 = 2\mu\mathbf{D}$, with $\mathbf{D}$ (Dirichlet like). The hypothesis of linearity for $\mathbf{T}_1$ in function of $\mathbf{D}$ is by no means needed, it is done for sake of simplicity, here we follow the method introduced in [7].

The third part provides a nonlinear instability result (Chetaev like).

2. Stability in the mean

Theory of elasticity accounts for materials with capacity to store mechanical energy, here we consider continuous media that possess a capacity to both store and dissipate mechanical energy.

Nonlinear stability properties will be analyzed in the following four cases:

$\mathbf{T} = \mathbf{T}_0(\mathbf{F})$; $\mathbf{T} = \mathbf{T}_0(\mathbf{F}, \theta)$; $\mathbf{T} = \mathbf{T}_0(\mathbf{F}) + \mathbf{T}_1(\mathbf{D})$; $\mathbf{T} = \mathbf{T}_0(\mathbf{F}, \theta) + \mathbf{T}_1(\mathbf{D})$,

where $\mathbf{D} = \frac{1}{2}(\nabla\mathbf{v} + (\nabla\mathbf{v})^T)$ is the velocity deformation tensor; θ is the temperature. Dissipative effect in the stress constitutive relation is inserted, in order to recover asymptotic stability for certain rest states of the elastic body.

The objective of this paper is to adapt and extend Lyapunov ideas to nonlinear theory of elastic-viscous bodies.

For sake of simplicity, we consider hyperelastic materials, under the action of **dead loads**, and with **displacement condition at boundary**.

We study stability of *equilibrium solutions*, say C_i, $i = 1, \dots, n$, in the correspondence of given boundary data and given forces.

The stability of an equilibrium configuration $C_i = C_*$, is determined through a precise rule. Here, we prove stability of C_* in the class of regular unsteady motions, with initial data belonging to a finite (**not infinitesimal!**) neighborhood I_* of C_*.

Basic equations

Set $\frac{d}{dt} = \frac{\partial}{\partial t} + \mathbf{v} \cdot \nabla$, ρ is the density, $\mathbf{u}$ the displacement, $\mathbf{v}$ the velocity. An elasto-viscous material admits a stress tensor of the form $\mathbf{T} = \mathbf{T}_0 + \mathbf{T}_1$.

To make assumptions on the elastic behavior of the body we need the Lagrangian representation of the stress which is furnished by the first Piola-Kirchoff tensor $\mathbf{K}(\mathbf{F})$. The link between the Cauchy and Piola-Kirchoff tensor is given by (1.6). We adopt the hypothesis that the material is hyperelastic, that means $K_{rl} = \frac{\rho_k \epsilon}{\partial F_{rl}}$, $\rho = \frac{\rho_k}{J}$, where ρ_k denotes the density in the reference configuration, and $\mathcal{E}$ denotes the elastic energy. As boundary and initial conditions, for sake of simplicity, we suppose $\partial C = \partial C_*$ is a rigid fixed boundary where the displacement is prescribed. Hence, the motion is given by the displacement, and velocity fields solutions to the system

$$
\begin{aligned}
&\rho\frac{d\mathbf{v}}{dt} = \frac{1}{J}\nabla_y \cdot \mathbf{K} + \nabla_x \cdot \mathbf{T}_1 + \rho\mathbf{f}, \qquad x, t \in C \times (0,T)\\
&\frac{d\mathbf{w}}{dt} = \mathbf{v}, \qquad x, t \in C \times (0,T)\\
&\mathbf{w}(x,t) = 0, \qquad \mathbf{v}(x,t) = 0, \qquad \text{on} \quad \partial C,\\
&\mathbf{v}(x,0) = \mathbf{v}_0(x) \qquad \mathbf{w}(x,0) = \mathbf{w}_0(x), \qquad x \in C.
\end{aligned}
\tag{2.1}
$$

The rest state

In order to have as basic state the rest we consider an external force satisfying the following equation

$$
0 = \frac{1}{J_*}\nabla_y \cdot \mathbf{K}(\mathbf{F}_*) + \frac{\rho_k}{J_*}\mathbf{f}, \qquad x \in C_*. \tag{2.2}
$$

Problem (2.2) may admit several equilibrium solutions, we assume that $\mathbf{x}_*(\mathbf{y})$ is a proper minimum for the energy E. The **Dirichlet-Lagrange theorem** reads now

Theorem 2.1 (Conservative force). *Assume that the total energy $\mathcal{E}(\mathbf{F})$ has a strict minimum at C_*, then C_* is a stable equilibrium position.*

Our condition should be compared with the Hadamard criterion (1903), Langebach (1959). Alive conservative forces can be considered by following Beju estimates (1971). The result is quite well established for isothermal elastic bodies, it becomes original for heat conducting elastic bodies.

Theorem 2.2 (Non-conservative force). *Assume that the elastic energy $\mathcal{E}(\mathbf{F})$ has a strict minimum at C_*, then for C_* it holds continuous dependence on the data.*

Our result should be compared with the test proposed by Beatty (1965–68).

We omit the proof for conservative forces and consider only **nonconservative** force. In this case, we must make a smallness assumption on the non-conservative part of the force. Let us prove that the rest state is stable.

Subtracting (2.2) multiplied by $(J_*/J)\mathbf{v}$ from $(2.1)_1$ multiplied by $\mathbf{v}$, and integrating over the actual domain C, we obtain

$$\frac{d}{dt}\int_C \rho\frac{v^2}{2}dc+\mathcal{D} = \int_C \frac{1}{J}\nabla_y(\mathbf{K}(\mathbf{F})-\mathbf{K}_*(\mathbf{F}_*))\frac{d\mathbf{x}}{dt}dc+\int_C \rho(\mathbf{f}(\mathbf{x})-\mathbf{f}(\mathbf{x}_*))\mathbf{v}dc, \quad (2.3)$$

with the same definition for $\mathcal{D}$.

Adding and subtracting the term

$$\int_{C_k} \frac{\partial(\rho_k\mathcal{E})}{\partial\mathbf{F}_*}\frac{d\mathbf{F}(\mathbf{x}_*)}{dt}dc_k,$$

in the second term at the right-hand side of (2.3), we deduce

$$\begin{aligned}
&\int_{C_k} \frac{\partial(\rho_k\mathcal{E})}{\partial\mathbf{F}_*}\frac{d\mathbf{F}(\mathbf{x})}{dt}dc_k + \int_C \rho(\mathbf{f}(\mathbf{x})-\mathbf{f}(\mathbf{x}_*))\mathbf{v}dc \\
&= \int_{C_k} \frac{\partial(\rho_k\mathcal{E})}{\partial F_{rl}}\Big|_* \frac{dF_{rl}(\mathbf{x}_*)}{dt}dc_k + \int_{C_k} \frac{\partial(\rho_k\mathcal{E})}{\partial F_{rl}}\Big|_* \frac{dF_{rl}(\mathbf{u})}{dt}dc_k + \int_C \rho(\mathbf{f}(\mathbf{x})-\mathbf{f}(\mathbf{x}_*))\mathbf{v}dc \\
&= \frac{d}{dt}\int_{C_k} \rho_k\mathcal{E}(\mathbf{F}_*)dc_k + \frac{d}{dt}\int_{C_k} \frac{\partial(\rho_k\mathcal{E})}{\partial F_{rl}}\Big|_* F_{rl}(\mathbf{u})dc_k + r_0(\mathbf{v},\mathbf{u}),
\end{aligned} \quad (2.4)$$

where

$$r_0(\mathbf{v},\mathbf{u}) = -\int_{C_k} \mathbf{v}\cdot\nabla_x\frac{\partial\rho_k\mathcal{E}}{\partial F_{rl}}\Big|_* F_{rl}(\mathbf{u})dc_k + \int_C \rho(\mathbf{f}(\mathbf{x})-(\mathbf{x}_*))\mathbf{v}dc.$$

In this way, we obtain the energy identity

$$\frac{d}{dt}\Big\{\int_C \rho\frac{v^2}{2}dc + \int_{C_k}\Big[\rho_k(\mathcal{E}(\mathbf{F})-\mathcal{E}(\mathbf{F}_*)) - \frac{\partial(\rho_k\mathcal{E})}{\partial F_{rl}}\Big|_* F_{rl}(\mathbf{u})\Big]dc_k\Big\} + \mathcal{D} = r(\mathbf{v},\mathbf{u}). \quad (2.5)$$

Furthermore, from the hypothesis that C_* is a proper minimum for $\mathcal{E}$, by the Taylor expansion we have

$$\begin{aligned}\int_{C_k} \rho_k(\mathcal{E}(\mathbf{F}) - \mathcal{E}(\mathbf{F}_*)dc_k - \frac{\partial(\rho_k\mathcal{E})}{\partial F_{rl}}\Big|_* F_{rl}(\mathbf{u}))dc_k \\ = \int_{C_k} \frac{\partial^2 \rho_k \mathcal{E}}{\partial F_{rl}\partial F_{sm}}(\bar{\mathbf{F}})F_{rl}(\mathbf{u})F_{sm}(\mathbf{u})dc_k \geq a\|\mathbf{F}(\mathbf{u})\|^2 dc_k,\end{aligned} \tag{2.6}$$

with a positive constant, also $\|.\|$ mean $L^2(C)$ norm. Therefore, from (2.5) we get

$$\frac{d}{dt}\mathcal{V} + \mathcal{D} = r(\mathbf{v}, \mathbf{u}), \tag{2.7}$$

where

$$\mathcal{V} = \int_C \rho\frac{v^2}{2}dc + \int_{C_k} \left\{\rho_k(\mathcal{E}(\mathbf{F}) - \mathcal{E}(\mathbf{F}_*)) - \frac{\partial \rho_k \mathcal{E}}{\partial F_{rl}}\Big|_* F_{rl}(\mathbf{u})\right\}dc_k$$

is a good Lyapunov functional, if C_* is a minimum for $\mathcal{E}$, that is

$$\mathcal{V} \geq b(\|\mathbf{v}\|^2 + \|\mathbf{F}(\mathbf{u})\|^2)$$

with b positive constant. However, since $\mathcal{D}$ increases only the norm of velocity, and $r(\mathbf{F}, \mathbf{v})$ is increased by the product of the L^2-norms of $\mathbf{v}$, and of $\mathbf{F}$). Hence, if the forces are not conservative, we cannot prove even stability in the mean, if the term $\mathbf{T}_1$ is zero. Nevertheless, we are still able to prove continuous dependence on initial data, in particular uniqueness from inequality

$$\frac{d}{dt}\mathcal{V} + \mathcal{D} \leq c(\|\mathbf{v}\|^2 + \|\mathbf{F}\|^2) \leq c_1\mathcal{V}. \tag{2.8}$$

3. Asymptotic decay for hyperelastic, viscous materials

In this section we prove the following

Theorem 3.1 (Conservative force). *Assume and that the total energy $\mathcal{E}(\mathbf{F})$ has a strict minimum at C_*, and $\mathbf{T}_1 = \mathbf{T}_1(\mathbf{D})$, then C_* is a asymptotically stable equilibrium position.*

We prove exponential decay to zero for a suitable norm of each perturbation (L^2 for $\mathbf{v}$, W_2^1 for $\mathbf{u}$). The result will be achieved by using any external force $\mathbf{f}$. To this end, we multiply equation $(2.3)_1$ by $\mathbf{u}$, integrating over C, and integrating by parts at the left-hand side, we have the pivot equation

$$\frac{d}{dt}\int_C \rho\mathbf{v}\cdot\mathbf{u}dc - \int_C \rho\mathbf{v}^2 dc = -\int_C \mathbf{T}_1 : \nabla_x \mathbf{u}dc + \int_C \frac{1}{J}\nabla_y \cdot [\mathbf{K}(\mathbf{F}) - \mathbf{K}(\mathbf{F}_*)]\cdot\mathbf{u}dc. \tag{3.1}$$

Derivating by parts, and employing *the minimum hypothesis*, we get

$$\int_C \frac{1}{J}\nabla_y \cdot [\mathbf{K}(\mathbf{F}) - \mathbf{K}(\mathbf{F}_*)] \cdot \mathbf{u} = -\int_{C_k} \frac{\partial^2(\rho_k \mathcal{E})}{\partial F_{rl}\partial F_{sm}}\Big|_{\mathbf{F}(\bar{\mathbf{x}})} F_{rl}(\mathbf{u}) F_{sm}(\mathbf{u}) dc_k$$
$$\leq -a\|\mathbf{F}(\mathbf{u})\|^2. \tag{3.2}$$

Hence, back into (3.1) we get

$$\frac{d}{dt}\int_C \rho\mathbf{v}\cdot\mathbf{u}dc + a\|\mathbf{F}(\mathbf{u})\|^2 \leq \int_C \rho\mathbf{v}^2 dc - \int_C \mathbf{T}_1 : \nabla_x \mathbf{u} dc. \tag{3.3}$$

Adding (2.5) to (3.3) multiplied by a small positive constant γ, we deduce

$$\begin{aligned}&\frac{d}{dt}\Big[\mathcal{V} + \gamma\int_C \rho\mathbf{v}\cdot\mathbf{u}dc\Big] + \int_C \mathbf{T}_1 : \mathbf{D}dc + \gamma a\|\mathbf{F}(\mathbf{u})\|^2 \\ &+\gamma\Big(\int_C \mathbf{T}_1 : \nabla_x \mathbf{u}dc - \int_C \rho\mathbf{v}^2 dc\Big) - r(\mathbf{v},\mathbf{u}) \leq 0.\end{aligned} \tag{3.4}$$

Since $\mathcal{V}$ is a **positive definite quadratic form** in the variables $\mathbf{v}$, $\mathbf{F}(\mathbf{u})$, if γ is small enough (*but finite!*), also $\mathcal{V}_1 = \mathcal{V} + \int_C \gamma\rho\mathbf{v}\cdot\mathbf{u}dc$ will be **positive definite**, that is

$$\mathcal{V} + \gamma\int_C \rho\mathbf{v}\cdot\mathbf{u}dc \geq b_1(\|\mathbf{v}\|^2 + \|\mathbf{F}(\mathbf{u})\|^2). \tag{3.5}$$

This states that $\mathcal{V}_1$ is equivalent to the L^2 norms of $\mathbf{v}$, $\mathbf{F}$. Furthermore, since $\mathbf{T}_1$ is a dissipative term, by Clausius-Duhem inequality we know that $\int_C \mathbf{T}_1 : \mathbf{D}dc$ increases the norm of the gradient of velocity. In particular, we have

$$\begin{aligned}&\int_C \mathbf{T}_1 : \mathbf{D}dc + \gamma a\|\mathbf{F}(\mathbf{u})\|^2 + \gamma\Big(\int_C \mathbf{T}_1 : \nabla_x \mathbf{u}dc - \int_C \rho\mathbf{v}^2 dc\Big) \\ &- r(\mathbf{v},\mathbf{u}) \geq c_1(\|\mathbf{v}\|^2 + \|\mathbf{F}(\mathbf{u})\|^2),\end{aligned} \tag{3.6}$$

with c_1 function of γ, and the maximum in C of the derivatives of $\mathbf{K}(\mathbf{F}_*)$.

It is worth of remarking that the on perturbations are done hypotheses of regularity!

From (3.5), (3.6) we deduce the following differential inequality

$$\frac{d}{dt}\mathcal{V} + c_2\mathcal{V} \leq 0, \tag{3.7}$$

with c_2 function of γ, and the maximum in C of the derivatives of $\mathbf{K}(\mathbf{F}_*)$. For γ small enough and under smallness assumption on $\mathbf{f}$, it can be proved that $c_2 > 0$, this implies exponential nonlinear stability.

4. Nonlinear instability for hyperelastic bodies

We limit ourselves to internal energies of the forms $\mathcal{E}(\mathbf{F})$, and give the following definition

The Hessian of total energy $\mathcal{E}(\mathbf{F})$ enjoys property (HI) at C_* iff there exists a matrix $\mathbf{L}$*, such that*

$$\mathbf{L} \cdot \nabla \otimes \nabla \mathcal{E}(\mathbf{F}_*) \mathbf{L} < 0 \tag{4.1}$$

In this section we prove the following two instability theorems.

Theorem 4.1 (Conservative force). Assume and that the Hessian of total energy $\mathcal{E}(\mathbf{F})$ satisfies property (HI) at C_*, and $\mathbf{T}_1 = 0$, then the equilibrium position C_* is instable.

We use the methods of [1]. Assume, by absurdum, that the motion is stable in $W^{3,2}$ norm, this allows us to choose initial data $\mathbf{u}_0$ such small that $\mathbf{u}$, in the norm $W^{3,2}$ is less of a arbitrarily small constant ϵ for all times t. Furthermore, we choose initial data such that $\nabla_y \mathbf{u}_0 = \epsilon \mathbf{L}$. First we remind that the conservation of total energy holds for the motion $x(y,t)$ in the form

$$\int_C \frac{1}{2}\rho \mathbf{v}^2 dc + \int_{C_k} E dc_k = \int_{C_0} \frac{1}{2}\rho_0 \mathbf{v}_0^2 dC_0 + \int_{C_k} E_0 dc_k. \tag{4.2}$$

Since the first derivative of E at C_* is zero, we deduce also

$$\int_C \frac{1}{2}\rho \mathbf{v}^2 dc + \frac{1}{2}\int_{C_k} \frac{\partial^2 E}{\partial F_{rl}\partial F_{sm}}\Big|_{x_*} F_{rl}(\mathbf{u})F_{sm}(\mathbf{u})dc_k + 0(\epsilon^3)$$
$$= \int_{C_0} \frac{1}{2}\rho_0 \mathbf{v}_0^2 dC_0 + \int_{C_k} (E_0 - E_b)dc_k. \tag{4.3}$$

The right-hand side can be chosen negative, because the Hessian satisfies property (HI). Therefore we have

$$\int_{C_0} \frac{1}{2}\rho_0 \mathbf{v}_0^2 dC_0 + \int_{C_k} (E_0 - E_b)dc_k = -A\epsilon^2,$$

and

$$\int_C \frac{1}{2}\rho \mathbf{v}^2 dc + A\epsilon^2 = -\frac{1}{2}\int_{C_k} \frac{\partial^2 E}{\partial F_{rl}\partial F_{sm}}\Big|_{x_*} F_{rl}(\mathbf{u})F_{sm}(\mathbf{u})dc_k + 0(\epsilon^3). \tag{4.4}$$

Let us multiply equation $(2.3)_1$ by $\mathbf{u}$, and integrate over C, integrating by parts at the left-hand side, we get

$$\frac{d}{dt}\int_C \rho \mathbf{v} \cdot \mathbf{u} dc - \int_C \rho \mathbf{v}^2 dc = -\int_C \frac{1}{J}[\mathbf{K}(\mathbf{F}) - \mathbf{K}(\mathbf{F}_*)] \cdot \mathbf{F}(\mathbf{u})dc_t \tag{4.5}$$
$$= -\int_{C_k} \mathbf{F}(\mathbf{u}) \cdot \nabla \otimes \nabla \mathcal{E}(\mathbf{F}(\mathbf{x}_*))\mathbf{F}(\mathbf{u})dc_k - O(\epsilon^3) = \int_C \rho \mathbf{v}^2 dc + 2A\epsilon^2 - O(\epsilon^3),$$

where we have employed (4.4).

We deduce, integrating in time

$$\frac{d}{dt}\int_C \frac{1}{2}\rho \mathbf{u}^2 dc \geq \frac{d}{dt}\int_C \frac{1}{2}\rho \mathbf{u}^2 dc\Big|_0. \tag{4.6}$$

Finally, by taking as initial data $\mathbf{v}_0 \cdot \mathbf{u}_0 = b_0 > 0$, integrating again over t, we get the growing to infinity of this solution.

Theorem 4.2 (Dissipative force). *Let $\mathbf{T}_1 = \mathbf{T}_1(\mathbf{v})$ be not too large, then the equilibrium position C_* is nonlinear unstable if $\mathcal{E}$ assumes a maximum, for suitable norms of perturbations (L^2 for $\mathbf{v}$, W_2^1 for $\mathbf{u}$).*

The result will be achieved by using any external force $\mathbf{f}$, dead loads. To this end, we multiply equation $(2.3)_1$ by $\mathbf{u}$, integrating over C, and integrating by parts at the left-hand side, we have

$$\frac{d}{dt}\int_C \rho\mathbf{v}\cdot\mathbf{u}dc - \int_C \rho\mathbf{v}^2 dc = -\int_C \mathbf{T}_1 : \nabla_x \mathbf{u}dc + \int_C \frac{1}{J}\nabla_y\cdot[\mathbf{K}(\mathbf{F}) - \mathbf{K}(\mathbf{F}_*)]\cdot\mathbf{u}dc, \tag{4.7}$$

Assume, by absurdum, that the motion is stable in $W^{3,2}$ norm, this allows us to choose initial data $\mathbf{u}_0$ such small that $\mathbf{u}$, $\mathbf{v}$ in the norm $W^{3,2}$ are less of a arbitrarily small constant ϵ for all times t. Integrating by parts in (4.7), we get

$$-\int_C \frac{1}{J}[\mathbf{K}(\mathbf{F}) - \mathbf{K}(\mathbf{F}_*)]\cdot\mathbf{F}(\mathbf{u}) = -\mathbf{F}(\mathbf{u})\cdot\nabla\otimes\nabla\mathcal{E}(\mathbf{F}(\mathbf{x}_*))\mathbf{F}(\mathbf{u}) - o(\epsilon^3) \geq a\|\mathbf{F}(\mathbf{u})\|^2, \tag{4.8}$$

where we have employed *the maximum hypothesis* for the energy, where a is a positive constant.

Integrating in time

$$\frac{d}{dt}\int_C \frac{1}{2}\rho\mathbf{u}^2 dc \geq \frac{d}{dt}\int_C \frac{1}{2}\rho\mathbf{u}^2 dc\Big|_0 + \int_0^t\int_C \Big[\rho\mathbf{v}^2 + a\|\mathbf{F}(\mathbf{u})\|^2 - \mathbf{T}_1 : \nabla_x\mathbf{u}dC_s\Big]ds. \tag{4.9}$$

In (4.9) the last integral at the r.h.s. can be taken positive if $\mathbf{T}_1$ is not too large. Finally, by taking as initial data $\mathbf{v}_0\cdot\mathbf{u}_0 = b_0 > 0$, integrating again over t, we get the growing to infinity of this solution.

Acknowledgment

The author wishes to thank professors Finn and Solonnikov for helpful and stimulating discussions. The author also acknowledges the Italian GNFM-INDAM and 60% and 40% Murst.

References

[1] R. Abeyaratne & J.K. Knowles, *Elastic materials with two stress-free configurations*, J. of Elasticity **67** (2002), 61–69.

[2] R. Benabidallah & M. Padula, *Sulla stabilità di un fluido politropico, viscoso, pesante in un contenitore limitato da pareti perfette conduttrici di calore*, Annali Univ. Ferrara, sez VII, Sc. Mat. **XLV** (1999), 127–161.

[3] J.L. Ericksen [E], *Introduction to the thermodynamics of solids*, Applied Math. and Math. Comput., **1** (1991).

[4] G. Guidoboni & B.J. Jin, *On the nonlinear stability of Marangoni-Benard problem with free surface in Boussinesq approximation*, to appear.

[5] G. Grioli, *Lezioni di Meccanica Razionale, Ed. Libreria*, Cortina Padova, (1991).

[6] B.J. Jin & M. Padula, *On existence of nonsteady compressible viscous flows in a horizontal layer with free upper surface*, Comm. Pure Appl. Anal. **1**(3) (2002), 370–415.

[7] M. Padula, *On the exponential decay to the rest state for a viscous isothermal fluid*, J. Math. Fluid Mech., **1** (1999), 1–16.

[8] M. Padula, *On the exponential stability of the rest of a viscous compressible fluid*, Proc. of Society of Trends in Application of Mathematics to Mechanics, Nice 25–29 may (1998).

[9] M. Padula & M. Pokorny, *Stability and decay to zero of the L^2 norms of perturbations to a viscous compressible heat conductive fluid motion exterior to a ball*, J. Math. Fluid Mech., **3** (2001), 317–408.

[10] M. Padula & V.A. Solonnikov, *On the Rayleigh-Taylor Stability*, Annali dell'Università di Ferrara, (sez.VIII, Sci. Mat.) **46** (2000), 307–336.

[11] M. Padula & V.A. Solonnikov, *On the global existence of nonsteady motions of a fluid drop and their exponential decay to a rigid rotation*, to appear on Quaderni di Matematica.

[12] M. Padula & V.A. Solonnikov, *Existence of non-steady flows of an incompressible, viscous drop of fluid in a frame rotating with finite angular velocity*, Proceeding Fourth European Conference on Elliptic and Parabolic Problems, Rolduc and Gaeta 2001, World Scientific, New Jersey-London-Singapore-Hong Kong, 180–203.

[13] M. Padula, *On direct Ljapunov method in continuum theories*, Nonlinear problems in mathematical physics and related topics I, in honor of prof. Ladyzhenskaya, eds. Sh. Birman, S. Hildebrandt, V.A. Solonnikov, O.A. Uralsteva, Kluwer Academic/Plenum Publisher, New York, Boston, Dordrecht, London, Moscow, 289–302, Novosibirsk (2002), 271–283.

[14] L. Truskinovsky & G. Zanzotto, *Ericksen bar revisited: energy wiggles*, J. Mech.Phys. Solids, **44** (1996), 1371–1408.

[15] C.C. Wang and C. Truesdell, *Introduction to rational elasticity*, Nordhoff Int. Publ. 1973.

M. Padula
Department of Mathematics
Via Machiavelli 35
I-44100 Ferrara, Italy
e-mail: `pad@dns.unife.it`

Progress in Nonlinear Differential Equations
and Their Applications, Vol. 61, 225–241

Stochastic Geometry Approach to the Kinematic Dynamo Equation of Magnetohydrodynamics

Diego L. Rapoport

Abstract. We review the geometry of diffusion processes of differential forms on smooth compact manifolds, as a basis for the random representations of the kinematic dynamo equations on these manifolds. We realize these representations in terms of sequences of ordinary (for almost all times) differential equations. We construct the random symplectic geometry and the random Hamiltonian structure for these equations, and derive a new class of Poincaré-Cartan invariants of magnetohydrodynamics. We obtain a random Liouville invariant. We work out in detail the case of R^3.

1. Introduction

Geometrical and topological invariants in hydrodynamics and magnetohydrodynamics have been extensively considered by several authors (see [10, 16, 18]). Thus, for the Euler equation for perfect fluids, an infinite-dimensional symplectic geometry theory was constructed by V.I. Arnold (see refs. in [10]), followed by work by D. Ebin and J. Marsden [17], which is widely perceived as a beautiful example of the differential-geometrical methods in fluid-dynamics, while a theory for the case of viscous fluids and for magnetohydrodynamics has only been recently constructed by the present author. This theory stems from stochastic differential geometry, i.e., a geometrical theory of diffusion processes or still, a stochastic theory of gauge-theoretical structures (linear connections of Riemann-Cartan-Weyl), so that geometrical and probabilistic structures become unified in a single theory which has been applied to several areas of mathematical and theoretical physics (see [5] and references therein); in particular, this theory has yielded a new class of random symplectic invariants for the invariant Navier-Stokes equations (NS, for short in the following), from which in the case of vanishing kinematical viscosity, we retrieve the Arnold-Ebin-Marsden theory. Furthermore, this approach has yielded analytical representations for NS on smooth compact manifolds with or without smooth boundaries, and still on Euclidean spaces and semispaces. [3,7]. The subject of this article is in giving for a start a rather sketchy (unfortunately due to page limitations) review of the fundamental elements of stochastic differential geometry [1, 2, 11, 12], to further extend this methodology previously used to construct the theory for NS, to the kinematic dynamo equation of magnetohydrodynamics (KDE, for short, in the following). Thus, in this article, we shall present a realization by sequences of ordinary differential equations, of the random representations of KDE on smooth compact connected manifolds without boundary, M, which are further isometrically embedded in Euclidean space. Furthermore, we shall extend these constructions to the cotangent manifold, T^*M, which provided with the canonical symplectic structure, will allow us to construct a random symplectic theory for KDE, and a new class of random invariants of passive magnetohydrodynamics. Our constructions will follow the formulation of stochastic differential geometry that stems from the developing method due to E. Cartan, for which a smooth curve lying on a Euclidean n-space is roled (keeping first-order contact) on a smooth n-manifold, extending it to the random development of Wiener processes on the same Euclidean space, as the geometrical construction of the most general diffusion processes on general manifolds [1, 2, 11]. This is the transfer method from which stemmed our representations for NS and KDE for smooth boundary compact manifolds [5], and will now be applied to KDE for boundaryless manifolds. We shall finally present in detail, as an example of these constructions, the Euclidean case R^3.

2. Riemann-Cartan-Weyl geometry of diffusions

In this section we follow [3, 5]. In this article M denotes a smooth connected compact orientable n-dimensional manifold (without boundary). We shall further provide M with a linear connection described by a covariant derivative operator ∇ which we assume to be compatible with a given metric g on M, i.e., $\nabla g = 0$. Given a coordinate chart (x^α) $(\alpha = 1, \ldots, n)$ of M, a system of functions on M (the Christoffel symbols of ∇) are defined by $\nabla_{\frac{\partial}{\partial x^\beta}} \frac{\partial}{\partial x^\gamma} = \Gamma(x)^\alpha_{\beta\gamma} \frac{\partial}{\partial x^\alpha}$. The Christoffel coefficients of ∇ can be decomposed as:

$$\Gamma^\alpha_{\beta\gamma} = \left\{ \begin{matrix} \alpha \\ \beta\gamma \end{matrix} \right\} + \frac{1}{2} K^\alpha_{\beta\gamma}. \tag{1}$$

The first term in (1) stands for the metric Christoffel coefficients of the Levi-Civita connection ∇^g associated to g, i.e., $\left\{ \begin{smallmatrix} \alpha \\ \beta\gamma \end{smallmatrix} \right\} = \frac{1}{2}(\frac{\partial}{\partial x^\beta} g_{\nu\gamma} + \frac{\partial}{\partial x^\gamma} g_{\beta\nu} - \frac{\partial}{\partial x^\nu} g_{\beta\gamma}) g^{\alpha\nu}$, and

$$K^\alpha_{\beta\gamma} = T^\alpha_{\beta\gamma} + S^\alpha_{\beta\gamma} + S^\alpha_{\gamma\beta}, \tag{2}$$

is the cotorsion tensor, with $S^\alpha_{\beta\gamma} = g^{\alpha\nu} g_{\beta\kappa} T^\kappa_{\nu\gamma}$, and $T^\alpha_{\beta\gamma} = (\Gamma^\alpha_{\beta\gamma} - \Gamma^\alpha_{\gamma\beta})$ the skew-symmetric torsion tensor. We are interested in (one-half) the Laplacian operator associated to ∇, i.e., the operator acting on smooth functions on M defined as

$$H(\nabla) := 1/2\nabla^2 = 1/2 g^{\alpha\beta} \nabla_\alpha \nabla_\beta. \tag{3}$$

A straightforward computation shows that $H(\nabla)$ only depends in the trace of the torsion tensor and g, since it is

$$H(\nabla) = 1/2\triangle_g + \hat{Q}, \tag{4}$$

with $Q := Q_\beta dx^\beta = T^\nu_{\nu\beta}dx^\beta$ the trace-torsion one-form and where $\hat{Q}$ is the vector field associated to Q via g: $\hat{Q}(f) = g(Q, df)$, for any smooth function f defined on M. Finally, $\triangle_g$ is the Laplace-Beltrami operator of g: $\triangle_g f = \mathrm{div_g}\ \mathrm{grad} f$, $f \in C^\infty(M)$, with $\mathrm{div_g}$ the Riemannian divergence. Thus for any smooth function, we have $\triangle_g f = 1/[det(g)]^{\frac{1}{2}} g^{\alpha\beta}\frac{\partial}{\partial x^\beta}([det(g)]^{\frac{1}{2}}\frac{\partial}{\partial x^\alpha} f)$. Consider the family of zeroth-order differential operators acting on smooth k-forms, i.e., differential forms of degree k ($k = 0, \ldots, n$) defined on M:

$$H_k(g, Q) := 1/2\triangle_k + L_{\hat{Q}}, \tag{5}$$

In the first summand of the r.h.s. of (5) we have the Hodge operator acting on k-forms:

$$\triangle_k = (d - \delta)^2 = -(d\delta + \delta d), \tag{6}$$

with d and δ the exterior differential and codifferential operators respectively, i.e., δ is the adjoint operator of d defined through the pairing of k-forms on M: $(\omega_1, \omega_2) := \int \otimes^k g(\omega_1, \omega_2)\mathrm{vol}_g$, for arbitrary k-forms ω_1, ω_2, where $vol_g(x) = \det(g(x))^{\frac{1}{2}}dx$ is the volume density. The last identity in (6) follows from the fact that $d^2 = 0$ so that $\delta^2 = 0$. Furthermore, the second term in (5) denotes the Lie-derivative with respect to the vector field $\hat{Q}$: $L_{\hat{Q}} = i_{\hat{Q}}d + di_{\hat{Q}}$, where $i_{\hat{Q}}$ is the interior product with respect to $\hat{Q}$: for arbitrary vector fields $X_1, \ldots, X_{k-1}$ and ϕ a k-form defined on M, we have $(i_{\hat{Q}}\phi)(X_1, \ldots, X_{k-1}) = \phi(\hat{Q}, X_1, \ldots, X_{k-1})$. Then, for f a scalar field, $i_{\hat{Q}}f = 0$ and

$$L_{\hat{Q}}f = (i_{\hat{Q}}d + di_{\hat{Q}})f = i_{\hat{Q}}df = g(Q, df) = \hat{Q}(f). \tag{7}$$

Since $\triangle_0 = (\nabla^g)^2 = \triangle_g$, we see that from the family defined in (5) we retrieve for scalar fields ($k = 0$) the operator $H(\nabla)$ defined in (3 & 4). The Hodge Laplacian can be further written expliciting the Weitzenbock metric curvature term, so that when dealing with $M = R^n$ provided with the Euclidean metric, $\triangle_k$ is the standard Euclidean Laplacian acting on the components of a k-form defined on R^n ($0 \leq k \leq n$).

Proposition 1. Assume that g is non-degenerate. There is a one-to-one mapping

$$\nabla \rightsquigarrow H_k(g, Q) = 1/2\triangle_k + L_{\hat{Q}}$$

between the space of g-compatible linear connections ∇ with Christoffel coefficients of the form

$$\Gamma^\alpha_{\beta\gamma} = \begin{Bmatrix} \alpha \\ \beta\gamma \end{Bmatrix} + \frac{2}{(n-1)}\left\{\delta^\alpha_\beta\, Q_\gamma\ -\ g_{\beta\gamma}\, Q^\alpha\right\}, n \neq 1 \tag{8}$$

and the space of elliptic second-order differential operators on k-forms (k=0,...,n).

3. Riemann-Cartan-Weyl diffusions on the tangent manifold

In this section we shall present the setting for the extension of the correspondence of Proposition 1 to a correspondence between RCW connections defined by (8) and diffusion processes of k-forms ($k = 0, \dots, n$) having $H_k(g, Q)$ as infinitesimal generators (i.g. for short, in the following). For this, we shall see this correspondence in the case of scalars, and then prepare the extension by defining diffusion processes on the tangent manifold.[1] We have already seen that introduction of more general covariant derivative operators (or still, of linear connections) than the Levi-Civita connection, is naturally associated with the appearance of an interaction term in the generalized Laplacians, which is the vector field given by the g-conjugate of a trace-torsion 1- form and thus with a RCW connection. We shall further see that in introducing the Wiener processes (white noise) and the rules of stochastic analysis [1], the present approach will lead us to associate the noise tensor of a generalized diffusion process with the Riemannian metric and the trace-torsion interaction term with the drift of a diffusion process.

For the sake of generality, in the following we shall further assume that $Q = Q(\tau, x)$ is a time-dependent 1-form. The stochastic flow associated to the diffusion generated by $H_0(g, Q)$ has for sample paths the continuous curves $\tau \mapsto x(\tau) \in M$ satisfying the Ito invariant non-degenerate s.d.e. (stochastic differential equation)

$$dx(\tau) = X(x(\tau))dW(\tau) + \hat{Q}(\tau, x(\tau))d\tau. \tag{9}$$

In this expression, $X : M \times R^m \to TM$ is such that $X(x) : R^m \to TM$ is linear for any $x \in M$, so that we write $X(x) = (X_i^\alpha(x))$ $(1 \leq \alpha \leq n,\ 1 \leq i \leq m)$ which satisfies $X_i^\alpha X_i^\beta = g^{\alpha\beta}$, where $g = (g^{\alpha\beta})$, and $\{W(\tau), \tau \geq 0\}$ is a standard Wiener process on R^m. Here τ denotes the time-evolution parameter of the diffusion (in a relativistic setting it should not be confused with the time variable), and for simplicity we shall assume always that $\tau \geq 0$. Indeed, taking in account the rules of stochastic analysis for which $dW^\alpha(\tau)dW^\beta(\tau) = \delta_\beta^\alpha d\tau$ (the Kronecker tensor), $d\tau dW(\tau) = 0$ and $(d\tau)^2 = 0$, we find that if $f : R \times M \to R$ is a C^2 function on the M-variables and C^1 in the τ-variable, then a Taylor expansion yields

$$\begin{aligned} f(\tau, x(\tau)) = f(0, x(0)) + [\frac{\partial f}{\partial \tau} + H_0(g, Q)f](\tau, x(\tau))d\tau \\ + \frac{\partial f}{\partial x^\alpha}(\tau, x(\tau))X_i^\alpha(x(\tau))dW^i(\tau) \end{aligned}$$

and thus $\frac{\partial}{\partial \tau} + H_0(g, Q)$ is the infinitesimal generator of the diffusion represented by integrating the s.d.e. (9). Furthermore, this identity sets up the so-called martingale problem approach to the random integration of linear evolution equations for scalar fields [1], and further, for differential forms as we shall see next. Note, that if we start with Eq. (9), we can reconstruct the associated RCW connection.

Our next step, is to extend the above results to differential forms. Consider the canonical Wiener space Ω of continuous maps $\omega : R \to R^m, \omega(0) = 0$, with the

[1] Thus, naturally we shall call these processes as *RCW diffusion processes*.

canonical realization of the Wiener process $W(\tau)(\omega) = \omega(\tau)$. The (stochastic) flow of the s.d.e. (9) is a mapping $F_\tau : M \times \Omega \to M$, $\tau \geq 0$, such that for each $\omega \in \Omega$, the mapping $F_.(. , \omega) : [0, \infty) \times M \to M$, is continuous and such that $\{F_\tau(x) : \tau \geq 0\}$ is a solution of equation (9) with $F_0(x) = x$, for any $x \in M$. Let us assume in the following that the components X_i^α, $\hat{Q}^\alpha$, $\alpha, \beta = 1, \ldots, n$ of the vector fields X and $\hat{Q}$ on M in Eq. (9) are predictable functions which further belong to $C_b^{m,\epsilon}$ ($0 < \epsilon < 1$, m a non-negative integer), the space of Hölder bounded continuous functions of degree $m \geq 1$ and exponent ϵ, and also that $\hat{Q}^\alpha(\tau) \in L^1(R)$, for any $\alpha = 1, \ldots, n$. With these regularity conditions, if we further assume that $x(\tau)$ is a semimartingale on a probability space $(\Omega, \mathcal{F}, P)$, then it follows that the flow of Eq. (9) has a modification (which with abuse of notation we denote as) $F_\tau(\omega) : M \to M$, $F_\tau(\omega)(x) = F_\tau(x, \omega)$, which is a diffeomorphism of class C^m, almost surely for $\tau \geq 0$ and $\omega \in \Omega$ [8]. We would like to point out that a similar result follows from working with Sobolev space regularity conditions instead of Hölder continuity. Indeed, assume that the components of X and $\hat{Q}$, $X_i^\alpha \in H^{s+2}(M)$ and $\hat{Q}^\beta \in H^{s+1}(M)$, $1 \leq i \leq m$, $1 \leq \beta \leq n$, where the Sobolev space $H^s(M) = W^{2,s}(M)$ with $s > \frac{n}{2} + m$, $m \geq 1$. Then, the flow of Eq. (9) for fixed ω defines a diffeomorphism in $H^s(M, M)$, and hence by the Sobolev embedding theorem, a diffeomorphism in $C^m(M, M)$ [9]. Let us describe the (first) derivative (or *jacobian*) flow of Eq. (9), i.e., the stochastic process $\{v(\tau) := T_{x_0}F_\tau(v(0)) \in T_{F_\tau(x_0)}M, v(0) \in T_{x_0}M\}$; here T_zM denotes the tangent space to M at z and $T_{x_0}F_\tau$ is the linear derivative of F_τ at x_0. The process $\{v_\tau, \tau \geq 0\}$ can be described [12] as the solution of the invariant Ito s.d.e. on TM:

$$dv(\tau) = \nabla^g \hat{Q}(\tau, v(\tau))d\tau + \nabla^g X(v(\tau))dW(\tau) \tag{10}$$

If we take U to be an open neighborhood in M so that the tangent space on U is $TU = U \times R^n$, then $v(\tau) = (x(\tau), \tilde{v}(\tau))$ is described by the system given by integrating Eq. (9) and the invariant Ito s.d.e.

$$d\tilde{v}(\tau)(x(\tau)) = \nabla^g X(x(\tau))(\tilde{v}(\tau))dW(\tau) + \nabla^g \hat{Q}(\tau, x(\tau))(\tilde{v}(\tau))d\tau, \tag{11}$$

with initial condition $\tilde{v}(0) = v_0$. Thus, $\{v(\tau) = (x(\tau), \tilde{v}(\tau)), \tau \geq 0\}$ defines a random flow on TM.

4. Realization of the RCW diffusions by ODE's

To realize the s.d.e's by o.d.e's it is mandatory to pass to the Stratonovich mid-point prescription, which are well known to have the same transformation rules in stochastic analysis that those of classical flows [1,2]. The need for such approximations is obvious whenever the noise tensor is not trivial, and thus the random integration may be extremely difficult; in the trivial noise case it becomes superfluous, as we shall see in the last section of this article. Thus, instead of Eq. (9) we consider the Stratonovich s.d.e. (here denoted, as usual, by the symbol $\circ$) for it given by :

$$dx(\tau) = X(x(\tau)) \circ dW(\tau) + b^{Q,X}(\tau, x(\tau))d\tau, \\ \text{where } b^{Q,X}(\tau, x(\tau)) = \hat{Q}(\tau, x(\tau)) + S(\nabla^g, X)(x(\tau)), \quad (12)$$

where the drift now contains an additional term, the Stratonovich correction term, given by $S(\nabla^g, X) = \frac{1}{2}\mathrm{tr}(\nabla^g_X X)$, where $\nabla^g_X X$, the Levi-Civita covariant derivative of X in the same direction and thus it is an element of TM, so that in local coordinates we have $S(\nabla^g, X)^\beta = \frac{1}{2}X_i^\beta \nabla^g_{\frac{\partial}{\partial x^\alpha}} X_i^\alpha$. Now we also represent the Jacobian flow using the Stratonovich prescription

$$d\tilde{v}(\tau) = \nabla^g X(x(\tau))(\tilde{v}(\tau)) \circ dW(\tau) + \nabla^g b^{Q,X}(\tau, x(\tau))(\tilde{v}(\tau))d\tau. \quad (13)$$

Now we shall construct classical flows to approximate the random flow $\{x(\tau) : \tau \geq 0\}$. We start by constructing a piecewise linear approximation of the Wiener process. Thus, we set for each $k = 1, 2, \ldots,$

$$W_k(\tau) = k[(\frac{j+1}{k} - \tau)W(\frac{j}{k}) + (\tau - \frac{j}{k})W(\frac{j+1}{k})], \\ \text{if } \frac{j}{k} \leq \tau \leq \frac{j+1}{k}, \quad j = 0, 1, \ldots \quad (14)$$

and we further consider the sequence $\{x_k(\tau)\}_{k\in N}$ satisfying

$$\frac{dx_k(\tau)}{d\tau} = X(x_k(\tau))\frac{dW_k}{d\tau}(\tau) + b^{Q,X}(\tau, x_k(\tau)), \quad (15)$$

$$\frac{d\tilde{v}_k(\tau)}{d\tau} = \nabla^g X(x_k(\tau))(\tilde{v}_k(\tau))\frac{dW_k}{d\tau}(\tau) + \nabla^g b^{Q,X}(\tau, x_k(\tau))(\tilde{v}_k(\tau)), \quad (16)$$

$$\frac{dW_k}{d\tau}(\tau) = k[W(\frac{j+1}{k}) - W(\frac{j}{k})] \text{ for} \frac{j}{k} < \tau < \frac{j+1}{k}, \quad (17)$$

(otherwise, it is undefined) so that $\frac{dW_k}{d\tau}(\tau)$ exists for almost all values of τ. Since $\{W_k(\tau)\}_{k\in N}$ is differentiable a.e., thus $\{x_k(\tau) : x_k(0) = x(0)\}_{k\in N}$ is a sequence of flows obtained by integration of well-defined o.d.e's on M, almost everywhere (a.e.) on τ, for all $W \in \Omega$. We remark that $\{x_k(\tau)\}_{k\in n}$ depends on the (here chosen canonical) realization of $W \in \Omega$ so that in rigour, we should write $\{x_k(\tau, W, x_0)\}_{k\in N}$ to describe the flow; the same observation is valid for the approximation of the derivative flow below. With the additional assumption that X and Q are smooth, then the previous sequence defines for almost all τ and for all $W \in \Omega$, a flow of smooth diffeomorphisms of M, and thus, the flow $\{v_k(\tau) = (x_k(\tau), \tilde{v}_k(\tau)) : v_k(0) = (x(0), v(0))\}$ defines a flow of smooth diffeomorphisms of TM. In this case, this flow converges uniformly in probability, in the group of smooth diffeomorphisms of TM, to the the flow of random diffeomorphisms on TM defined by Eqs. (12–13) [1, 2, 11].

5. RCW gradient diffusions of differential forms

Assume that there is an isometric immersion of an n-dimensional manifold M into a Euclidean space R^m given by the mapping $f:M\to R^m, f(x)=(f^1(x),\ldots,f^m(x))$. For example, $M = S^n, T^n$, the n-dimensional sphere or torus respectively, and f is an isometric embedding into R^{n+1}, or still $M = R^m$ with f given by the identity map. The existence of such a smooth immersion is proved by the Nash theorem in the compact manifold case, yet the result is known to be valid as well for non compact manifolds [15]. Assume further that $X(x) : R^m \to T_xM$, is the orthogonal projection of R^m onto T_xM the tangent space at x to M, considered as a subset of R^m. Then, if $e_1,\ldots,e_m$ denotes the standard basis of R^m, we have

$$X(x) = X^i(x)e_i, \text{ with } X^i(x) = \text{grad } f^i(x), i = 1,\ldots,m. \tag{18}$$

We should remark for the benefit of the reader, that although the noise term is provided by the isometric immersion and thus associated as in the general case with the Levi-Civita covariant derivative operator, we still have a more general covariant derivative, in fact a RCW connection, since the drift of the diffusion process will continue to be associated with the g-conjugate of the trace-torsion of this connection, which together with the metric, yields the RCW connection.

So we are interested in the RCW *gradient* diffusion processes on compact manifolds isometrically immersed in Euclidean space, given by (9) with the diffusion tensor X given by (18). We shall now give the Ito-Elworthy formula for k-forms ($0 \le k \le n$) on compact manifolds which are isometrically immersed in Euclidean space. Recall that the kth exterior product of k time-dependant vector fields $v_1,\ldots,v_k$ is written as $v_1 \wedge v_2 \wedge \ldots \wedge v_k$ and $\Lambda^k(R \times TM)$ is the vector space generated by them. We further denote by $C^{1,2}_c(\Lambda^k(R \times M))$ the space of time-dependant k-forms on M continuously differentiable with respect to the time variable and of class C^2 with respect to the M variable and of compact support with its derivatives.

Theorem 1 (Ito-Elworthy Formula for k-forms [12]). *Let M be isometrically immersed in R^m as above. Let $V_0 \in \Lambda^k T_{x_0}M$, $0 \le k \le n$. Set $V_\tau = \Lambda^k(TF_\tau)(V_0)$, the kth Grassmann product of the Jacobian flow of the RCW gradient diffusion with noise tensor $X = \nabla f$. Then $\partial_\tau + H_k(g,\hat{Q})$ is the i.g. (with domain of definition the differential forms of degree k in $C^{1,2}_c(\Lambda^k(R \times M))$) of $\{V_\tau : \tau \ge 0\}$.*

Remarks 2. Therefore, starting from the flow $\{F_\tau : \tau \ge 0\}$ of the s.d.e. (9) (or its Stratonovich version given by Eq. (12)) with i.g. given by $\partial_\tau + H_0(g,Q)$, we construct (fibered on it) the derived velocity process $\{v(\tau) : \tau \ge 0\}$ given by (10) (or (9 & 11), with the diffusion tensor given by (18), or still, its Stratonovich version given by Eqs. (12–13)) which has $\partial_\tau + H_1(g,Q)$ for i.g. Finally, if we consider the diffusion processes of differential forms of degree $k \ge 1$, we further get that $\partial_\tau + H_k(g,Q)$ is the i.g. of the process $\{\Lambda^k v(\tau) : \tau \ge 0\}$, on the Grassmannian bundle $\Lambda^k(R \times TM)$, $(k = 0,\ldots,n)$. Note that consistent with our notation, and since $\Lambda^0(TM) = M$ we have that $\Lambda^0 v(\tau) \equiv x(\tau), \forall\tau \ge 0$. In particular, $\partial_\tau + H_2(g,Q)$ is the i.g. of the stochastic process $\{v(\tau) \wedge v(\tau) : \tau \ge 0\}$ on $(R \times TM) \wedge (R \times TM)$.

Consider on a smooth manifold M isometrically immersed in Euclidean space, the following initial value problem: We want to solve

$$\frac{\partial}{\partial \tau}\beta = H_k(g,Q)\beta_\tau, \text{with } \beta(0,x) = \beta_0(x), 0 \leq k \leq n, \tag{19}$$

for an arbitrary time-dependant k-form $\beta = \beta_\tau(x) = \beta(\tau, x)$ defined on M which belongs to $C_c^{1,2}(\Lambda^k(R \times M))$. Then, the formal solution of this problem is as follows [13]: Consider the stochastic differential equation given by running backwards in time Eq. (14)[2]:

$$dx^{\tau,s,x} = X(x^{\tau,s,x}) \circ dW(s) + b^{Q,X}(\tau - s, x^{\tau,s,x})ds, x^{\tau,0,x} = x \in M, \tag{20}$$

and the derived velocity process $\{v^{\tau,s,v(x)}, v^{\tau,0,v(x)} = v(x) \in T_xM, 0 \leq s \leq \tau\}$ which in a coordinate system we write as $v^{\tau,s,v(x)} = (x^{\tau,s,x}, \tilde{v}^{\tau,s,v(x)})$ verifying (20) and the s.d.e.

$$\begin{aligned} d\tilde{v}^{\tau,s,v(x)} &= \nabla^g X(x^{\tau,s,x})(\tilde{v}^{\tau,s,v(x)}) \circ dW(s) \\ &\quad + \nabla^g b^{Q,X}(\tau - s, x^{\tau,s,x})(\tilde{v}^{\tau,s,v(x)})ds, \tilde{v}^{\tau,0,v(x)} = v(x). \end{aligned} \tag{21}$$

Notice that this system is nothing else than the Jacobian process running backwards in time until the beginning.

Theorem 2 [12]. *The formal solution of the initial value problem* (19) *is*

$$\beta(\tau,x)(\Lambda^k v(x)) = E_x[\beta_0(x^{\tau,\tau,x})(\Lambda^k \tilde{v}^{\tau,\tau,v(x)})], \tag{22}$$

where the l.h.s. $\Lambda^k v(x)$ denotes the exterior product of k linearly independant tangent vectors at x, and in the r.h.s. $\Lambda^k v^{\tau,\tau,v(x)}$ denotes the exterior product of the flows having initial condition given by $\Lambda^k v(x)$.

Proof. It follows from the Ito-Elworthy formula.

6. KDE and RCW gradient diffusions

The kinematic dynamo equation for a passive magnetic field transported by an incompressible fluid, is the system of equations [10] for the time-dependant magnetic vector field $B(\tau, x) = B_\tau(x)$ on M defined by $i_{B_\tau}\mu(x) = \omega_\tau(x)$ (for $\tau \geq 0$), satisfying

$$\partial_\tau \omega + (L_{\hat{u}_\tau} - \nu^m \triangle_{n-1})\omega_\tau = 0, \omega(0,x) = \omega(x), 0 \leq \tau, \tag{23}$$

where ν^m is the magnetic diffusivity, and we recall that $\mu = vol(g) = det(g)^{\frac{1}{2}} dx^1 \wedge \dots \wedge dx^n$ is the Riemannian volume; here the *velocity* 1-form $u_\tau(x) = u(\tau, x)$ satisfies the incompressibility condition $\delta u_\tau = - \div (\hat{u}_\tau) = 0, \forall \tau$ and the invariant Navier-Stokes equations (NS, on the following)

$$\frac{\partial u}{\partial \tau} = [\nu \triangle_1 - L_{\hat{u}_\tau}]u_\tau - dp_\tau, \tag{24}$$

[2]We can, of course, solve this problem by running the Ito form [12]

where p_τ is a time-dependant function, the pressure, ν is the kinematical viscosity, or either, the Euler equations obtained by setting $\nu = 0$. Note that we can rewrite KDE as

$$\partial_\tau \omega = H_{n-1}(2\nu^m g, -\frac{1}{2\nu^m} u_\tau)\omega_\tau, \omega(0,x) = \omega(x), 0 \leq \tau \tag{25}$$

while NS can be written as

$$\frac{\partial u}{\partial \tau} = H_1(2\nu g, \frac{-1}{2\nu} u_\tau)u_\tau, \delta u_\tau = 0,$$

which by considering the vorticity time-dependant 2-form $\Omega_\tau := du_\tau$ we have the equivalent system of equations

$$\frac{\partial \Omega_\tau}{\partial \tau} = H_2(2\nu g, \frac{-1}{2\nu} u_\tau)\Omega_\tau, H_1(g,0)u_\tau = -\delta\Omega_\tau,$$

the first one being NS for the vorticity obtained by applying d to Eq. (24) and the second one is the Poisson-de Rham equation, obtained by applying δ to the definition of Ω. In [3–7], the geometrical theory of diffusion processes was applied to give exact implicit representations for this system, in terms of stochastic differential equations, and further realize these representations in terms of systems of ordinary differential equations, and still to construct the random symplectic structure. In this article, we shall follow the same line of approach but for KDE, which for $n = 3$ is identical to NS for the vorticity, with ν^m instead of ν, yet we must keep in mind that for KDE we are after B_τ.

In the following we assume additional conditions on M, namely that it is isometrically immersed in an Euclidean space, so that the diffusion tensor is given in terms of the immersion f by $X = \nabla f$. Let u denote a solution of Eq. (24) (or still, of the Euler equation with $\nu = 0$) and consider the flow $\{F_\tau : \tau \geq 0\}$) of the s.d.e. whose i.g. is $\frac{\partial}{\partial \tau} + H_0(2\nu^m g, \frac{-1}{2\nu^m} u)$; from Eq. (9) and Theorem 1 we know that this is the flow defined by integrating the non-autonomous Ito s.d.e.

$$dx(\tau) = [2\nu^m]^{\frac{1}{2}} X(x(\tau))dW(\tau) - \hat{u}(\tau, x(\tau))d\tau, x(0) = x, 0 \leq \tau. \tag{26}$$

We shall assume in the following that X and $\hat{u}$ have the regularity conditions stated in Section 3 so that the random flow of Eq. (26) is a diffeomorphism of M of class C^m. Now if we express the random Lagrangian flow in Stratonovich form

$$dx(\tau) \quad = \quad [2\nu^m]^{\frac{1}{2}} X(x(\tau)) \circ dW(\tau) + b^{-u,X}(\tau, x(\tau))d\tau, \tag{27}$$

with

$$b^{-u,X}(\tau, x(\tau)) \quad = \quad \nu^m \text{tr } (\nabla^g_X X)(x(\tau)) - \hat{u}(\tau, x(\tau))), \tag{28}$$

we can approximate in the group of diffeomorphisms of M this flow by considering the sequence of a.e. o.d.e's

$$\frac{dx_k}{d\tau}(\tau) = [2\nu^m]^{\frac{1}{2}} X(x_k(\tau)) \frac{dW_k}{d\tau}(\tau) + b^{-u,X}(\tau, x_k(\tau)), k \in N, \tag{29}$$

with $\frac{dW_k}{d\tau}$ defined in Eq. (17), and we consider as well the Jacobian flow on TM, $\{v(\tau) = (x(\tau), \tilde{v}(\tau))\}$ with $\tilde{v}(\tau)$ satisfying the Stratonovich equations

$$d\tilde{v}(\tau)(x(\tau)) = [2\nu^m]^{\frac{1}{2}} \nabla^g X(x(\tau))(\tilde{v}(\tau)) \circ dW(\tau) + \nabla^g b^{-u,X}(\tau, x(\tau))(\tilde{v}(\tau))d\tau, \tag{30}$$

which can be approximated by $\{x_k(\tau), \tilde{v}_k(\tau))\}_{k\in N}$ given by integrating the a.e. o.d.e.

$$\frac{d\tilde{v}_k(\tau)}{d\tau} = [2\nu^m]^{\frac{1}{2}} \nabla^g X(x_k(\tau))(\tilde{v}_k(\tau)) \frac{dW_k}{d\tau}(\tau) + \nabla^g b^{-u,X}(\tau, x_k(\tau))(\tilde{v}_k(\tau)). \tag{31}$$

Thus, from [11] follows that the flow of the system of a.e. o.d.e's given by Eqs. (29, 31), and under the assumption that u is of class C^m ($m \geq 1$), converges uniformly in probability, in the group of diffeomorphisms of TM of class C^{m-1} to the random diffeomorphism flow, of the same class, that integrates KDE, as we shall see next.

Let us find the form of the strong solution (whenever it exists) of the initial value problem for $\omega(\tau, x)$ satisfying (23) with initial condition $\omega(0, x) = \omega_0(x)$ which we assume to be of class C^2. For this, we run backwards in time the random Lagrangian flow Eq. (26): For each $\tau \geq 0$ consider the s.d.e. (with $s \in [0, \tau]$):

$$dx^{\tau,s,x} = [2\nu^m]^{\frac{1}{2}} X(x^{\tau,s,x}) \circ dW(s) + b^{-u,X}(\tau - s, x^{\tau,s,x})ds, x^{\tau,0,x} = x. \tag{32}$$

and the derived velocity process $\{v^{\tau,s,v(x)} : v^{\tau,0,v(x)} = v(x) \in T_xM, 0 \leq s \leq \tau\}$ which in a coordinate system we write as $v^{\tau,s,v(x)} = (x^{\tau,s,x}, \tilde{v}^{\tau,s,v(x)})$ verifying (32) and the s.d.e.

$$\begin{aligned} d\tilde{v}^{\tau,s,v(x)} &= [2\nu^m]^{\frac{1}{2}} \nabla^g X(x^{\tau,s,x})(\tilde{v}^{\tau,s,v(x)}) \circ dW(s) \\ &\quad + \nabla^g b^{-u,X}(\tau - s, x^{\tau,s,x})(\tilde{v}^{\tau,s,v(x)})ds, \tilde{v}_0^{\tau,0,v(x)} = v(x) \in T_xM. \end{aligned} \tag{33}$$

Let $v^1(x), \ldots, v^{n-1}(x)$ linearly independent vectors in T_xM, be initial conditions for the flow $\tilde{v}^{\tau,x,v(x)}$.

Theorem 3. *If there is a $C^{1,2}$ (i.e., continuously differentiable in the time variable $\tau \in [0, T)$, and of class C^2 in the space variable) solution $\tilde{\omega}_\tau(x)$ of the initial value problem, it is*

$$\tilde{\omega}_\tau(v^1(x) \wedge \ldots v^{n-1}(x)) = E_x[\omega_0(x^{\tau,\tau,x})(\tilde{v}^{\tau,\tau,v^1(x)} \wedge \cdots \wedge \tilde{v}^{\tau,\tau,v^{n-1}(x)})], \tag{34}$$

where E_x denotes the expectation value with respect to the measure on $\{x^{\tau,\tau,x} : \tau \geq 0\}$.

Proof. It is evident from Theorems 1 and 2.

Now we can approximate Equation (34) by taking the Jacobian flow $\{(x_k^{\tau,s,x},\tilde{v}_k^{\tau,s,v(x)})\}_{k\in N}$ on TM given by

$$\begin{aligned}
\frac{dx_k^{\tau,s,x}}{ds}(s) &= [2\nu^m]^{\frac{1}{2}}X(x_k^{\tau,s,x})\frac{dW_k(s)}{ds}+b^{-u,X}(\tau-s,x_k^{\tau,s,x}),x_k^{\tau,0,x}=x,\\
\frac{d\tilde{v}_k^{\tau,s,v(x)}}{ds}(s) &= [2\nu^m]^{\frac{1}{2}}\nabla^g X(x_k^{\tau,s,x})(\tilde{v}_k^{\tau,s,v(x)})\frac{dW_k(s)}{ds}\\
&+ \nabla^g b^{-u,X}(\tau-s,x_k^{\tau,s,x})(\tilde{v}_k^{\tau,s,v(x)})ds,\tilde{v}_k^{\tau,0,v(x)}=v(x)\in T_xM\\
\text{where } \frac{dW_k(s)}{ds} &= 2^k\{W(\frac{[2^ks/\tau]+1}{2^k})-W(\frac{[2^ks/\tau]}{2^k})\}, s\in[0,\tau],(\tau>0), \quad (35)
\end{aligned}$$

with $[z]$ the integer part of $z\in(0,1]$, is the Stroock & Varadhan polygonal approximation [11]. Thus, we can write the expression:

$$\tilde{\omega}_\tau(v^1(x)\wedge\dots v^{n-1}(x)) = lim_{k\to\infty}E_x[\omega_0(x_k^{\tau,\tau,x})(\tilde{v}_k^{\tau,\tau,v^1(x)}\wedge\dots\wedge\tilde{v}_k^{\tau,\tau,v^{n-1}(x)})]. \quad (36)$$

7. KDE and random symplectic diffusions

Starting with a general RCW diffusion of 1-forms generated by $H_1(g,Q)$, we introduce a family of Hamiltonian functions, $\mathcal{H}_k(k\in N)$ defined on the cotangent manifold $T^*M=\{(x,p)/p:T_xM\to R \text{ linear}\}$ by

$$\mathcal{H}_k=\mathcal{H}_{X,k}+\mathcal{H}_Q, \quad (37)$$

with (in the following $\langle -,-\rangle$ denotes the natural pairing between vectors and covectors)

$$\mathcal{H}_{X,k}(x,p)=\langle\langle p,X(x)\rangle,\frac{dW_k}{d\tau}\rangle, \quad (38)$$

where the derivatives of W_k are given in (17), and

$$\mathcal{H}_{\hat{Q}}(x,p)=\langle p,b^{Q,X}(x)\rangle. \quad (39)$$

Now, we have a sequence of a.a. classical Hamiltonian flow, defined by integrating for each $k\in N$ the a.a. system of o.d.e.'s

$$\begin{aligned}
\frac{dx_k(\tau)}{d\tau} &= X(x_k(\tau))\frac{dW_k}{d\tau}+b^{Q,X}(x_k(\tau)), \quad (40)\\
\frac{dp_k(\tau)}{d\tau} &= -\langle\langle p_k(\tau),\nabla^g X(x_k(\tau))\rangle,\frac{dW_k(\tau)}{d\tau}\rangle-\langle p_k(\tau),\nabla^g b^{Q,X}(\tau,x_k(\tau))\rangle \quad (41)
\end{aligned}$$

which preserves the canonical 1-form $p_kdx_k=(p_k)_\alpha d(x_k)^\alpha$ (no summation on k!), and then preserves its exterior differential, the canonical symplectic form $S_k=dp_k\wedge dx_k$. We shall denote this flow as $\phi_.^k(\omega,.)$; thus $\phi_\tau^k(\omega,.):T^*_{x_k(0)}M\to T^*_{x_k(\tau)}M$, is a symplectic diffeomorphism, for any $\tau\in R_+$ and $\omega\in\Omega$. Furthermore, if we consider the contact 1-form [14] on $R\times T^*M$ given by $\gamma_k:=p_kdx_k-\mathcal{H}_{X,k}d\tau-\mathcal{H}_{\hat{Q}}d\tau,\forall k\in N$, we obtain a classical Poincaré-Cartan integral invariant: Let two smooth *closed* curves σ_1 and σ_2 in $T^*M \times \{\tau = \text{constant}\}$ encircle the same tube of trajectories of the Hamiltonian equations for $\mathcal{H}_k$, i.e., Eqs. (40, 41); then $\int_{\sigma_1} \gamma_k = \int_{\sigma_2} \gamma_k$. Furthermore, if $\sigma_1 - \sigma_2 = \partial\rho$, where ρ is a piece of the vortex tube determined by the trajectories of the classical Hamilton's equations, then it follows from the Stokes theorem [14] that

$$\int_{\sigma_1} \gamma_k - \int_{\sigma_2} \gamma_k = \int_{\sigma_1} p_k dx_k - \int_{\sigma_2} p_k dx_k = \int_{\rho} d\gamma_k = 0. \tag{42}$$

Returning to our construction of the random Hamiltonian system, we know already that for X and $\hat{Q}$ smooth, the Hamiltonian sequence of flows described by Eqs. (40, 41) converges uniformly in probability in the group of diffeomorphisms of T^*M, to the random flow of the system given by Eqs. (27, 28) and

$$dp(\tau) \quad = \quad -\langle\langle p(\tau), \nabla^g X(x(\tau))\rangle, \circ dW(\tau)\rangle - \langle p(\tau), \nabla^g b^{Q,X}(\tau, x(\tau))d\tau\rangle. \tag{43}$$

Furthermore this flow of diffeomorphisms is the mapping:

$$\phi_\tau(\omega, ., .)(x, p) = (F_\tau(\omega, x), F_\tau^*(\omega, x)p),$$

where $F_\tau^*(\omega, x)$ is the adjoint mapping of the Jacobian transformation. This map preserves the canonical 1-form pdx, and consequently preserves the canonical symplectic 2-form $S = d(pdx) = dp \wedge dx$, and thus $\phi_\tau(\omega, .) : T^*_{x(0)}M \to T^*_{x(\tau)}M$ is a flow of symplectic diffeomorphisms on T^*M for each $\omega \in \Omega$ [11]. Consequently, $\Lambda^n S$ is preserved by this flow, and thus we have obtained the Liouville measure invariant by a random symplectic diffeomorphism. We shall write onwards, the formal Hamiltonian function on T^*M defined by this approximation scheme as

$$\mathcal{H}(x, p) := \langle\langle p, X(x)\rangle, \frac{dW_\tau}{d\tau}\rangle + \mathcal{H}_{\hat{Q}}(x, p). \tag{44}$$

We proceed now to introduce the random Poincaré-Cartan integral invariant for this flow. Define the formal 1-form by the expression

$$\gamma := pdx - \mathcal{H}_{\hat{Q}} d\tau - \langle p, X\rangle \circ dW(\tau), \tag{45}$$

and its formal exterior differential (with respect to the $\mathcal{N} = T^*M$ variables only)

$$d_{\mathcal{N}}\gamma = dp \wedge dx - d_{\mathcal{N}}\mathcal{H}_{\hat{Q}} \wedge d\tau - d_{\mathcal{N}}\langle p, X\rangle \circ dW(\tau). \tag{46}$$

Clearly, we have a random differential form whose definition was given by Bismut [11,7]. Let a smooth r-simplex with values in $R_+ \times T^*M$ be given as

$$\sigma : s \in S_r \to (\tau_s, x_s, p_s),$$

where

$$S_r = \{s = (s_1, \ldots, s_r) \in [0, \infty)^r, s_1 + \ldots + s_r \leq 1\}, \tag{47}$$

with boundary $\partial\sigma$ the $(r-1)$-chain $\partial\sigma = \sum_{i=1}^{r+1} (-1)^{i-1}\sigma^i$, where σ^i are the $(r-1)$ singular simplexes given by the faces of σ. σ can be extended by linearity to any smooth singular r-chains. We shall now consider the *random continuous* r-simplex,

c, the image of σ by the flow of symplectic diffeomorphisms ϕ, i.e., the image in $R \times T^*M$

$$\phi(\tau_s, \omega, x_s, p_s) = (\tau_s, F_\tau(\omega, x_s), F^*_\tau(\omega, x_s)p_s), \text{ for fixed } \omega \in \Omega, \tag{48}$$

where $F_\tau(\omega, x)$ and $F^*_\tau(\omega, x)p$ are defined by Eqs. (27, 28 & 43), respectively.

Then, given α_0 a time-dependant 1-form on $\mathcal{N}$, $\beta_0, \ldots, \beta_m$ functions defined on $R \times \mathcal{N}$, the meaning of a random differential 1-form

$$\gamma = \alpha_0 + \beta_0 d\tau + \beta_i \circ dW^i(\tau), i = 1, \ldots, m, \tag{49}$$

is expressed by its integration on a *continuous* 1-simplex

$$c : s \to (\tau_s, \phi_{\tau_s}(\omega, n_s)), \text{ where } n_s = (x_s, p_s) \in T^*M, \tag{50}$$

the image by $\phi_.(\omega, .)$, $(\omega \in \Omega)$ the random flow of symplectomorphisms on T^*M, of the smooth 1-simplex $\sigma : s \in S_1 \to (\tau_s, (x_s, p_s))$. Then, $\int_c \gamma$ is a measurable real-valued function defined on the probability space Ω in [11,7]. Now we shall review the random differential 2-forms. Let now $\tilde{\alpha}_0$ be a time-dependant 2-form on $\mathcal{N}$, thus $\tilde{\alpha}_0(\tau, n)$ which we further assume to be smooth. Furthermore, let $\tilde{\beta}_0(\tau, n), \ldots, \tilde{\beta}_m(\tau, n)$ be smooth time-dependant 1-forms on $\mathcal{N}$ and we wish to give a meaning to the random differential 2-form

$$\gamma = \tilde{\alpha}_0 + d\tau \wedge \tilde{\beta}_0 + dW^1(\tau) \wedge \tilde{\beta}_1 + \ldots + dW^m(\tau) \wedge \tilde{\beta}_m. \tag{51}$$

on integrating it on a continuous 2-simplex $c : s \to (\tau_s, \phi_{\tau_s}(\omega, n_s))$, or which we define it as a measurable real-valued function on Ω in [11,7]. To obtain the random Poincaré-Cartan invariant we need the following results on the approximations of random differential 1- and 2-forms by classical differential forms. Given as before $\tilde{\alpha}_0$ a time-dependant smooth 2-form on $\mathcal{N}$ and time-dependant smooth 1-forms $\tilde{\beta}_1, \ldots, \tilde{\beta}_m$ on $\mathcal{N}$, there exists a subsequence k_i and a zero-measure $\hat{\Omega}$ subset of Ω dependant on $\tilde{\alpha}_0, \tilde{\beta}_1, \ldots, \tilde{\beta}_m$ such that for all $\omega \notin \hat{\Omega}$, $\phi^{k_i}_.(\omega, .)$ converges uniformly on any compact subset of $R_+ \times R^{2n}$ to $\phi_.(\omega, .)$ as well as all its derivatives $\frac{\partial^l \phi^{k_i}_.}{\partial n^l}(\omega, .)$ with $|l| \leq m$, converges to $\frac{\partial^l \phi_.}{\partial n^l}(\omega, .)$, and for any smooth 2-simplex, $\sigma : s \to (\tau_s, n_s)$ valued on $R_+ \times \mathcal{N}$, if

$$\gamma_k = \tilde{\alpha}_0 + d\tau \wedge (\tilde{\beta}_0 + \tilde{\beta}_1 \frac{dW^1_k}{d\tau} + \ldots \tilde{\beta}_m \frac{dW^m_k}{d\tau}) \tag{52}$$

and if c^k is the 2-simplex given by the image of a smooth 2-chain by the a.a. smooth diffeomorphism $\phi^k_.(\omega, .)$ defined by integration of Eqs. (40, 41): $c^k : s \to (\tau_s, \phi^k_{\tau_s}(\omega, n_s))$, and c is the continuous 2-chain $s \to (\phi_{\tau_s}(\omega, n_s))$, then $\int_{c^{k_i}} \gamma^{k_i}$ converges to $\int_c \gamma$. If instead we take a time-dependant 1-forms α_0 and time-dependant functions $\beta_0, \ldots, \beta_m$ on $\mathcal{N}$ and consider the time-dependant 1-form on $\mathcal{N}$ given by

$$\gamma_k = \alpha_0 + (\beta_0 + \beta_1 \frac{dW^1_k}{d\tau} + \ldots \beta_m \frac{dW^m_k}{d\tau}) d\tau \tag{53}$$

and for any a.e. smooth 1-simplex $c^k : s \to (\tau_s, \phi^k_{\tau_s}(\omega, n_s))$ then there exists a subsequence k_i and a zero-measure set $\hat{\Omega}$, dependant of $\alpha_0, \beta_0, \ldots, \beta_m$, such that for all $\omega \notin \hat{\Omega}$, $\phi^{k_i}_.(\omega, .)$ converges uniformly over all compacts of $R^+ \times R^{2n}$ with all its derivatives of order up to m to those of $\phi_{.}(\omega,.)$, and if c is the continuous 1-simplex $c: s \to (\tau_s, \phi_s(\omega, n_s))$, then $\int_{c^{k_i}} \gamma^{k_i}$ converges to $\int_c \gamma$, with γ defined in Eq. (49).

Then, we can state the fundamental theorem of Stokes for this random setting, which is due to Bismut [11] (Theorem 3.4). Let c be a random continuous 2-simplex image of an arbitrary smooth 2-simplex by the flow $\phi_{.}(\omega,.)$. There exists a zero-measure set $\tilde{\Omega} \subset \Omega$ such that for any $\omega \notin \Omega$, then $\int_c d\gamma = \int_{\partial c} \gamma$, for any differential random 1-form γ.

In the following in the case defined by KDE, for which $\hat{Q} = -\hat{u}$ with u a solution of NS or Euler equations, so that from Eqs. (45 & 49) we set

$$\alpha_0 = pdx, \beta_0 = -\mathcal{H}_{-\hat{u}} \equiv \mathcal{H}_{\hat{u}}, \beta_i = -(2\nu^m)^{\frac{1}{2}} \langle p, X_i \rangle \equiv p_\alpha X_i^\alpha, i = 1, \dots, m, \quad (54)$$

where $X : R^m \to TM$ with $X(x) = \text{grad}f$ with $f : M \to R^d$ is an isometric immersion of M, then

$$\begin{aligned} \gamma &= pdx + \mathcal{H}_{\hat{u}} d\tau - (2\nu^m)^{\frac{1}{2}} \langle p, X \rangle_i \circ dW^i(\tau) \\ &\equiv p_\alpha (dx^\alpha + (b^{u,X})^\alpha d\tau - (2\nu^m)^{\frac{1}{2}} X_i^\alpha \circ dW^i(\tau)), \end{aligned} \quad (55)$$

is the random Poincaré-Cartan 1-form defined on $R^+ \times \mathcal{N}$. The Hamiltonian function for KDE is

$$\mathcal{H}(x,p) := [2\nu^m]^{\frac{1}{2}} \langle \langle p, X(x) \rangle, \frac{dW_\tau}{d\tau} + \mathcal{H}_{-\hat{u}}(x,p), \quad (56)$$

with

$$\mathcal{H}_{-\hat{u}}(x,p) = p_\alpha (b^{-u,X})^\alpha = g^{\alpha\beta} p_\alpha (-u_\beta + \nu^m X_i^\alpha \nabla^g_{\frac{\partial}{\partial x^\beta}} X_i^\beta) \quad (57)$$

so that the Hamiltonian system is given by Eqs. (27, 28 & 43). As in the general case, we then obtain a Liouville invariant measure produced from the nth exterior product of the canonical symplectic form.

8. The Euclidean case

To illustrate with an example, consider $M = R^3$, $f(x) = x$, $\forall x \in M$, and then $X = \nabla f \equiv I$, the identity matrix, as well as $g = XX^\dagger = I$ the Euclidean metric, and $\nabla^g = \nabla$, is the gradient operator acting on the components of differential forms. Consequently, the Stratonovich correction term vanishes since $\nabla_X X = 0$ and thus the drift in the Stratonovich s.d.e's. is the vector field $b^{-u,X} = -\hat{u} = -u$ (we recall that $\hat{u}$ is the g-conjugate of the 1-form u, but here $g = I$). In this case the time-dependant 'magnetic' $(n-1)$-form in KDE, $\omega_\tau(x)$ is a 2-form on R^3 (or still, an R^3-valued function defined on R^3). The stochastic flow which integrates KDE is given by integrating the system of equations ($s \in [0, \tau]$)

$$\begin{aligned} dx^{\tau,s,x} &= [2\nu^m]^{\frac{1}{2}} \circ dW(s) - u(\tau - s, x^{\tau,s,x}) ds, x^{\tau,0,x} = x, \\ d\tilde{v}^{\tau,s,v(x))} &= -\nabla u(\tau - s, x^{\tau,s,x})(\tilde{v}^{\tau,s,v(x)}) ds, \tilde{v}^{\tau,0,v(x)} = v(x) \end{aligned} \quad (58)$$

the second being an ordinary differential equation (here, in the canonical basis of R^3 provided with Cartesian coordinates (x^1, x^2, x^3), ∇u is the matrix $(\frac{\partial u^i}{\partial x^j})$ for $u(\tau, x) = (u^1(\tau, x), u^2(\tau, x), u^3(\tau, x))$. Since $\int_0^\tau \circ dW(s) = W(\tau) - W(0) = W(\tau)$, we obtain

$$x^{\tau,s,x} = x + [2\nu]^{\frac{1}{2}} W(s) - \int_0^s u(\tau - r, x^{\tau,r,x}) dr, s \in [0, \tau], \tag{59}$$

and

$$\tilde{v}^{\tau,s,v(x)} = e^{-s\nabla u(\tau - s, x^{\tau,s,x})} v(x). \tag{60}$$

Finally, from Theorem 3 we have (with $v^i(x), i = 1, 2$ as before)

$$\tilde{\omega}_\tau(v^1(x) \wedge v^2(x)) = E_x[\omega_0(x^{\tau,\tau,x})(\tilde{v}^{\tau,\tau,v^1(x)} \wedge \tilde{v}^{\tau,\tau,v^2(x)})], \tag{61}$$

where the expectation value is taken with respect to the standard Gaussian function defined on R^3, albeit not centered on the origin of R^3 due to the last term in Eq. (59). We would like to remark that we can still follow [3, 4] to give in closed form, the implicit representation for $u(\tau, x)$ obeying NS on R^3.

We finally proceed to present the random symplectic theory for KDE on R^3. In account of (44) with the above choices, the Hamiltonian function is

$$\mathcal{H}(x, p) := [2\nu^m]^{\frac{1}{2}} \langle p, \frac{dW(\tau)}{d\tau} \rangle + \mathcal{H}_{-\hat{u}}(x, p), \tag{62}$$

with

$$\mathcal{H}_{-\hat{u}}(x, p) = -\langle p, u \rangle. \tag{63}$$

The Hamiltonian system is described by the Stratonovich s.d.e. for $x(\tau) \in R^3, \forall \tau \geq 0$:

$$dx(\tau) = [2\nu^m]^{\frac{1}{2}} \circ dW(\tau) - u(\tau, x(\tau)) d\tau, \tag{64}$$

and the o.d.e

$$dp(\tau) = -\langle p(\tau), \nabla u(\tau, x(\tau)) \rangle d\tau. \tag{65}$$

If we further set $x(0) = x$ and $p(0) = p$, the Hamiltonian flow preserving the canonical symplectic form $S = dp \wedge dx$ on R^6 is given by

$$\begin{aligned}\phi_\tau(.,.)(x, p) &= (x(\tau), p(\tau)) \\ &= (x + [2\nu^m]^{\frac{1}{2}} W(\tau) - \int_0^\tau u(r, x(r)) dr, e^{-\tau \nabla u(\tau, x(\tau))} p).\end{aligned} \tag{66}$$

Finally, the Poincaré -Cartan 1-form takes the form

$$\gamma = \langle p, dx - u d\tau - (2\nu^m)^{\frac{1}{2}} \circ dW(\tau) \rangle, \tag{67}$$

and the Liouville invariant is $S \wedge S \wedge S$. This completes the implementation of the general construction on $3D$.

Final Remarks. Geometrical-topological invariants in magnetohydrodynamics and hydrodynamics have been widely studied [10,16,18]. Following the presentation in [7] which lead to the random symplectic invariants of NS, in this article we have studied their extension to passive magnetohydrodynamics. The invariants produced in this approach are new to the best understanding of this author. A similar theory can be produced for the smooth boundary case.

Acknowledgements

The author would like to express his gratitude to Prof. José Francisco Rodrigues and the Organizing Committee of the Conference, for their kind support and a most pleasant stay at Obidos, and to Prof. Ana Bela Cruzeiro for helpful comments and her kind interest in the author's work. Finally, our gratitude to the reviewer for suggesting the presentation of the Euclidean case and several improvements to the original draft.

References

[1] N. Ikeda and S. Watanabe, *Stochastic Differential Equations and Diffusion Processes*, North-Holland/Kodansha, 1981.

[2] P. Malliavin, Géométrie Différentielle Stochastique, Les Presses Univ., Montreal, 1978.

[3] D.L. Rapoport, Random diffeomorphisms and integration of the classical Navier-Stokes equations, *Rep. Math. Phys.* **49**, no. 1, 1–27 (2002).

[4] D.L. Rapoport, Random representations for viscous fluids and the passive magnetic fields transported by them, in *Proceedings, Third International Conference on Dynamical Systems and Differential Equations, Kennesaw, May 2000*, Discrete & Cont. Dyn. Sys. special issue, S. Hu (ed.), 2000.

[5] D.L. Rapoport, On the geometry of the random representations for viscous fluids and a remarkable pure noise representation, *Rep. Math. Phys.* **50**, no. 2, 211–250 (2002).

[6] D.L. Rapoport, Martingale problem approach to the representations of the Navier-Stokes equations on smooth boundary manifolds and semispace, *Rand. Oper. Stoch. Eqts.* **11**, no. 2, 109–136 (2003).

[7] D.L. Rapoport, Random symplectic geometry and the realization of the random representations of the Navier-Stokes equations by ordinary differential equations, *Rand. Oper. Stoch. Eqts.* **vol. 11**, no. 2, 359–380 (2003).

[8] K. Kunita, *Stochastic Flows and Stochastic Differential Equations*, Cambridge Univ. Press, Cambridge, 1994.

[9] P. Baxendale and K.D. Elworthy, Flows of Stochastic Systems, Z. Wahrschein. verw. Gebiete 65, 245–267 (1983).

[10] V.I. Arnold and B.A. Khesin, *Topological Methods in Hydrodynamics*, Springer, New York, 1999.

[11] J.M. Bismut, *Mécanique Analytique*, Springer LNM **866**, 1982.

[12] K.D. Elworthy, Stochastic Flows on Riemannian Manifolds, in *Diffusion Processes and Related Problems in Analysis*, M. Pinsky et al. (eds.), vol. II, Birkhäuser, 1992.

[13] Elworthy, K.D. and Li, X.M., Formula for the derivative of heat semigroups, *J. Func. Anal.* **125**, 252–286 (1994).

[14] V.I. Arnold, *Mathematical Methods in Classical Mechanics*, Springer Verlag, 1982.

[15] M. Taylor, *Partial Differential Equations, vol. I*, Springer Verlag,1995.

[16] M. Ghill & S. Childress, *Stretch, Twist and Fold*, Springer Verlag, 1995

[17] D. Ebin & J. Marsden,Groups of diffeomorphisms and the motion of incompressible fluids, *Ann. Math.* **92**, 102–163 (1971).

[18] H.K. Moffatt & R.L. Ricca, Helicity and Calugareanu invariants, *Procds. Roy. London* **A 439** (1992), 411–429.

Diego L. Rapoport
Applied Mechanics, Faculty of Engineering
Univ. of Buenos Aires, Buenos Aires & Department of Sciences
Universidad Macional de Quilmes
Bernal, Argentina
e-mail: `drapo@unq.edi.ar`

Progress in Nonlinear Differential Equations
and Their Applications, Vol. 61, 243–256

Quasi-Lipschitz Conditions in Euler Flows

Reimund Rautmann

Abstract. In mathematical models of incompressible flow problems, quasi-Lipschitz conditions present a useful link between a class of singular integrals and systems of ordinary differential equations. Such a condition, established in suitable form for the first-order derivatives of Newtonian potentials in $\mathbb{R}^n$ (Section 2) gives the main tool for the proof (in Sections 3–6) of the existence of a unique classical solution to Cauchy's problem of Helmholtz's vorticity transport equation with partial discretization in $\mathbb{R}^3$ for each bounded time interval. The solution depends continuously on its initial value and, in addition, fulfills a discretized form of Cauchy's vorticity equation.

Mathematics Subject Classification (2000). 35 A 05, 35 A 35, 34 A 40, 34 A 12, 76 C 05.

Keywords. Helmholtz's and Cauchy's vorticity equations with a discretization in Hölder spaces. Quasi-Lipschitz conditions.

1. Introduction

The aim of this note is point out the rôle of quasi-Lipschitz conditions as a useful link between a class of singular integrals, representing solutions of partial differential equations, and systems of ordinary differential equations modelling incompressible flow problems. Namely after having shown the quasi-Lipschitz condition for gradients of Newtonian potentials in $\mathbb{R}^n$ (Section 2), in the following Sections 3–6 we will use it in order to prove the existence of a unique classical solution to the Cauchy problem of Helmholtz's vorticity transport equation with partial discretization in $\mathbb{R}^3$ for each bounded time interval.

Let v denote a continuous function defined for $(t, x) \in J \times \mathbb{R}^n$, $v(t, x) \in \mathbb{R}^n$, $J = [0, a]$, $n \geq 2$. The global existence of a unique flow $X = Lv$, $X = X(t, s, x) \in \mathbb{R}^n$ solving the initial value problem of the differential equation

$$\frac{\partial}{\partial t} X = v(t, X)\,, \quad t \in J, \tag{1.1}$$
$$X(s, s, x) = x\,, \quad s \in J,$$

is guaranteed by a uniform Lipschitz condition for $v(t,x)$ with respect to $x \in \mathbb{R}^n$, but also by the more general condition

$$[v(t,\cdot)]_\ell = \sup_{\substack{x,y\in\mathbb{R}^n \\ x\neq y}} \frac{|v(t,y)-v(t,x)|}{\ell(|y-x|)} \leq \lambda < \infty \tag{1.2}$$

with the function

$$\ell(r) = \begin{cases} 0 & , \quad r = 0, \\ -r\ln r & , \quad r \in (0, e^{-1}), \\ r & , \quad r \geq e^{-1}, \end{cases}$$

and a constant λ. The uniqueness of (1.1) under condition (1.2) goes back to Osgood [13], the global existence of the solution $X(t,s,x)$ for all $(t,s,x) \in J \times J \times \mathbb{R}^n$ results easily from (1.2) by Wintner's criterium [5]. Following Kato [7], the requirement (1.2) is referred to as quasi-Lipschitz condition for v.

In many nonlinear problems of fluid dynamics the direction field v (i.e., the flow velocity) is represented by first-order derivatives of a Newtonian potential V in which the mass density depends again on flow variables, but in such a way that bounds for the supremum norm of the density are available. However, as shown in potential theory [4], in order to get a Lipschitz bound for ∇V, stronger Hölder norm estimates for the mass density in V would be required.

This shows the useful strength of quasi-Lipschitz conditions: Namely, such a condition already holds for *a* direction field $v = \nabla V$ if the continuous mass density in the Newtonian potential V has compact support in $\mathbb{R}^n$.

In the case of a 2-dimensional Euler flow, representing the velocity $v(t,x)$ in terms of its vorticity $w(t,x) = \operatorname{rot} v(t,x)$, Wolibner [22], Hölder [6] and Kato [7] have used condition (1.2) as a decisive tool for proving global existence of unique classical solutions to the initial value problem of the Euler equations.

The analogous result for the Cauchy problem of the nonstationary vorticity transport-diffusion equation in $\mathbb{R}^2$ has been established in [15] also by making essential use of condition (1.2).

Notations. Besides the usual Banach space $C^0(\Omega)$ of all uniformly bounded, continuous functions f, $C^0(\Omega)$ being equipped with the norm $|f|_0\Omega = \sup_{z\in\Omega}|f(z)|$, $f(z) \in \mathbb{R}^n$, $n = 1,3$, where Ω denotes any one of the sets $\mathbb{R}^3$, $J = [0,a]$, $J \times \mathbb{R}^3$, $B = \{x \in \mathbb{R}^3 \,|\, |x| \leq R\}$, we will work with the space $C^1(\Omega)$, or $C^{0,1}(J \times \Omega)$ of all functions $f \in C^0(\Omega)$, or $f \in C^0(J \times \Omega)$, which also have continuous and uniformly bounded first-order derivatives with respect to all coordinates $(z_i) \in \Omega$, or in $(t,(z_i)) \in J \times \Omega$ only with respect to the coordinates $(z_i) \in \Omega$, respectively.

For any Hölder exponent $\alpha \in (0,1)$, boundedness of the Hölder seminorm

$$[g]_\alpha^\Omega = \sup_{\substack{x,y\in\Omega \\ 0<|y-x|<1}} \frac{|g(y)-g(x)|}{|y-x|^\alpha} \tag{1.3}$$

is required in the Hölder spaces $C^\alpha(\Omega) = \{g \in C^0(\Omega) | [g]_\alpha^\Omega < \infty\}$, $C^\alpha(\Omega)$ being equipped with the Hölder norm $|g|_\alpha\Omega = |g|_0\Omega + [g]_\alpha\Omega$, $C^{0,\alpha}(J\times\Omega) = \{f \in C^0(J\times$

$\Omega)|\sup_{t\in J}[f(t,\cdot)]_\alpha^\Omega < \infty\}$. We will omit the index Ω, if no confusion is possible. By $c_0, c_1, c', \ldots$ we will denote constants which may have different values at different places.

2. A quasi-Lipschitz condition for first-order derivatives of Newtonian potentials in $\mathbb{R}^n$.

With points $x = (x_i)$, $y = (y_i) \in \mathbb{R}^n$, $r = |x - y|$, these derivatives have the form

$$(K_i f)(x) = \int_{\mathbb{R}^n} (x_i - y_i) \cdot r^{-n} \cdot f(y)dy\,, \quad i = 1, \ldots, n, \tag{2.1}$$

where f denotes any given continuous real-valued function, f having compact support. Thus $K_i f$ is continuous on $\mathbb{R}^n$. Estimates of the type below go back to [2,8].

Proposition 2.1. *For all continuous functions $f : \mathbb{R}^n \to \mathbb{R}, n \geq 2, f$ having compact support* $\operatorname{supp} f \subset \mathbb{R}^n$, *the estimates*

(i) $|K_i f|_0 \leq c \cdot |f|_0, \quad c = \omega_n + |\overset{\circ}{\operatorname{supp}} f|,$
(ii) $[K_i f]_\ell \leq c_1 \cdot |f|_0, \quad c_1 = c_0 \cdot c,$
(iii) $|\nabla K_i f|_{\alpha^\cdot} \leq c_\alpha \cdot |f|_\alpha$ *hold, where ω_n denotes the measure of the $n-1$-dimensional unit sphere in $\mathbb{R}^n, c_\alpha$ depending on $\alpha \in (0,1), n$, and $|\overset{\circ}{\operatorname{supp}} f|$ only.*

Proof. The proof given in [7] for 2-dimensional bounded domains easily can be extended to $\mathbb{R}^n$, but for completeness of our estimates in Section 5 and 6 below we will stress the fact that the value of the bound c in (i) depends only on n and the Lebesgue measure $|\overset{\circ}{\operatorname{supp}}\ f|$ of the open support of f.

Writing $r = |x - y|$, we find (i) from

$$|(K_i f)(x)| \leq |f|_0 \cdot \left\{ \int_{r\leq 1} r^{1-n} dy + \int_{1<r, y\in \overset{\circ}{\operatorname{supp}} f} dy \right\}.$$

To prove (ii), we consider two different points

$$x_m = (x_{mi}) \in \mathbb{R}^n, \text{ with } r_m = |x_m - y|, \quad m = 0, 1\,, \quad d = |x_1 - x_0|.$$

The difference $k = |(x_{1i} - y_i)r_1^{-n} - (x_{0i} - y_i)r_0^{-n}|\,|f(y)|$ firstly integrated over the ball $\{y \in \mathbb{R}^3 |\, r_0 \leq 2d\}$ gives

$$\begin{aligned}\int_{r_0\leq 2d} k\, dy &\leq |f|_0 \cdot \left\{ \int_{r_0\leq r_1, r_0\leq 2d} r_0^{1-n} dy \ + \int_{r_1\leq r_0, r_1\leq 2d} r_1^{1-n} dy \ + \int_{r_0\leq 2d} r_0^{1-n} dy \right\} \\ &\leq |f|_0 \cdot 6\omega_n d. \end{aligned} \tag{2.2}$$

On the complementary region $r_0 > 2d$ we rewrite the integrand in the form

$$k = \left|(x_{1i} - x_{0i})r_1^{-n} + (x_{0i} - y_i)(r_1^{-n} - r_0^{-n})\right|\left|f(y)\right|. \tag{2.3}$$

On the line segment $l = \{x_s\} = \{x_0 + s(x_1 - x_0)\big|s \in [0,1]\}$, with $r_s = |x_s - y|$ we find $\left|r_1^{-n} - r_0^{-n}\right| \le n \cdot d \cdot \left|\int_0^1 r_s^{-n-1} ds\right|$. An elementary geometric consideration in the plane through 3 linearly independent points x_0, x_1, y shows the relations

$$r_0 = c \cdot d, \quad r_s \ge r_1 \ge (c-1)d, \text{ thus } \frac{r_s}{r_0} \ge \frac{r_1}{r_0} \ge \frac{c-1}{c} \ge \frac{1}{2} \tag{2.4}$$

which hold for all points $y \in \mathbb{R}^n$ on the sphere $S = \{y|\, |y - x_0| = c \cdot d\}$ with $c \ge 2$. (i.e., on the line segment l the point x_1 with $r_1 = (c-1)d$ is the nearest one to the sphere S around x_0.) Introducing the latter bounds and $|x_{0i} - y_i| \le r_0$ in k from (2.3) we get

$$\int\limits_{2d<r_0} k\,dy \le |f|_0 \cdot d \cdot \Bigg\{ \int\limits_{2d \le r_0 \le 1} 2^n r_0^{-n} dy + \int\limits_{\overset{\circ}{\text{supp}} f} dy \tag{2.5}$$

$$+ n \cdot \Bigg(\int\limits_{2d \le r_0 \le 1} 2^{n+1} r_0^{-n} dy + \int\limits_{\overset{\circ}{\text{supp}} f} dy \Bigg) \Bigg\}$$

$$\le |f|_0 \cdot d \left\{ 2^n \omega_n (1 + 2n) |ln(2d)| + (1+n) |\overset{\circ}{\text{supp}}\ f| \right\},$$

where the first and the third integral above appear in case $2d < 1$ only. Adding (2.2) and (2.5) gives (ii) with suitable constant c_0. The last estimate (iii) is well known from potential theory, cf. [4] or [9 p. 46–49, 59–62]. □

3. The hydrodynamical equations of Euler and Helmholtz

The vorticity vector

$$w(t,y) = (w_j(t,x)) = \text{ rot } v(t,x)\,, \quad j = 1,2,3 \tag{3.1}$$

of any smooth solution $v(t,x)$ to the Cauchy problem of Euler's momentum transport equation

$$\frac{\partial}{\partial t} v + v \cdot \nabla v + \nabla p = 0, \qquad \text{div } v = 0 \quad \text{ in } J \times \mathbb{R}^3, \tag{3.2}$$

$$v = v_0, \quad t = 0,$$

containing the pressure function $p(t,x)$, fulfills Helmholtz's vorticity transport equation

$$\frac{\partial}{\partial t} w + v \cdot \nabla w = w \cdot \nabla v \quad \text{ in } J \times \mathbb{R}^3, \quad w = w_0 = \text{ rot } v_0, \quad t = 0, \tag{3.3}$$

where the velocity vector v is given by Biot-Savart's formula

$$v(t,x) = \frac{1}{4\pi} \cdot \int_{\mathbb{R}^3} \mathrm{rot}_x \frac{w(t,y)}{|x-y|}\, dy = (Kw(t,\cdot))(x) = (Kw)(t,x). \tag{3.4}$$

Remark 3.1. *Recalling the vector identity*

$$rot_x(|x-y|^{-1}w(t,y)) = -|x-y|^{-3}(x-y) \times w(t,y),$$

we see that the components of the vector Kw are differences of terms having the form $\frac{1}{4\pi}$ $K_i f$ in Proposition 2.1, *where $f = w_j$. Therefore if, e.g., $w(t,\cdot)$ is Hölder continuous in $x \in \mathbb{R}^3$ having always compact support in $\mathbb{R}^3$, the integral* (3.4) *exists and besides* (3.1) *also* $div\ v(t,\cdot) = 0$ *holds.*

We get Helmholtz's equation (3.3) by taking the rotation in (3.2). Thus in $\mathbb{R}^3$ (under the suitable smoothness assumption) equation (3.3) represents the necessary and sufficient condition for the term $\frac{\partial}{\partial t}v + v \cdot \nabla v$ being a gradient field.

Since in any 2-dimensional flow parallel to the (x_1, x_2)-plane in $\mathbb{R}^3$ the right-hand side in (3.3) vanishes, the resulting conservation law for $w(t,x)$ along particle's pathes opens the way to proofs for global existence of unique classical solutions to (3.3), (3.4) and (3.2). The quasi-Lipschitz condition (1.2) depending on a bound for $|w(t,\cdot)|_0$ only is the decisive tool, cf. [6, 7, 22]. Concerning 3-dimensional Euler flows a lot of work has been done on solutions to (3.2) which exist locally in time, and on blow-up criteria, cf. [10–12] and the citations there. Thus at least from the numerical point of view it seems to be a remarkable fact that a single discretization transforms (3.3) together with (3.4) into a Cauchy problem which has a unique classical solution on each bounded time interval.

4. Helmholtz and Cauchy's vorticity equation with a discretization

In order to fix a domain for the linear map $K : w \to Kw$, we introduce the linear subspace $C_B^0 \subset C^0(J \times \mathbb{R}^3)$, C_B^0 containing each function $f \in C^0(J \times \mathbb{R}^3)$ for which a ball $B_f \subset \mathbb{R}^3$ exists, B_f covering the support $\operatorname{supp} f(t,\cdot)$ for all $t \in J$:

$$C_B^0\{f \in C^0(J \times \mathbb{R}^3) | \exists B_f = \{x \in \mathbb{R}^3 | |x| \le R_f\}, \operatorname{supp} f(t,\cdot) \subset B_f\ \forall t \in J\}.$$

The point in our definition of C_B^0 is that for each single $f \in C_B^0$ we can find a ball B_f such that $f(t,x)$ is vanishing at all points x outside of B_f for all $t \in J$.

Theorem 4.1. *Let the prescribed initial value w_0 be one times continuously differentiable, w_0 having compact support, $\epsilon \neq 0$ denoting a real constant. Then the initial value problem*

$$\frac{\partial}{\partial t}w + v \cdot \nabla w = \frac{1}{\epsilon}\left\{v(t, x + \epsilon w(t,x)) - v(t,x)\right\}, \quad (t,x) \in J \times \mathbb{R}^3, \tag{4.1}$$

$$w(0,x) = w_0(x)$$

with $v = Kw$ from (3.4) has a unique global solution $w \in C^1(J \times \mathbb{R}^3) \cap C_B^0$. The function w can be approximated by iteration of a contracting map T and depends continuously on w_0.

For a proof of Theorem 4.1. which follows the ideas we have formulated in [14], in a preparatory step with prescribed $v \in C^{0,1}$, we will integrate (4.1), getting a discretized form of Cauchy's vorticity equation.

Proposition 4.2. *Assume $v \in C^{0.1}$, $w_0 \in C^1(\mathbb{R}^3)$ be given. Then*

(i) *each solution $w \in C^1(J \times \mathbb{R}^3)$ of (4.1.) with $w(0,\cdot) = w_0$ has the representation*

$$w(t,x) = \frac{1}{\epsilon}\{X(t,0,Z(\cdot)) - X(t,0,\cdot)\} \circ X(0,t,x) = \tag{4.2}$$

$$= (HX)(t,x), \quad (t,x) \in J \times \mathbb{R}^3 \quad \textit{with}$$

$$Z(x) = x + \epsilon w_0(x), \tag{4.3}$$

$X = Lv$ denoting the solution of (1.1).

(ii) *Conversely the function $w = HX$ in (4.2) belongs to $C^1(J \times \mathbb{R}^3)$ and solves* (4.1).

Proof. Recalling well-known facts from the theory of ordinary differential equations, in case $v \in C^{0,1}$ the unique solution $X(t,s,\widehat{x})$ of (1.1) belongs to $C^1(J \times J \times \mathbb{R}^3)$, $X(t,s,\cdot) = X^{-1}(s,t,\cdot)$ representing a C^1-diffeomorphism of $\mathbb{R}^3$. Thus the representation of any function $w(t,x)$ in Lagrangian coordinates $\widehat{x}$,

$$\widehat{w}(t,\widehat{x}) = w(t, X(t,0,\widehat{x})), \tag{4.4}$$

is well defined. Because of the equation

$$\left(\frac{\partial}{\partial t}\widehat{w}(t,\cdot)\right) \circ X^{-1}(t,0,x) = \left(\frac{\partial}{\partial t}w + v \cdot \nabla w\right)(t,x) \tag{4.5}$$

which results immediately from the chain rule, initial value problem (4.1) is equivalent to

$$\frac{\partial}{\partial t}\widehat{w} = \frac{1}{\epsilon}\{v(t, X + \epsilon w(t,X)) - v(t,X)\} \quad \text{or} \tag{4.6}$$

$$\frac{\partial}{\partial t}\{X + \epsilon\widehat{w}\} = v(t, X + \epsilon\widehat{w}), \quad t \in J, \text{ with} \tag{4.7}$$

$$X + \epsilon\widehat{w} = \widehat{x} + \epsilon w_0(\widehat{x}), \qquad t = 0,$$

where $v(t,X) = \frac{\partial}{\partial t}X$, $X = X(t,0,\widehat{x})$. Equation (4.7) expresses the initial value problem (1.1) for the function $X + \epsilon\widehat{w}$. Thus recalling again the unique solvability of (1.1) we find

$$(X + \epsilon\widehat{w})(t,\widehat{x}) = X(t,0,\widehat{x} + \epsilon w_0(\widehat{x})) \quad \text{or} \tag{4.8}$$

$$\widehat{w}(t,\widehat{x}) = \frac{1}{\epsilon}\{X(t,0,Z(\widehat{x})) - X(t,0,\widehat{x})\} \tag{4.9}$$

which shows (4.2) because of $\widehat{x} = X(0,t,x)$.

Finally recalling $X^{\pm 1}(t,0,\cdot) \in C^1(\mathbb{R}^3)$ in case $v \in C^{0,1}$, from our requirement in (ii) we conclude $w = HX \in C^1(J \times \mathbb{R}^3)$, equation (4.2) being equivalent to (4.9). Differentiating the latter with respect to t and using the representation (4.5) above of $\frac{\partial}{\partial t}\widehat{w}(t,\widehat{x})$ with $\widehat{x} = X^{-1}(t,0,x)$ we get (4.1). □

Remark 4.3. *Taking in* (4.9) *the limit* $\epsilon \to 0$ *we find Cauchy's vorticity equation, [19, p. 152].*

5. The fixpoint equation

In order from (4.2) to get a fixpoint equation for w, we only have to find appropriate domains for the maps K and L such that the composed map

$$T = HLK$$

is properly defined. For any given initial value $f_0 \in C^1(\mathbb{R}^3)$ with

$$\overset{\circ}{\operatorname{supp}} f_0 \subset B_{R_0},\ |\overset{\circ}{\operatorname{supp}} f_0| \le \tau \quad \text{and} \quad \text{fixed } R_1 \ge R_0,\ \tau > 0,$$

we consider the class

$$(5.0) \quad C_{f_0} = \{f \in C^0(J \times \mathbb{R}^3) | f(0,\cdot) = f_0,\ \operatorname{supp} f(t,\cdot) \subset B_{R_1}, |\overset{\circ}{\operatorname{supp}} f(t,\cdot)| \le \tau\}$$

of all continuous vector-valued functions $f, f(t,x) \in \mathbb{R}^3$, which always have their support $\operatorname{supp} f(t,\cdot)$ inside the *fixed* ball $B_{R_1} = \{x \in \mathbb{R}^3 | |x| \le R_1\}$, the Lebesgue measure of the open support $\overset{\circ}{\operatorname{supp}}\ f(t,\cdot)$ being uniformly bounded by τ. As we will see, for any $w \in C_{f_0}$ the quasi-Lipschitz condition for $v(t,x) = (Kw(t,\cdot)(x)$ in (3.4) ensures the existence of the homeomorphisms $X(t,s,\cdot) = X^{-1}(s,t,\cdot)$ of $\mathbb{R}^3$ which are uniquely defined by the solution $X = Lv$ of (1.1) for all $t, s \in J$. Thus we can calculate HX in (4.2), and T is well defined on C_{f_0}. More precisely we have

Proposition 5.1. *Assume* $w_0 \in C^1(\mathbb{R}^3)$, $\operatorname{supp} w_0 \subset B_{R_0}$, $|\overset{\circ}{\operatorname{supp}}\ w_0| \le \tau$, $R_1 \ge R_0$. *Then the composed map* $T = HKL$ *is defined for all* $w \in C_{w_0}$ *and fulfills*

(i) $T(C_{w_0}) \subset C^0(J \times \mathbb{R}^3)$, $\quad (Tw)(t,0) = w_0$ *for* $t = 0$, $w \in C_{w_0}$,

(ii) $\overset{\circ}{\operatorname{supp}}\ (Tw)(t,\cdot) = X(t,0,\cdot)(\overset{\circ}{\operatorname{supp}}\ w_0)$.

(iii) *In case* $w \in C_{w_0} \cap C^{0,\gamma}$ *for some* $\gamma \in (0,1)$, *there holds*

$$|\overset{\circ}{\operatorname{supp}}\ w(t,\cdot)| = \quad |\overset{\circ}{\operatorname{supp}}\ w_0|, \ \textit{and}\ Tw \in C^1(J \times \mathbb{R}^3).$$

Moreover, if in addition to (iii) we require $|w(t,\cdot)|_0 \le N(t) = |w_0|_0 e^{\frac{2c}{|\epsilon|}t}$, $\quad t \in J$, *we have*

(iv) $|(Tw)(t\cdot)|_0 \le N(t)$, *and with* $t^* = \max\{t,s\}$

(v) $|X(t,s,x) - x| \le c \cdot \frac{|\epsilon|}{2}|w_0|_0 \cdot |e^{\frac{2c}{|\epsilon|}t} - e^{\frac{2c}{|\epsilon|}s}| \le c \cdot \frac{|\epsilon|}{2}\{N(t^*) - |w_0|_0\} = \rho(t^*)$.

Proof. Recalling Proposition 2.1. and Remark 3.1 we see that the continuous function $v(t,x) = (Kw(t,\cdot))(x) = (Kw)(t,x)$ is defined for all $w \in C_{w_0}$, and for

$w \in C_{w_0} \cap C^{0,\gamma}$ with some $\gamma \in (0,1)$, there exists even $\nabla_x v(t,x)$ and div $v(t,x) = 0$ holds. Moreover, from Proposition 2.1. we find the estimates

$$|Kw(t,\cdot)|_0 \le c|w(t,\cdot)|_0 \,, \tag{5.1}$$

$$[Kw(t,\cdot)]_\ell \le c|w(t,\cdot)|_0 \,, \quad w \in C_{w_0}, \tag{5.2}$$

$$|\nabla Kw(t,\cdot)|_\gamma \le c_1|w(t,\cdot)|_\gamma \,, \quad w \in C_{w_0} \cap C^{0,\gamma},\ \gamma \in (0,1), \text{ with} \tag{5.3}$$

c, c_1 depending on the fixed $\tau \ge |\overset{\circ}{\text{supp}}\, w(t,\cdot)|$, but not on $w \in C_{w_0}$.

The continuity of $v(t,x) = (Kw(t,\cdot))(x)$ or of $\nabla v(t,x)$ with respect to $t \in J$ results from the continuity of $|w(t,\cdot)|_0$ or, for some $\gamma \in (0,1)$, of $|w(t,\cdot)|_\gamma$ in t, respectively, which will be ensured by the following

Remark 5.2. (a) *Since $J = [0,a]$ and $B_R \subset \mathbb{R}^3$ are compact, for any $R > 0$ from* supp $w(t,\cdot) \subset B_R$ *for all $t \in J$ and the continuity of $w(t,x)$ in $J \times \mathbb{R}^3$ we conclude that the function $W(t) = \sup_{x \in B_R} |w(t,x)| = |w(t,\cdot)|_0$ is uniformly continuous on J.*
(b) *In case $w \in C_{w_0} \cap C^{0,\gamma}$, for any $\gamma' \in (0,\gamma)$ the Hölder quotients*

$$H_{\gamma'}w(t,x,y) = \begin{cases} \frac{|w(t,y)-w(t,x)|}{|y-x|^{\gamma'}} & \text{for } x \ne y, \\ 0 & \text{for } x = y \end{cases}$$

are continuous in $(t,x,y) \in J \times \mathbb{R}^3 \times \mathbb{R}^3$. Therefore by (a) (with $B_R \times B_R$ instead of B_R) the Hölder seminorm $[w(t,\cdot)]_{\gamma'} = \sup_{(x,y) \in B_R \times B_R, |y-x|<1} H_{\gamma'}w(t,x,y)$ is uniformly continuous in $t \in J$ for each fixed $\gamma' \in (0,\gamma)$.

The quasi-Lipschitz condition (5.2) for the continuous direction field $v(t,x) = (Kw)(t,\cdot))(x)$ guarantees uniqueness and global existence of the flow $X = Lv$ of (1.1) on $J \times J \times \mathbb{R}^3$, $X(t,s,\cdot) = X^{-1}(s,t,\cdot)$ being a homeomorphism of $\mathbb{R}^3$ for all $t,s \in J$, [1, 5]. Therefore with H from (4.2), $X = Lv$, the composed map $T = HLK$ is well defined, and we see $T : C_{w_0} \to C^0(J \times \mathbb{R}^3)$, which shows (i), since $(Tw)(0,\cdot) = w_0$ is clear from (4.2) and $X(s,s,x) \equiv x$.
Equations (4.2), (4.3) show that $w(t,x) = (HX)(t,x) \ne 0$ holds if and only if $w_0(\widehat{x}) \ne 0$, where $\widehat{x} = X(0,t,x)$. This proves (ii).

In the subspace $C_{w_0} \cap C^{0,\gamma}$, from (3.4), from the inequalities (5.1), (5.3) and Remark 5.2., we get $v = Kw \in C^{0,1+\gamma}$, div $v = 0$, which gives the first statement in (iii) by (ii), since then $X(t,s,\cdot)$ is measure preserving. The second statement $Tw \in C^1(J \times \mathbb{R}^3)$ follows from $X = Lv \in C^1(J \times J \times \mathbb{R}^3)$, $w_0 \in C^1(\mathbb{R}^3)$ by the chain rule.

For proving (iv), (v), we assume $w \in C_{w_0} \cap C^{0,\gamma}$ and

$$|w(t,\cdot)|_0 \le N(t) \tag{5.4}$$

with some continuous function N, $t \in J$. Since (5.1) gives $|v(t,\cdot)|_0 \le c \cdot N(t)$, from the differential equation (1.1) we see

$$|X(t,s,\widehat{x}) - \widehat{x}| \le c\left|\int_s^t N(t')dt'\right| \qquad \text{for all } \widehat{x} \in \mathbb{R}^3, \tag{5.5}$$

which because of (4.2) implies

$$|(Tw)(t,x)| \leq \frac{1}{|\epsilon|}\left\{\begin{array}{l} |X(t,0,Z(X(0,t,x)) - Z(X(0,t,x)|+ \\ |Z(X(0,t,x)) - X(0,t,x)|+ \\ |X(0,t,x) - x| \end{array}\right\}$$

$\leq \frac{2c}{\epsilon} \cdot \int_0^t N(t')dt' + |w_0|_0$.

Therefore T will preserve (5.4), if $N(t)$ is positive solution of the linear Volterra integral inequality

$$|w_0|_0 + \frac{2c}{|\epsilon|}\int_0^t N(t')dt' \leq N(t), \tag{5.6}$$

having the minimal solution

$$N(t) = |w_0|_0 e^{\frac{2c}{|\epsilon|}t}, \qquad t \in J. \tag{5.7}$$

Thus (iv) holds. Integrating in (5.5) with N from (5.7) gives (v), too.

Immediate consequence of Proposition 5.1 is

Corollary 5.3. *Under the assumptions of Proposition* 5.1 (iii) *and* (iv), *we have*

(i) $X(t,s,\cdot)(B_R) \subset B_{R'}$ *with any* $0 < R$ *and* $R' = R + \rho(t^*)$,

(ii) $\text{supp}(Tw)(t,\cdot) \subset B_{R_1}$ *with* $R_1 = R_0 + \rho(a)$, *and*

(iii) $T(C_{w_0} \cap C^{0,\gamma}) \subset C_{w_0}$ *if we require* $R_1 = R_0 + \rho(a)$.

Proof. From Proposition 5.1 (v) we see (i), which together with (ii) in Proposition 5.1. implies (ii) and (iii). □

6. Application of the contracting mapping principle

In order to find closed subsets of $C_{f_0} \cap C^{0,\gamma}$ in which the map T is contracting, we need additional bounds in Hölder norms. For any function $N = N(t)$ from (5.7), any $\beta \in (0,1)$ and any constant $M_1 > 0$, we define the bounded subsets

$$\begin{array}{rcl} C_N^{0,\beta} & = & \{f \in C^{0,\beta}(J \times \mathbb{R}^3) | |f(t,\cdot)|_0 \leq N(t)\}, \\ C_{N,M_1}^{0,\beta} & = & \{f \in C^{0,\beta}(J \times \mathbb{R}^3) | |f(t,\cdot)_0| \leq N(t),\ [f(t,\cdot)]_\beta \leq M_1\} \end{array}$$

of $C^{0,\beta}$ and write $A_{w_0} = C_{w_0} \cap C_{N,M_1}^{0,\beta}$.

Proposition 6.1. *Assume* $w_0 \in C^1(\mathbb{R}^3)$,

$$\text{supp}\, w_0 \subset B_{R_0}, |\overset{\circ}{\text{supp}}\ w_0| \leq \tau. \tag{6.1}$$

(i) *Then the composed map* $T = HLK$ *fulfills*

$$T : C_{w_0} \cap C_N^{0,\gamma} \longrightarrow C_{N,M_1}^{0,\beta} \cap C_{w_0} \cap C^1 = A_{w_0} \cap C_1, \quad \textit{where} \tag{6.2}$$

$$N = N(t) = |w_0| \in^{\frac{2c}{|\epsilon|}t},\ R_1 = R_0 + \rho(a),\ (\rho \textit{ from Proposition } 5.1.\ (\text{v}).), \tag{6.3}$$

$$\alpha = e^{-cN(a)a}, \quad \beta = \alpha^2, \tag{6.4}$$

$$M_1 = \frac{1}{|\epsilon|}\left\{c'(1+|\epsilon||\nabla w_0|_0)^\alpha \cdot (\max\{2eR_4, 1\})^{1+\alpha} + (2R_4)^{1-\beta}\right\}. \tag{6.5}$$

(ii) *In case of two vector valued functions* $w_m \in A_{w_{m0}}$ *having the initial values* $w_m(0,\cdot) = w_{m0} \in C^1(\mathbb{R}^3)$ *which both fulfill* (6.1), *the inequality*

$$|Tw_2 - Tw_1|_* \le c \cdot |w_{20} - w_{10}|_0 + \frac{c_1}{b}|w_2 - w_1|_* \tag{6.6}$$

holds in the norm

$$|f|_* = \sup_{t\in J} e^{-(b+cM)t} \cdot |f(t,\cdot)|_0 \tag{6.7}$$

which is equivalent to the norm $|f|_0$ *for* $f \in C^0(J \times \mathbb{R}^3)$, *with* $M = c(N(a) + M_1)$, $c_1 = \frac{c}{|\epsilon|}(3 + ce^{cMa} \cdot (1 + |\epsilon||\nabla w_{10}|_0))$, *and arbitrary* $b \in (0,\infty)$.

Proof. In Proposition 5.1. (iii) and Corollary 5.3. we have already stated the inclusion $T(C_{w_0} \cap C^{0,\gamma}) \subset C_{w_0} \cap C^1$. Moreover in Corollary 5.3. (ii) we have seen that for all $w \in C_{w_0} \cap C^{0,\gamma}$ with $|w(t,\cdot)|_0 \le N(t)$ the function $(Tw)(t,x)$ is vanishing at all points $x \in \mathbb{R}^3$ outside of the ball $B_{R_1}, R_1 = R_0 + \rho(a)$. Consequently we have to perform the estimates of $(Tw)(t,x)$ only for points $x \in B_{R_1}$, where, by Corollary 5.3. (i), the values $|X(t,s,x)|$ or $|X(t,s,Z(\widehat{x}))|$, with $\widehat{x} = X(0,t,x)$, are uniformly bounded by $R_2 = R_1 + \rho(a)$ or $R_3 = R_2 + \rho(a) + |\epsilon||w_0|_0$, respectively, and similarly for $x \in B_{R_3}$ we find $|X(t,s,x)| \le R_4 = R_3 + \rho(a)$. Therefore the Hölder estimates for $X = Lv$ resulting on B_{R_3} from the quasi-Lipschitz condition (5.2) give

$$[X(t,s,\cdot)]_\alpha B_{R_3} \le \max\{2eR_4, 1\} \text{ with } \alpha = e^{-cN(a)a}, \text{ [15,17]}. \tag{6.8}$$

In addition, if $w_m \in C_{w_{m0}} \cap C^{0,\gamma}$, thus $v_m = Kw_m \in C^{0,1+\gamma}$, the Lipschitz bound for $X_m = Lv_m$ on B_{R_3} reads

$$|X_2(t,s,\cdot) - X_1(t,s,\cdot)|_0 B_{R_3} \le ce^{cM|t-s|} \cdot \Big|\int_s^t e^{-cM|t'-s|} \cdot \delta(t')dt'\Big|, \tag{6.9}$$

where $|\nabla v_m(t,\cdot)|_0 \le M, \quad \delta(t) = |v_2(t,\cdot) - v_1(t,\cdot)|_0 B_{R_4}$, [15, 17].
($\delta(t)$ being continuous in $t \in J$ as shown in Remark 5.2. The supremum in the definition of $\delta(t)$ is required over B_{R_4}, since in the differential inequality resulting from (1.1) for $|X_2(t,s,x) - X_1(t,s,x)|$, $x \in B_{R_3}$, even the values $X_m(t,s,x) \in B_{R_4}$ enter the spatial argument of the direction field v.) In [17] we have proved that statement (i) in Proposition 6.1. follows from (6.8), while (ii) results from (6.9), since we have $|\nabla v_m(t,\cdot)|_0 = |\nabla K w_m(t,\cdot)|_0 \le c\{|w_m(t,\cdot)|_0 + [w_m(t,\cdot)]_\beta\}$ due to (5.1), (5.3). □

Proposition 6.2. *Assume* $w_0 \in C^1(\mathbb{R}^3)$ *fulfills* (6.1), *and* $N = N(t), \alpha, \beta, M_1$ *are given by* (6.3)–(6.5). *Then*

(i) *the class* $C_{w_0} \cap C^{0,\beta}_{N,M_1} = A_{w_0}$ *constitutes a closed subset in* $C^0(J \times \mathbb{R}^3)$ *with respect to the norm* $|\cdot|_*$.

(ii) *There holds* $TA_{w_0} \subset A_{w_0}$, T *being in case* $b > c_1$ *a contraction of* A_{w_0} *with respect to* $|\cdot|_*$.

(iii) *The fixpoint equation* $w = Tw$ *has a unique solution* $w \in A_{w_0}$. *The fixpoint* w *belongs even to* $C^1(J \times \mathbb{R}^3) \cap A_{w_0}$, w *being there the unique solution of the initial value problem* (4.1) *with* $v = Kw$.

(iv) *In the norm* $|\cdot|_0$, *the solution* $w = Tw \in A_{w_0}$ *depends continuously on its initial value* $w_0 \in C^1(\mathbb{R}^3)$, w_0 *fulfilling* (6.1).

Proof. The norms $|\cdot|_*$ and $|\cdot|_0$ being equivalent, we easily see the closeness of A_{w_0} with respect to $|\cdot|_*$ in the Banach space $C^0(J \times \mathbb{R}^3)$: Namely, in case of uniform convergence $|f_k - f_1|_0 \longrightarrow 0$ in A_{w_0} with $k \to \infty$, each uniform Hölder estimate $|f_k(t,\cdot)| \leq N(t), [f_k(t,\cdot)]_\beta \leq M_1$ remains valid for the limit f_1, too, and the same holds for the requirements $\operatorname{supp} f_k(t,\cdot) \subset B_{R_1}, f_k(0,\cdot) = w_0$, since B_{R_1} and w_0 are fixed independently of k.

In order also to verify the third requirement $|\overset{\circ}{\operatorname{supp}}\, f_1(t,\cdot)| \leq \tau$, we consider the measurable sets $S_{m,k} = \{x \in \mathbb{R}^3 || f_k(t,x)| > \frac{1}{m}\}$, $m = 1, 2, \dots$. Because of the uniform convergence $f_k \to f_1$, for each $m = 1, 2 \cdots$ there exists some k_m with $S_{m,1} \subset S_{2m,k}$ for all $k \geq k_m$, thus there holds $|S_{m,1}| \leq |S_{2m,k}| \leq \tau$, since $f_k \in A_{w_0}$. The sequence $(S_{m,1})$ being increasing with $\overset{\circ}{\operatorname{supp}}\, f_1(t,\cdot) = \bigcup_{m=1}^\infty S_{m,1}$, we conclude $|\overset{\circ}{\operatorname{supp}}\, f_1(t,\cdot)| = \lim_{m \to \infty} |S_{m,1}| \leq \tau$. This proves (i).

The first statement in (ii) follows from Proposition 6.1. (i). If we take $b > c_1$, the contracting property of T on A_{w_0} follows from Proposition 6.1. (ii), since there the first term on the right-hand side vanishes.

Because of (i) and (ii) the contraction mapping principle [21] ensures the existence of a unique fixpoint $w = Tw \in A_{w_0}$, which can be approximated (with respect to the norm $|\cdot|_*$) by iteration of T.

For any fixpoint $w = Tw \in A_{w_0}$, from Proposition 2.1. (iii), Remark 3.1., 5.2., Proposition 4.2. (ii) and Proposition 6.1. (i) we see that $w \in C^1$ fulfills (4.1). Conversely for each solution $w \in C^1 \cap A_{w_0} \subset C^{0,\gamma} \cap A_{w_0}$ of the initial value problem (4.1) we find $v = Kw \in C^{0,1}$ from Proposition 2.1. and Remarks 3.1., 5.2. Therefore, as stated in Proposition 4.2. (i), w has the representation $w = HX$ in (4.2), thus $w = Tw$ holds because of $X = Lv, v = Kw$.

Finally, if $w_m = Tw_m \in A_{w_{m0}}$ with initial values $w_{m0} \in C^1$ fulfilling (6.1), and if we take $b > c_1$, setting $q = \frac{c_1}{b}$, from Proposition 6.1. (ii) we get

$$|w_2 - w_1|_* \leq \frac{c}{1-q} |w_{20} - w_{10}|_0, \tag{6.10}$$

which shows the continuous dependence of w on its initial value w_0 even in the norm of $C^0(J \times \mathbb{R}^3$.) □

With Proposition 6.2. (iii), (iv) we have proved the statement of Theorem 4.1., with the only exception that until now we have shown uniqueness of solutions w to (4.1) and their continuous dependence on w_0 in the classes $A_{w_0} \cap C^1(J \times \mathbb{R}^3)$, but not yet in the class $C_B^0 \cap C^1(J \times \mathbb{R}^3)$. Therefore the proof of Theorem 4.1. will be completed by

Corollary 6.3. *For two arbitrary vector-valued functions* $w_m \in C^0_B \cap C^1\ (J \times \mathbb{R}^3)$ *having the same initial value* $w_m(0,\cdot) = w_0$*, there always exists a class* $A_{w_0} = C_{w_0} \cap C^{0,\beta}_{N,M_1}$ *as we had defined it in Proposition* 6.1.*,* A_{w_0} *containing both* w_m*,* $m = 1,2$*. Thus, in particular, each solution* $w \in C^0_B \cap C^1$ *of* (4.1) *belongs to such a class* A_{w_0}*, where existence, uniqueness and continuous dependence on the initial value of solutions to* (4.1) *holds by Proposition* 6.2.

Proof. Let $w_m \in C^0_B \cap C^1(J \times \mathbb{R}^3)$, with $w_m(0,\cdot) = w_0$ denote two vector-valued functions which, by definition of C^0_B, for all $t \in J$ have their supports $\operatorname{supp} w_m(t,\cdot)$ in a suitable ball $B = \{x \in \mathbb{R}^3 \,|\, |x| \le R\}$. Taking

$$R_1 \ge R_0 = R, \qquad \tau \ge |B| = \frac{4}{3}\pi R^3, \tag{6.11}$$

we find $\operatorname{supp} w_0 \subset B_{R_0}$, $|\overset{\circ}{\operatorname{supp}}\, w_0| \le \tau$, and consequently $w_m \in C_{w_0}$ from (5.0), $m = 1,2$. Moreover, a class $C^{0,\beta}_{N,M_1}$ certainly contains both w_m, if there holds

$$|w_m(t,\cdot)|_0 \le N(t)\,, \quad M_1 \ge [w_m(t,\cdot)]_\beta \quad t \in J, \quad m = 1,2. \tag{6.12}$$

Recalling Proposition 2.1. (i) we see that the bound c in (5.1), (5.7) considered in dependence on τ can be written in the form

$$c = c_1 \cdot (1 + \tau) \tag{6.13}$$

with constant $c_1 \ge 1$. Therefore the function $N(t)$ as well as the bound $R_1 = R_0 + \rho(a)$ in (6.3) are with $\tau \to \infty$ strictly increasing to ∞, whereas α in (6.4) is strictly decreasing to 0. Nevertheless the bound M_1 in (6.5) for $\alpha \le 1/2$ has the minorant

$$M_* = \frac{1}{|\epsilon|}\left\{c' \cdot \max\{2\,|\in|R_4, 1\} + (2R_4)^{3/4}\right\} \le M_1, \tag{6.14}$$

M_* being again strictly increasing to ∞ with $\tau \to \infty$. In addition we note that the finite value $M = c'(|\nabla w_1|_0 + |\nabla w_2|_0)$ (where the constant $c' \ge 1$ depends only on the special norm we use for 3×3-matrices) gives a uniform upper bound of the values $[w_m(t,\cdot)]_\beta$, which are decreasing with $\beta \to 0$ due to our definition (1.3), $m = 1,2$, $t \in J$, $\beta \in (0,1)$. Finally we set $M' = \max_{m=1,2} \left|\frac{\partial}{\partial t} w_m\right|_0$, M' being finite because of $w_m \in C^0_B \cap C^1$.

In order that $N(t)$ in (6.3) becomes upper bound of both $|w_m(t,\cdot)|_0$, in a first step for any fixed $x \in B$ we consider the continuous function $\varphi_m(t) = |w_m(t,x)|$ which has the Dini derivative

$$D^+_t \varphi_m(t) \in [-\infty, \infty] \text{ with } \quad D^+_t \varphi_m(t) \le \left|\frac{\partial}{\partial t} w_m(t,x),\right|, \tag{6.15}$$

[20]. Due to a basic Lemma in differential inequalities [20, p. 64, 70], for any constant $\delta > 0$ from the differential inequality

$$D^+_t \{N(t) - \varphi_m(t) - \delta \cdot t\} > 0 \text{ for } t \in J \tag{6.16}$$

we can conclude that the function $N(t) - \varphi_m(t)$ is monotone increasing and that

$$N(t) \ge \varphi_m(t) \tag{6.17}$$

holds for $t \in J, m = 1, 2$, because of

$$N(0) = |w_0|_0 \geq |w_0(x)| = |w_{m0}(x)|.$$

Recalling (6.3) and our definition of M' above, (6.16) results from the requirement

$$M' + \delta < \frac{2c}{|\epsilon|}|w_0|_0. \tag{6.18}$$

Using (6.13) we easily can fulfill (6.18) together with (6.11) for $R_1 = R_0 + \rho(a)$ and also $M_* \geq M$ by taking $\tau \geq |B|$ sufficiently large, which proves $w_m \in A_{w_0}$, $m = 1, 2$. □

References

[1] E.A. Coddington, N. Levinson, *Theory of Ordinary Differential Equations,* McGraw-Hill, New York, 1955.

[2] U. Dini, *Sur la méthode des approximations successives pour les équations aux derivées partielles du deuxième ordre,* Acta Math. 25 (1902), 185–230.

[3] N.M. Günter, *Theory of potentials and its application to basic problems of mathematical physics,* Moscow, 1953.

[4] N.M. Günter, *Die Potentialtheorie und ihre Anwendung auf Grundaufgaben der Mathematischen Physik,* Teubner Leipzig, 1957.

[5] Ph. Hartman, *Ordinary Differential Equations,* Wiley New York, 1964.

[6] E. Hölder, *Über die unbeschränkte Fortsetzbarkeit einer stetigen ebenen Bewegung in einer unbegrenzten Flüssigkeit,* Math. Z. 37 (1933), 727–738.

[7] T. Kato, *On classical solutions of the two-dimensional nonstationary Euler equations,* Arch. Rat. Mech. Anal. 25 (1967), 188–200.

[8] L. Lichtenstein, *Grundlagen der Hydromechanik,* Springer Berlin, 1929.

[9] S.G. Mikhlin, *Multidimensional Singular Integrals and Integral Equations,* New York, Pergamon-Press, 1965.

[10] T. Ogawa, Y. Taniuchi, *Remarks on uniqueness and blow-up criterion to the Euler equations in the generalized Besov spaces, and J. Korean* Math. Soc. 37 (2000), 6, 1007–1019.

[11] T.Ogawa, Y. Taniuchi, *A note on blow-up criterion to the 3-D-Euler equations in a bounded domain,* J. Math. Fluid Mech. 5 (2003), 17–23.

[12] T. Ogawa, Y. Taniuchi, *On blow-up criteria of smooth solutions to the 3-D-Euler equations in a bounded domain,* J. Differential Equations 190 (2003), 39–63.

[13] W.F. Osgood, *Beweis der Existenz einer Lösung der Differentialgleichung* $\frac{dy}{dx} = f(x, y)$ *ohne Hinzunahme der Cauchy-Lipschitzschen Bedingung,* Monatshefte Math. Phys. 9 (1898), 331–345.

[14] R. Rautmann, *Das Cauchyproblem der Helmholtzschen Wirbelgleichung mit einer Differenzennäherung. Recent developments in theoretical and experimental fluid mechanics,* U. Müller, K.G. Roesner, B. Schmidt (eds.), Springer (1979), 295–308.

[15] R. Rautmann, *Ein Näherungsverfahren für spezielle parabolische Anfangswertaufgaben mit Operatoren, in: Numerische Lösung nichtlinearer partieller Differential- und Integralgleichungen,*R. Ansorge, W. Törnig (eds.), Springer, Lecture Notes in Math. 267 (1972), 187–231.

[16] R. Rautmann, V.A. Solonnikov, *Quasi-Lipschitz condition in potential theory,* to appear.

[17] R. Rautmann, V.A. Solonnikov, *Helmholtz's vorticity transport equation with partial discretization in bounded 3-dimensional domains,* to appear.

[18] J. Schauder, *Potentialtheoretische Untersuchungen,* Math. Z. 33 (1931), 602–640.

[19] J. Serrin, *Mathematical principles of classical fluid mechanics, in: Flügge (ed),* Handbuch der Physik 8, 125–263, Springer Berlin 1959.

[20] W. Walter, *Differential and Integral Inequalities,* Springer Berlin 1970.

[21] J. Weissinger, *Zur Theorie und Anwendung des Iterationsverfahrens,* Math. Nachrichten 8 (1952), 193–212.

[22] W. Wolibner, *Un théorème sur l'existence du mouvement plan d'un fluide parfait, homogène, incompressible, pendant un temps infiniment longue,* Math. Z., 37 (1933), 698–726.

[23] W.I. Yudovič, *Nonstationary flows of an ideal incompressible fluid,* Zh. Vyc. Math. 3 (1963), 1032–1066.

Reimund Rautmann
Institut für Mathematik der Universität Paderborn
D-33095 Paderborn, Germany
e-mail: `rautmann@upb.de`

Progress in Nonlinear Differential Equations
and Their Applications, Vol. 61, 257–273

Interfaces in Solutions of Diffusion-absorption Equations in Arbitrary Space Dimension

Sergei Shmarev

Abstract. We study the Cauchy-Dirichlet problem for the degenerate parabolic equation

$$u_t = \Delta u^m - a\, u^p \quad \text{in } \mathcal{E}$$

with the parameters $a \in \mathbb{R}$, $m > 1$, $p > 0$, satisfying the condition $m + p \geq 2$. The problem domain $\mathcal{E}$ is the exterior of the cylinder bounded by a simple-connected surface S, $\operatorname{supp} u_0$ is an annular domain $\mathbb{R}^n$. We show that the velocity of the outer interface $\Gamma = \partial\{\operatorname{supp} u(x,t)\}$ is given by the formula

$$\mathbf{v} = \left[-\frac{m}{m-1} \nabla u^{m-1} + \nabla \Pi \right]\Bigg|_{\Gamma},$$

where $\Pi(x,t)$ is a solution of the degenerate elliptic equation

$$\operatorname{div}(u \nabla \Pi) = a\, u^p, \qquad \Pi = 0 \text{ on } \Gamma,$$

depending on t as a parameter. It is proved that the solution and its interface Γ preserve their initial regularity with respect to the space variables, and that they are real analytic functions of time t. We also show that the regularity of the velocity $\mathbf{v}$ is better than it was at the initial instant. For the space dimensions $n = 1, 2, 3$, these results were established in [8]. We propose a modification of the method of [8] that makes it applicable to equations with an arbitrary number of independent variables.

Mathematics Subject Classification (2000). 35K55, 35K57, 35K65, 35B65, 35R35.

Keywords. diffusion-absorption equation, degenerate parabolic equation, free boundary problem, regularity of interface.

The author was supported by the Spanish Research Project BFM2000-1324 and by the European RTN Programme HPRN-CT-2002-00274.

1. Introduction

We study the regularity properties of weak, nonnegative, continuous solutions of the Cauchy-Dirichlet problem

$$\begin{cases} u_t = \Delta\, u^m - a\, u^p & \text{in } \mathcal{E}, \\ u = \psi > 0 \text{ on } S, & u(x,0) = u_0(x) \geq 0 \text{ in } E(0) \end{cases} \tag{1.1}$$

in the range of the parameters $a \in \mathbb{R}$, $m > 1$, $p > 0$,

$$m + p \geq 2. \tag{1.2}$$

The domain $\mathcal{E}$ is the exterior of the cylinder bounded by a simple-connected surface S.

Equation (1.1) is strictly parabolic for $u > 0$ and degenerates at the level $u = 0$. In this paper we study the properties of the a priori unknown surface of degeneracy alias *the free boundary* or *the interface.* A weak solution of problem (1.1) is understood in the following sense.

Definition 1.1. *A function $u(x,t)$ is said to be a weak solution of problem* (1.1) *if*

1. *$u(x,t)$ is bounded, nonnegative, and continuous in $\overline{\mathcal{E}}$; $u = \psi$ on S, $u(x,0) = u_0$ in $E(0) = \overline{\mathcal{E}} \cap \{t = 0\}$;*
2. *for every test-function $\eta(x,t) \in C^1(\mathbb{R}^n \times [0,T])$ vanishing when $t = T$, on S, and for $|x| > R$ with some $R > 0$, the following identity holds:*

$$\int_{E(0)} u_0 \eta(x,0)\, dx + \int_{\mathcal{E}} (\eta_t u - \nabla_x \eta \cdot \nabla_x u^m - a\eta\, u^p)\, dx\, dt = 0.$$

The second term on the right-hand side of equation (1.1) models the process of absorption ($a > 0$) or reaction ($a < 0$). It is known that in the range of the exponents $m > 1$, $p > 1$, $a < 0$ the solutions may blowup in finite time, while for $0 < p < 1$ and $a > 0$ the support of the solution may shrink. As is proved in [1, Chapter 3, Section 4], in the latter case even if the solution of equation (1.1) is strictly positive on the parabolic boundary of a cylinder $\{x \in \mathbb{R}^n : |x - x_0| < R\} \times \{t \in (0,T)\}$, there always appears a zero cavern inside the cylinder, provided that R and T are chosen sufficiently large. This is true for solutions of a nonlinear equation which contains equation (1.1) as a partial case.

For $n = 1$, the study of the behavior and regularity of interfaces in solutions of equation (1.1) was performed in the series of papers [2, 3, 4, 5]. It was shown that in the range of parameters $m > 1$, $p \in (0,1)$, $m + p \geq 2$, ($a > 0$), the interface is governed by the first-order equation

$$\eta'(t) = -\frac{m}{m-1}\left(u^{m-1}\right)_x (\eta(t),t) + \frac{a(1-p)}{(u^{1-p})_x\, (\eta(t),t)}, \tag{1.3}$$

which generalizes the Darcy law. It is proved in [2, 3] that the both terms on the right-hand side of (1.3) exist (finite or infinite) until the moment when the solution vanishes, and that at each instant t only one of the two terms is distinct from zero. In this range of the exponents the interfaces are proved to be Lipschitz-continuous [2, 3]. The behavior of interfaces in the multi-dimensional case is studied in [8]. It was shown that in the case $n = 1, 2, 3$ the velocity of the outer interface $\Gamma = \partial\{\operatorname{supp} u(x,t)\}$ in the solutions of the Cauchy problem for (1.1) is given by the formula

$$\mathbf{v} = \left[-\frac{m}{m-1} \nabla u^{m-1} + \nabla \Pi \right] \Bigg|_{\Gamma(t)}, \tag{1.4}$$

where $\Pi(x,t)$ is the solution of the degenerate elliptic equation

$$\operatorname{div}(u \nabla \Pi) = a\, u^p, \qquad \Pi = 0 \text{ on } \Gamma,$$

depending on t as a parameter. It is proved that the solution $u(x,t)$ and its interface Γ preserve their initial regularity with respect to the space variables, that they are real analytic functions of time t, and that the regularity of the velocity $\mathbf{v}$ is better than it was at the initial instant. Formula (1.4) generalizes the interface equation (1.3) to the cases $n = 2, 3$, for $n = 1$ (1.4) coincides with (1.3). These assertions were proved in [8] by means of the method of Lagrangian coordinates. A drawback of the method is that it is not applicable to the case $n \geq 4$. This is the aim of the present work to get rid of the restriction on the spatial dimension and to show that the same results are true for any natural n.

Let us denote

$$P_0 = \frac{m}{m-1} u_0^{m-1}, \qquad p = \frac{m}{m-1} u^{m-1},$$

and assume that the data of problem (1.1) satisfy the following properties:

$$\begin{cases} \operatorname{supp} P_0 \text{ is an annular domain in } \mathbb{R}^n \\ \text{with the interior boundary } s(0) = \overline{S} \cap \{t = 0\}, \\ P_0 \in C^1(\overline{\operatorname{supp} P_0}), \quad P_0 \in V(2k+1, \operatorname{supp} P_0) \text{ with } k \geq 1 \\ |\nabla P_0| + P_0 \geq \delta > 0 \text{ in } \overline{\operatorname{supp} P_0}, \quad \delta = \text{const}, \\ P_0 \text{ and } \psi \text{ satisfy the first-order compatibility conditions on } s(0). \end{cases} \tag{1.5}$$

(The definition of the weighted Hölder spaces $V(2k+1, G)$ is given in Section 3.)

$$\begin{cases} S \in C^{2+\alpha} \text{ in the sense of the definition given in [7, Chapter IV, Section 7],} \\ \psi > 0 \text{ on } \overline{S},\ u_0 \text{ and } \psi \text{ satisfy the compatibility conditions:} \\ u_0 = \psi, \qquad \psi_t - \Delta u_0^m + a\, u_0^p = 0 \text{ on } \overline{S} \cap \{t = 0\}. \end{cases} \tag{1.6}$$

The following is the main result of this paper.

Theorem 1.1. *Let us assume that the data of problem* (1.1) *satisfy conditions* (1.5) *and* (1.6). *There exist* $a^* > 0$, $\epsilon^* > 0$, $M > 0$, *and* $T > 0$ *such that for every* $|a| < a^*$ *and* $\|P_0\|_{V(2k+1, \operatorname{supp} P_0)} < \epsilon^*$ *the Cauchy-Dirichlet problem* (1.1) *has a weak continuous nonnegative solution* $u(x,t)$ *satisfying the properties:*

1. *for all* $t \in (0, T]$ *the set* $\operatorname{supp} u(x,t) \cap \{t = \text{const}\} \subset \mathbb{R}^n$ *is a finite annular domain with the exterior boundary* $\gamma(t)$*;*

2. *the set $\gamma(t)$ is obtained as a one-to-one mapping from the set*
$$\gamma(0) = \partial\{\operatorname{supp} P_0\} \cap \{P_0 = 0\};$$
3. *the function $p(x,t)$ and the surface $\gamma(t)$ preserve their initial regularity with respect to the spatial variables;*
4. *the solution $u(x,t)$ and the interface $\Gamma = \bigcup_{t\in[0,T]} \gamma(t)$ are real analytic with respect to the variable t;*
5. *the interface velocity is given by formula (1.4); for all $t > 0$ the regularity of $\mathbf{v}$ with respect to te spatial variables in one order higher than it was at the initial instant.*

Assume that problem (1.1) has a weak continuous solution in the sense of Definition 1.1. Let us take a smooth closed surface $\Sigma = \bigcup_{t\in(0,T)} \sigma(t)$ such that $S \subset \Sigma$ and $S \cap \Sigma = \emptyset$, denote by $\mathcal{D}$ the exterior of the cylinder bounded by Σ, and then set $D(t) = \mathcal{D} \cap \{t = \text{const}\}$.

By the continuity of u, the surface Σ can be chosen in the special way:

$$u > 0 \text{ on } \Sigma, \qquad \forall\, t \in (0,T) \int_{D(t)} u(x,t)\,dx = \int_{D(0)} u_0(x)\,dx = \text{const}\,. \tag{1.7}$$

The function u can be formally viewed now as a solution of the following problem:

$$\begin{cases} u_t = \Delta\, u^m - a\, u^p, \quad \text{in } \mathcal{D}, \\ u(x,0) = u_0(x) \text{ in } D(0), \\ \text{the domain } D(t) \text{ and the solution } u(x,t) \text{ satisfy condition (1.7).} \end{cases} \tag{1.8}$$

The solution of problem (1.8) is understood as follows:

Definition 1.2. *A pair (u, Σ) is said to be a weak solution of problem (1.8) if $\Sigma \in C^{2+\alpha}$, u is nonnegative in $\mathcal{D}$, continuous in $\overline{\mathcal{D}}$, and for every test-function $\eta \in C^1(\overline{\mathcal{D}})$, vanishing on Σ, for $t = T$, and for all sufficiently large $|x|$, the following identity holds:*

$$\int_{D(0)} \eta(x,0) u_0\,dx + \int_{\mathcal{D}} [\eta_t\, u - \nabla\, \eta \cdot \nabla\, u^m - a\, \eta\, u^p]\,dxdt = 0. \tag{1.9}$$

Unlike the original problem (1.1) problem (1.8) contains two free boundaries: the zero-level surface $\Gamma = \{(x,t) : u = 0\}$ where the equation degenerates, and a surface Σ that has to be chosen according to condition (1.7). Of course, such a choice of Σ can be done in many different ways and the solution of problem (1.8) need be unique. It will be sufficient for our purposes to construct *any* of the possible solutions of problem (1.8). Given the initial surface $\sigma(0)$, we explicitly construct the corresponding surface $\sigma(t)$ and the function $u(x,t)$. The most of attention is focused on the study of problem (1.8).

In Section 2 we introduce a system of Lagrangian coordinates generated by a function u satisfying (1.8). Problem (1.8) is considered as the mathematical model of motion of a continuum bounded by the free boundaries and preserving the mass. The Lagrangian description of such a motion leads to a new problem posed in a cylinder with the annular base and vertical lateral boundaries. Instead of the free-boundary problem for a single degenerate parabolic equation we arrive at a system of nonlinear equations posed in a time-independent domain. We show that a classical solution of the new problem exists and allows one to define a solution of (1.8) (u, Σ).

In Section 3 we define the weighted Hölder spaces $V(2k+1, G)$ and present a decomposition of the space of vector-valued functions into the direct sum of two orthogonal subspaces. In Section 4 we revisit the Lagrangian counterpart of problem (1.8) and show that the particle velocity can be thought as a potential vector on the plane of Lagrangian coordinates. This is the main novelty: instead of searching the velocity in the form $\nabla v + \mathbf{rot}\,\mathbf{w}$ [8], which is only possible if $n \leq 3$ and requires solving the Poisson equation for the components of $\mathbf{w}$, we deal with the projection of the problem on the subspace of potential vectors. This allows us to reduce the system of nonlinear equations to a system of scalar degenerate elliptic and parabolic equations without any restriction on the space dimension.

The analysis of the problem posed in Lagrangian coordinates follows [8]. When problem (1.8) is solved, we recover the solution of the original problem (1.1) by "pasting" the solution of (1.8) to the solution of the Dirichlet problem for equation (1.8) in the annular cylinder $\mathcal{E} \setminus \mathcal{D}$ with the natural boundary condition on Σ.

The choice of the exterior Cauchy-Dirichlet problem is more the question of presentation of the method. In the same way a set of local Lagrangian coordinates can be introduced in any problem where the appearing free boundary is a simple connected surface.

2. Lagrangian coordinates

Let $u(x,t)$ be a solution of problem (1.8). Throughout the rest of the paper we use the following notation:

$$\Omega(t) = D(t) \cap \operatorname{supp} u(x,t), \qquad \partial\Omega(t) = \sigma(t) \cup \gamma(t),$$

where $\sigma(t)$ and $\gamma(t)$ are the interior and exterior boundaries of $\Omega(t)$,

$$\begin{cases} \Omega_0 = D(0) \cap \operatorname{supp} P_0, \qquad Q = \Omega_0 \times (0,T], \\ \Gamma_0 = \gamma(0) \times [0,T],\ \Sigma_0 = \sigma(0) \times [0,T] \\ \text{are the exterior and interior lateral boundaries of } Q. \end{cases}$$

2.1. Auxiliary mechanical problem

Let $u(x,t)$ be a solution of the free-boundary problem (1.8). Throughout the section we assume that the solution is as smooth as is needed to perform all the requested transformations.

Let us start with reformulation of problem (1.8) viewed as the mathematical description of the process of propagation of a polytropic gas in a porous medium. There are two possible ways to describe the motions of continua. The first one, usually referred to as *the Euler method,* consists in viewing the characteristics of motion (such as velocity, density, etc.) as functions of time t and some Cartesian coordinate system $x_1, \dots, x_n$ not connected with the medium. The alternative description is due to *Lagrange.* In this method all magnitudes describing the motion are considered as functions of time and the initial state of the continuum.

Let Ω_0 and $\Omega(t) \subset \mathbb{R}^n$ be the domains occupied by a polytropic gas at the moments $t = 0$ and $t > 0$. This correspondence defines the mapping

$$x = X(\xi, t), \quad \xi \in \Omega_0,$$

which assigns the position $X(\xi, t)$ to the particle initially located at the point $\xi \in \Omega_0$. Given the velocity field $\mathbf{v}(x, t)$, the motion of this particle is controlled by *the trajectory equation*

$$\begin{cases} X_t(\xi, t) & = \mathbf{v}\left[X(\xi, t), t\right], \quad t > 0, \\ X(\xi, 0) & = \xi, \quad \xi \in \Omega_0. \end{cases} \tag{2.1}$$

Another ingredient of the description is the mass conservation law. We assume that any volume $\omega(t)$ constituted by the same particles at every instant t preserves its mass in time:

$$\frac{d}{dt}\left\{\int_{\omega(t)} u(x, t)\, dx\right\} = 0,$$

where $u(x, t)$ denotes the density at the point (x, t). Let $U(\xi, t) = u[X(\xi, t), t]$ will be the density, $P(\xi, t) = p[X(\xi, t), t]$ be the pressure, and

$$J = [J_{ij}], \quad J_{ij} = \frac{\partial X_i}{\partial \xi_j}, \quad i, j = 1, \dots, n$$

be the Jacobi matrix of the mapping $\xi \mapsto X$ and $|J| \equiv \det(\partial X/\partial \xi)$. Using the trajectory equation (2.1) and applying the rule of differentiation of determinants, it is easy to verify the validity of the relation

$$\frac{d|J|}{dt} = |J| \operatorname{div}_x \mathbf{v},$$

called *the Cauchy identity.* Formally passing to the coordinates ξ, we have now:

$$\begin{aligned} 0 = & \frac{d}{dt}\left\{\int_{\omega(t)} u(x, t)\, dx\right\} = \int_{\omega(0)} \frac{d}{dt}\left\{u\,|J|\right\}\, d\xi \\ = & \int_{\omega(0)} \left[u_t + \nabla_x u \cdot \mathbf{v}[X(\xi, t), t] + u \operatorname{div}_x \mathbf{v}\right] |J|\, d\xi \\ = & \int_{\omega(0)} \left[u_t + \operatorname{div}_x(u\mathbf{v})\right] |J|\, d\xi = \int_{\omega(t)} \left[u_t + \operatorname{div}_x(u\mathbf{v})\right]\, dx. \end{aligned} \tag{2.2}$$

Since the volume $\omega(t)$ is assumed to be arbitrary, the mass conservation law in the Euler coordinates is given by the equation

$$u_t + \operatorname{div}_x(u\mathbf{v}) = 0 \tag{2.3}$$

and its Lagrangian counterpart has the form $\dfrac{d}{dt}(u\,|J|) = 0$. The mass conservation law can be written then in the form

$$U(\xi,t)\,|J| = u(\xi,0) \quad \text{in } \Omega_0, \quad t \geq 0. \tag{2.4}$$

We are able to state in this framework the Lagrangian analog of problem (1.8). Let us assume that the density of the gas is u, and that the gas velocity follows the law

$$\mathbf{v} = -\nabla p + \nabla \Pi \tag{2.5}$$

where $\Pi(x,t)$ is an unknown scalar function to be defined. We will assume that

the velocity $\mathbf{v}$ is uniformly bounded throughout the problem domain. (2.6)

Let (u,Σ) be a solution of problem (1.8). Multiplying equation (2.3) by an arbitrary test-function η satisfying the conditions of Definition 1.2 and integrating by parts we have

$$\int_{D(0)} \eta(x,0)\,u_0\,dx + \int_{\mathcal{D}} [\eta_t\,u + u\nabla\,\eta\cdot\mathbf{v}]\,dxdt - \int_\Gamma u\eta\,\mathbf{v}\,dS = 0$$

whence, using (2.5)–(2.6),

$$\int_{D(0)} \eta(x,0)\,u_0\,dx + \int_{\mathcal{D}} [\eta_t\,u - u\nabla\,\eta\cdot\nabla\,p + u\nabla\,\eta\cdot\nabla\,\Pi]\,dxdt = 0.$$

Comparing this relation with the integral identity from Definition 1.2 we may write: for every test-function η satisfying the conditions of Definition 1.2

$$\int_{\mathcal{D}} [u\nabla\,\eta\cdot\nabla\,\Pi - a\,\eta\,u^p]\,dxdt = 0. \tag{2.7}$$

This is true if we take for Π any function satisfying the degenerate elliptic equation end the boundary condition

$$\operatorname{div}(u\nabla\,\Pi) = a\,u^p \quad \text{in } \mathcal{D}, \qquad u\nabla\,\Pi\cdot\mathbf{n}|_\Gamma = 0. \tag{2.8}$$

Recalculating the derivatives in x by the rule

$$\nabla_x = \left(J^{-1}\right)^*\,\nabla_\xi, \quad (J^{-1})^* \text{ is the matrix transposed to } J^{-1},$$

we obtain the following system of equations posed in the annular cylinder $Q = \Omega_0 \times (0,T)$:

$$J^* X_t(\xi,t) = -\nabla_\xi P + \nabla_\xi \pi, \tag{2.9}$$

$$P\,|J|^{m-1} = P_0(\xi) \quad \text{in } Q, \tag{2.10}$$

$$\begin{cases} X(\xi,0) = \xi, \quad P(\xi,0) = P(\xi,0) \quad \text{in } \Omega_0; \\ P(\xi,t) = 0 \quad \text{on the exterior lateral boundary of } Q, \\ |\nabla_\xi P| + |\nabla_\xi \pi| \text{ is bounded in } \overline{Q}. \end{cases} \tag{2.11}$$

Passing to the variables ξ in the integral identity (2.7) and plugging (2.10), we have that the function $\pi(\xi,t) \equiv \Pi(X(\xi,t),t)$ satisfies the integral identity

$$\int_Q [u_0|J|J^{-1}(J^{-1})^*\nabla_\xi\widetilde{\eta}\cdot\nabla_\xi\pi - a\,\widetilde{\eta}\,u_0U^{p-1}]\,d\xi dt = 0, \quad \widetilde{\eta}(\xi,t) \equiv \eta(X,t),$$

which means that on the plane of Lagrangian coordinates the function π satisfies in the weak sense the degenerate elliptic equation

$$\operatorname{div}_\xi\left(u_0|J|J^{-1}(J^{-1})^*\nabla_\xi\pi\right) = a\,u_0U^{p-1} \quad \text{in } Q. \tag{2.12}$$

Let us notice that to formulate problem (2.9)–(2.11), (2.12) we only used a resemblance between an evolution equation and the mass balance law in the motion of a continuous medium, which is not a rigorous justification of the performed change of the independent variables. The next indispensable step is to answer the question whether the constructed solution of the auxiliary problem (2.9)–(2.11) allows one to recover the solution of the free-boundary problem (1.8) and, finally, of the original problem (1.1).

2.2. The inverse transformation

Let the triad (X, P, π) be a solution of problem (2.9)–(2.12). Let us assume that $P(\xi,t)$ is strictly positive in $\overline{Q}\setminus\Gamma_0$, bounded and continuous in $\overline{Q}$, $P=0$ on Γ_0. Define the mapping

$$x = X(\xi,t) \quad \text{in } Q, \qquad X(\xi,0) = \xi \quad \text{for } \xi\in\Omega_0,$$

and assume that for every $t\in(0,T)$ it is a bijection of $\overline{\Omega_0}$ onto $\overline{\Omega(t)}$, and that $|J|$ is separated away from zero and infinity in $\overline{Q}$. Set

$$X(\xi,t) = \xi - \int_0^t (J^{-1})^*\,(\nabla_\xi P - \nabla_\xi\pi)\,d\tau, \quad \xi\in\Omega_0, \tag{2.13}$$

and

$$p(x,t) = \frac{m}{m-1}u^{m-1}(x,t) = \begin{cases} P(\xi,t) & \text{if } x = X(\xi,t) \text{ with } \xi\in\Omega_0, \\ 0 & \text{in } \{D(t)\setminus\Omega(t)\}\times[0,T]. \end{cases} \tag{2.14}$$

If $X(\xi,t)$ is continuos in $\overline{Q}$, then the function $u(x,t)$ defined in this way is strictly positive in the image of $\overline{Q}\setminus\Gamma_0$ because of (2.10), vanishes on the image of Γ_0, is bounded and continuous in $\bigcup_{t\in[0,T]} D(t)$, assumes its initial values by continuity and satisfies equation (1.8) in the weak sense. Indeed, given an arbitrary test-function $\eta(x,t)\in C^1(\mathbb{R}^n\times[0,T])$, vanishing for large $|x|$, for $t=T$, and on the image of $\sigma(0)$ under the mapping (2.13), we may write:

$$
\begin{aligned}
-\int_{\Omega_0} u_0(x)\eta(x,0)dx &= \int_0^T \frac{d}{dt}\left\{\int_{\Omega(t)} u(x,t)\eta(x,t)dx\right\} dt \\
&= \int_Q \frac{d}{dt}\left(U|J|\eta[X(\xi,t),t]\right) dt\, d\xi \\
&= \int_Q \frac{d}{dt}\left(U|J|\right) \eta\, d\xi\, dt + \int_0^T \int_{\Omega_0} U|J| \frac{d\eta}{dt}\, d\xi\, dt \\
&= \int_Q \left[\eta_t + \nabla_x \eta \cdot \frac{dX}{dt}\right] U|J|\, d\xi\, dt \\
&= \int_Q \left[\eta_t - (\nabla_x p - \nabla_x \Pi)\cdot \nabla_x \eta\right] U|J|\, d\xi\, dt \\
&= \int_{\mathcal{D}} \left(u\, \eta_t - \nabla_x \eta \cdot \nabla_x u^m - a\eta\, u^p\right) dx\, dt,
\end{aligned}
\tag{2.15}
$$

where we made use of (2.1), (2.4) and (2.7) (alias (2.12)).

We thus obtain the parametric representations for both the domain $\mathcal{D}$ and the weak solution $u(x,t)$ of problem (1.8) via the solution of problem (2.9)–(2.12). If $X(\xi,t)$ is continuous in $\overline{Q}$, the free boundaries Σ and Γ in this solution are given by equation (2.9) (or by (2.13)) with $\xi \in \partial\Omega_0$, that is

$$
\sigma(t) = \{x : x = X(\xi,t),\ \xi \in \sigma(0)\}, \qquad \gamma(t) = \{x : x = X(\xi,t),\ \xi \in \gamma(0)\}.
$$

These arguments prove the following assertion:

Theorem 2.1. *Let the functions* (X, P, π) *satisfy conditions* (2.9)–(2.12). *If*

1. $P \in C\left(\overline{Q}\right)$, $P > 0$ *in* $\overline{Q} \setminus \Gamma_0$, $P_0 = 0$ *on* Γ_0, $|\nabla_\xi P| + |\nabla_\xi \pi| \le C$ *in* $\overline{Q}$,
2. $|J| \equiv \det\left[\partial X/\partial\xi\right]$ *is separated away from zero and infinity in* $\overline{Q}$,
3. $X(\xi,t) \in C(\overline{Q})$ *and for every* $t \in (0,T)$ *the mapping* $\xi \mapsto X(\xi,t)$ *is a bijection between* $\overline{\Omega}_0$ *and* $\overline{\Omega}(t)$,

then formulas (2.13)–(2.14) *define a weak continuous solution of problem* (1.8).

We may recover now the weak solution of problem (1.1) understood in the sense of Definition 1.1. Let (u,Σ) be the constructed solution of problem (1.8). Set $\phi = u|_\Sigma$ and consider the problem

$$
\begin{cases}
v_t = \Delta\, v^m - a\, v^p & \text{in the annular cylinder bounded by } S \text{ and } \Sigma, \\
v = \psi \text{ on } S, \quad v = \phi \text{ on } \Sigma, & \\
v(x,0) = u_0 \text{ in } E(0) \setminus D(0). &
\end{cases}
\tag{2.16}
$$

By assumption $\psi > 0$ on S, $u_0 > 0$ in $E(0)\setminus D(0)$, and by construction $\phi > 0$ on Σ. It follows from the maximum principle that any classical solution of problem (2.16) is separated away from zero, which allows us to treat the equation as nondegenerate parabolic. It is then standard to show the existence of a unique classical solution provided that the solution of problem (1.8) (the pair (u, Σ)) is sufficiently smooth. Moreover, one may show that the function

$$w(x,t) = \begin{cases} u(x,t) & \text{in } \mathcal{D}, \\ v(x,t) & \text{in } \mathcal{E} \setminus \mathcal{D} \end{cases} \tag{2.17}$$

is the desired weak continuous solution of the original problem (1.1). The rigorous proof is given in the end of the paper.

Since the assertion of Theorem 2.1 is true for *any* solution of problem (2.9)–(2.12), it will be convenient to specify the boundary and initial conditions for the function π. We will take for $\pi(\xi, t)$ the solution of the problem

$$\begin{cases} \operatorname{div}_\xi \left(u_0 |J| J^{-1} (J^{-1})^* \nabla_\xi \pi\right) = a\, u_0 U^{p-1} & \text{in } \Omega_0 \text{ for all } t \in [0,T], \\ \pi = 0 \text{ on } \partial\Omega_0, \qquad \pi(\xi, 0) = \pi_0(\xi), & |\nabla \pi| \le C \text{ in } \overline{Q}, \end{cases} \tag{2.18}$$

where π_0 is the solution of the problem

$$\operatorname{div}_\xi (u_0 \nabla_\xi \pi_0) = a\, u_0^p \quad \text{in } \Omega_0, \qquad \pi_0 = 0 \text{ on } \partial\Omega_0. \tag{2.19}$$

By Lagrangian counterpart of problem (1.8) *we will now mean the following problem: to find functions* (X, P, π) *satisfying conditions* (2.9), (2.10), (2.11), (2.18)–(2.19).

3. The weighted function spaces

3.1. Hölder spaces

Let $P_0 \in C^k(\overline{\Omega_0})$, $k \ge 1$, $P_0 = 0$ on $\gamma(0)$, and $|\nabla P_0| + P_0 \ge \kappa > 0$ in $\overline{\Omega_0}$. Then the $(n-1)$-dimensional manifold $\gamma(0)$ can be parametrized as follows: there exists $\epsilon > 0$ such that for every $\xi_0 \in \gamma(0)$ the set $B_\epsilon(\xi_0) \cap \gamma(0)$ is defined by the formulas

$$\begin{cases} \xi_i = y_i \text{ if } i \ne n, \\ y_n = P_0(y', \xi_n), \end{cases} \qquad y' = (y_1, \dots, y_{n-1}) \in B_\epsilon(\xi_0) \cap \{\xi_n = 0\}.$$

Adopt the notation

$$\left|D^k v\right| = \sum_{\beta = (\beta_1, \dots, \beta_N),\, |\beta| = k} \left|D^\beta v\right|.$$

Given a set $G \subseteq Q$ and a function P_0, we define the seminorms and norms: $|u|_{0,G} = \sup_G |u|$,

$$\{u\}_{\alpha,G} = \sup_{(x,t),(y,\tau) \in G,\, \delta((x,t),(y,\tau) \ne 0} \left\{ d^\alpha(x,y) \frac{|u(x,t) - u(y,t)|}{\delta^\alpha((x,t),(y,\tau))} \right\},$$

$$d(x,y) = \min\{P_0(x), P_0(y)\}, \quad \delta((x,t),(y,\tau)) = \sqrt{d(x,y)|t-\tau| + |x-y|^2},$$

$$\langle u\rangle_{0,G} = |u|_{0,G} + \{u\}_{\alpha,G},$$

$$\begin{aligned}\langle u\rangle_{2k+1,G} &= \sum_{2r+|\mu|=0}^{k} |D_t^r D^\mu u|_{0,G} + \sum_{2r+|\mu|=k+1}^{2k+1} \left| d^{|\mu|-k+r-\beta} D_t^r D^\mu u \right|_{0,G} \\ &+ \sum_{2r+|\mu|=2k+1} \{d^{k+1-\beta} D_t^r D^\mu u\}_{\alpha,G} \quad \text{for } k \geq 0, \quad \beta, \alpha \in (0,1).\end{aligned}$$

$$\langle\langle u\rangle\rangle_{k,G} = \sum_{2r+|\beta|=0}^{k} \left| d^{r+|\beta|} D_t^r D^\beta u \right|_{0,G} + \sum_{2r+|\beta|=k} \left\{ d^{r+|\beta|} D_t^r D^\beta u \right\}_{\alpha,G}, \quad k \geq 0.$$

The Banach spaces $V(2k+1, Q)$ with $k \geq 0$ and given parameters $\beta, \alpha \in (0,1)$ are defined as the completion of $C_0^\infty(\overline{Q})$ in the norm $\langle\cdot\rangle_{2k+1,Q}$.

If a function w does not depend on t we consider the function $\widetilde{w}(\xi,t) = w(\xi)$ with the dummy variable t and use the notation

$$\langle \widetilde{w}\rangle_{k,Q} = \langle w\rangle_{k,\Omega_0}, \qquad \|\widetilde{w}\|_{V(2k+1,Q)} = \|w\|_{V(2k+1,\Omega_0)}.$$

The Banach spaces Λ_i are defined as completion of $C_0^\infty\left(\overline{Q}\right)$ in the norms

$$\|u\|_{\Lambda_i} = \sum_{k=0}^{\infty} \frac{1}{k!M^k} \|(tD_t)^k u\|_{V(i,Q)}.$$

In these definition M is a finite number which will be specified later. It is easy to see that the elements of Λ_i, viewed as functions of the variable t and depending on $\xi \in \overline{\Omega_0}$ as a parameter, are real analytic. The radius of convergence of the corresponding power series is defined through M. Given a function $u \in \Lambda_i$, we introduce the new variable $\tau = \ln t$ and the function $U(\xi,\tau) = u(\xi,t)$. The Taylor expansion in τ of the function U has the form

$$U(\xi,\tau) = U(\xi,\tau_0) + \sum_{i=1}^{\infty} \frac{D_\tau^i U(\xi,\tau_0)}{M^i i!} \left[M(\tau-\tau_0)\right]^i.$$

This series is absolute and uniform convergent if $|\tau - \tau_0| < 1/M$.

3.2. Hilbert space $\mathbf{L}_2(\Omega_0, u_0)$ and its orthogonal subspaces.

Let (X, P, π) be a solution of problem (2.9)–(2.11), and U be defined by formulas (2.14). Assume that $|J|$ is separated away from zero and infinity so that the inverse matrix J^{-1} exists. Let us assume that the matrix J is symmetric. To be precise, let

$$J = I + \mathbf{D}(v), \quad \text{where } [\mathbf{D}(v)]_{ij} = D_{ij}^2 v \text{ with } v \in \Lambda_{2k+1}.$$

We define the space $\mathbf{L}_2(\Omega_0, u_0)$ as he closure of $C^\infty(\Omega_0)$ with respect to the weighted norm

$$\|\mathbf{v}\|_2^2 = \int_{\Omega_0} u_0 |J^{-1}\mathbf{v}|^2 \, d\xi.$$

The space $\mathbf{L}_2(\Omega_0, u_0)$ is the Hilbert space with the scalar product

$$\langle \mathbf{v}, \mathbf{u}\rangle = \int_{\Omega_0} u_0 \left(J^{-1}\mathbf{v} \cdot J^{-1}\mathbf{u}\right) d\xi.$$

The space $\mathbf{L}_2(\Omega_0, u_0)$ can be represented as the direct sum of the two subspaces orthogonal in the sense of the scalar product in $\mathbf{L}_2(\Omega_0, u_0)$:

$$\mathbf{G}(\Omega_0, u_0) = \{\mathbf{v} = \nabla w \in \mathbf{L}_2(\Omega_0, u_0),\ w = 0 \text{ on } \sigma(0)\},$$
$$\mathbf{J} = \{\mathbf{w} \in \mathbf{L}_2(\Omega_0, u_0) : \ \langle \mathbf{w}, \mathbf{v}\rangle = 0\ \forall\, \mathbf{v} \in \mathbf{G}(\Omega_0, u_0)\}.$$

Let us show that every vector field Φ admits the representation

$$\Phi = \nabla f + \Psi \quad \text{where } \Psi \in \mathbf{J}.$$

For f we take a function satisfying the conditions

$$\operatorname{div}\left(u_0\,(J^{-1})^2(\nabla f - \Phi)\right) = 0 \qquad \text{in } \Omega_0, \qquad f = 0 \text{ on } \partial\Omega_0. \tag{3.1}$$

Let us assume that equation (3.1) admits a classical solution f, and that Φ and ∇f are uniformly bounded in Ω_0. Set $\Psi = \Phi - \nabla f$ and let $\mathbf{w} \in \mathbf{G}(\Omega_0, u_0)$ be arbitrary. There exists η such that $\eta = 0$ on $\sigma(0)$, $\mathbf{w} = \nabla \eta$, and

$$\begin{aligned}\langle \Psi, \mathbf{w}\rangle &= \int_{\Omega_0} u_0\,(J^{-1}\Psi \cdot J^{-1}\nabla \eta)\,d\xi = \int_{\Omega_0} \eta \operatorname{div}\left(u_0\,(J^{-1})^2(\nabla f - \Phi)\right) d\xi \\ &+ \int_{\partial\Omega_0} \eta\, u_0\,(J^{-1})^2(\nabla f - \Phi, \mathbf{n})\, dS = 0.\end{aligned}$$

The desired decomposition is constructed.

Proposition 3.1. *Let* $u_0^{m-2}\operatorname{div}\left(u_0(J^{-1})^2\Phi\right) \in \Lambda_{2k-1}$ *with* $k \geq 1$, *and* $J_{ij} = \delta_{ij} + D^2_{ij}v$ *for some* $v \in \Lambda_{2k+3}$ *with the same* k. *If* $\operatorname{dist}(\sigma(0), \gamma(0))$ *is appropriately small, there holds the representation* $\Phi = \nabla f + \Psi$, *where* $f \in \Lambda_{2k+1}$ *is the solution of problem* (3.1) *and* $\Psi = \Phi - \nabla f \in \mathbf{J}$. *Moreover,*

$$\|f\|_{\Lambda_{2k+1}} \leq C\left\|u_0^{m-2}\operatorname{div}\left(u_0(J^{-1})^2\Phi\right)\right\|_{\Lambda_{2k-1}} \tag{3.2}$$

with a constant C *independent of* f.

The proof of solvability of problem (3.1) in the function spaces Λ_{2k+3} is an imitation of the proof given in [8, Section 5] for the case $J \equiv I$. The difference is that now $J = I + \mathbf{D}(v)$ with $v \in \Lambda_{2k+3}$.

3.3. Operators of orthogonal projection.

Let us define the operator $\mathcal{P}$ of the orthogonal projection onto the subspace $\mathbf{G}(\Omega_0, u_0)$. Given a vector field Φ and the matrix $J = I + \mathbf{D}(v)$ satisfying the conditions of Proposition 3.1, we set

$$\mathcal{P}\langle\Phi\rangle = \nabla f, \text{ where } f \in \Lambda_{2k+1} \text{ is the solution of problem (3.1).}$$

We also introduce the operator of orthogonal projection onto the subspace $\mathbf{J}$:

$$\mathcal{R}\langle\Phi\rangle = \Phi - \mathcal{P}\langle\Phi\rangle \equiv \Phi - \nabla f.$$

4. The gradient flow

It is shown in Section 2 that any solution of the problem formulated in Lagrangian coordinates generates a solution of problem (1.8) in the plane of Euler coordinates. Moreover, it is sufficient for our purposes to construct *any* of the possible solutions of the problem posed in Lagrangian coordinates. This allows us to limit ourselves to constructing special solutions that describe *the gradient flows on the plane of Lagrangian coordinates*: $X(\xi,t) = \xi + \nabla v(\xi,t)$.

Let us consider the new problem: to find a triad of scalar functions (v, P, π) such that (v, P, π) is a classical solution of the system of equations

$$\mathcal{P}\langle J\nabla v_t\rangle + \nabla_\xi (P - \pi) = 0 \quad \text{in } Q = \Omega_0 \times (0,T), \tag{4.1}$$

$$P\,|J|^{m-1} - P_0(\xi) = 0, \tag{4.2}$$

$$\operatorname{div}_\xi \left(u_0 |J| J^{-1}(J^{-1})^* \nabla_\xi \pi\right) = a\, u_0 U^{p-1}, \tag{4.3}$$

$$\begin{cases} v(\xi,0) = 0,\ \pi(\xi,0) = \pi_0(\xi),\ P(\xi,0) = P_0 \text{ in } \Omega_0, \\ \pi = v = 0 \text{ on } \partial\Omega_0,\ P = 0 \text{ on } \Gamma_0, \quad |\nabla\pi| + |\nabla P| \le C \text{ in } \overline{Q}. \end{cases} \tag{4.4}$$

Here π_0 is the solution of problem (2.19), J is the symmetric matrix with the entries $J_{ij} = \delta_{ij} + D^2_{ij} v$, so that the projection operator $\mathcal{P}$ is well defined.

Let us check that formulas (2.13)–(2.14) continue to define a weak solution of problem (1.8). Let the triad $(v, P, \pi) \in \Lambda_{2k+3} \times (\Lambda_{2k+1})^2$, $k \ge 1$, be a solution of problem (4.1), (4.2), (4.3), (4.4). Then for every $t > 0$

$$\mathcal{R}\langle J\nabla v_t\rangle = J\nabla v_t - \mathcal{P}\langle J\nabla v_t\rangle \in \mathbf{L}_2(\Omega_0, u_0).$$

Formulas (2.15) can be written in the form

$$-\int_{\Omega_0} u_0(x)\eta(x,0)dx = \int_0^T \frac{d}{dt}\left\{\int_{\Omega_0} u(x,t)\eta(x,t)dx\right\} dt$$

$$-\int_Q [\eta_t + \nabla_x\eta \cdot X_t]\; U|J|\, d\xi\, dt = \int_Q \eta_t\, U|J|\, d\xi\, dt + \Theta.$$

Since (v, P, π) is a solution of problem (4.1)–(4.4), then for every test-function $\eta \in C^1(\Omega_0)$ vanishing on $\sigma(0)$

$$\Theta = \int_Q u_0(X_t \cdot \nabla_x\eta)\, d\xi dt = \int_Q u_0 (J^{-1})^2 (JX_t \cdot \nabla_\xi\eta)\, d\xi dt = \int_0^T \Big\langle JX_t, \nabla_\xi\eta\Big\rangle dt$$

$$= \int_0^T \Big\langle \mathcal{P}\langle JX_t\rangle, \nabla_\xi\eta\Big\rangle dt + \int_0^T \Big\langle \mathcal{R}\langle JX_t\rangle, \nabla_\xi\eta\Big\rangle dt = -\int_0^T \Big\langle \nabla_\xi(P-\pi), \nabla_\xi\eta\Big\rangle dt$$

$$= -\int_Q u\nabla_x(p - \Pi)\cdot\nabla_x\eta |J|\, d\xi dt = -\int_D [\nabla_x u^m - a u^p \eta]\cdot\nabla_x\eta\, dx dt.$$

Notice that the condition $X = \xi + \nabla v \in C(\overline{Q})$, and items 1)–2) of the conditions of Theorem 2.1 are automatically fulfilled because of the inclusions $v \in \Lambda_{2k+3}$, $\pi \in \Lambda_{2k+1}$. The above arguments prove the following version of Theorem 2.1.

Theorem 4.1. *Let $P_0 \in V(2k+1, \Omega_0)$, $k \geq 1$, and $(v, P, \pi) \in \Lambda_{2k+3} \times (\Lambda_{2k+1})^2$ be a solution of problem (4.1)–(4.4). If*

1. *$|J| \equiv \det\left[I + D^2 v\right]$ is separated away from zero and infinity in $\overline{Q}$,*
2. *the mapping $\xi \mapsto X(\xi, t)$ is a bijection from $\overline{\Omega}_0$ to $\overline{\Omega}(t)$ for every $t \in (0, T]$,*

then formulas (2.13)–(2.14) define a weak continuous solution of problem (1.8).

4.1. Solution of problem (4.1)–(4.4). The linearized problem

Theorem 4.2. *Let conditions (1.5) be fulfilled and $\sigma(0) \in C^{2+\alpha}$. There exist $a^* > 0$, $\epsilon^* < 1$, M and T^* such that for every $\|P_0\|_{V(2k+1,\Omega_0)} < \epsilon^*$, $|a| < a^*$ problem (4.1)–(4.4) has in the cylinder Q with $T < T^*$ a unique solution (v, P, π). The function P is strictly positive in Q and $P = 0$ on Γ_0. The solution (v, P, π) satisfies the estimate*

$$\|\pi\|_{\Lambda_{2k+1}} + \|v\|_{\Lambda_{2k+3}} + \|P\|_{\Lambda_{2k+1}} \leq C\,\left(|a| + \|P_0\|_{V(2k+1,\Omega_0)}\right)$$

with a finite constant C independent of v, π, and P.

The proof literally repeats the proof of the analogous assertion given in [8], which is why we omit all the technical details.

The solution of nonlinear problem (4.1)–(4.4) is obtained by means of the modified Newton method. We consider problem (4.1)–(4.4) as the functional equation $\mathcal{F}(x) \equiv \{\mathcal{F}_1(x), \mathcal{F}_2(x), \mathcal{F}_3(x)\} = 0$ where $x = (v, P, \pi)$. The solution of the equation $\mathcal{F}(x) = 0$ is obtained as the limit of the sequence of solutions of the linear problems $\{x_n\}$ where

$$x_{n+1} = x_n - \mathcal{G}^{-1}\langle \mathcal{F}(x_n)\rangle \qquad \text{with } x_n = (v_n, P_n). \tag{4.5}$$

The linear operator $\mathcal{G}$ is the Frechét derivative of $\mathcal{F}$ at the initial state $x_0 = (0, P_0, \pi_0)$. To construct the operator $\mathcal{G}^{-1}$ we consider the problem $\mathcal{G}\langle x\rangle = g$, where $x = (w, P, \pi)$ and $g = (\nabla f, \Psi, H)$.

$$\mathcal{G}_1\langle (v, P)\rangle = \frac{d\mathcal{F}_1}{d\epsilon}(\epsilon\nabla v, P_0 + \epsilon P)\bigg|_{\epsilon=0} = \nabla(P - \pi) + \mathcal{P}\langle \nabla v_t\rangle \equiv \nabla(P - \pi + v_t),$$

To linearize $\mathcal{F}_2$ we make use of the Newton formulas:

$$\det[\lambda I - A] = \sum_{k=0}^{n} (-1)^k \alpha_k\, \lambda^{n-k},$$

$\alpha_0 = 1$, $k\,\alpha_k = \sum_{i=1}^{k} \alpha_{k-i}\,\mathrm{trace}\left(A^i\right)$ for $1 \leq k \leq n$. We have $|I + \epsilon\mathbf{D}(v)| = 1 + \epsilon\,\Delta\,v + \mathcal{O}(\epsilon^2)$, so that

$$\mathcal{G}_2\langle (v, P, \pi)\rangle = \frac{d\mathcal{F}_2}{d\epsilon}(\epsilon\nabla v, P_0 + \epsilon P, \pi_0 + \epsilon\pi)\bigg|_{\epsilon=0} = P + (m-1)P_0\Delta\,v.$$

Next,

$$\left((I + \epsilon\mathbf{D}(v))^{-1}\right)^2 = \left(\sum_{k=0}^{\infty} (-\epsilon)^k\,(\mathbf{D}(v))^k\right)^2 = I - 2\epsilon\,\mathbf{D}(v)) + \mathcal{O}\left(\epsilon^2\right),$$

whence

$$\mathcal{G}_3\langle(v,P,\pi)\rangle = \operatorname{div}(u_0\nabla\pi - 2u_0\,\mathbf{D}(v)\cdot\nabla\pi_0) + a(1-p)u_0^p\Delta v.$$

The linear problem problem $\{\mathcal{G}_1,\,\mathcal{G}_2,\mathcal{G}_3\}\langle(v,P,\pi)\rangle = (\nabla f, H)$ takes on the form

$$\begin{cases} \nabla v_t + \nabla(P-\pi) = \nabla f, \\ P + (m-1)P_0\,\Delta v = \Psi, \\ \operatorname{div}(u_0\nabla\pi - 2u_0\,\mathbf{D}(v)\cdot\nabla\pi_0) + a(1-p)u_0^p\Delta v = H. \end{cases} \tag{4.6}$$

Eliminating P we reduce this system to two scalar equations for the functions v and π:

$$\begin{cases} v_t - (m-1)P_0\Delta v = f + \pi - \Psi \quad \text{in } Q, \\ v = 0 \text{ on the parabolic boundary of } Q, \end{cases} \tag{4.7}$$

$$\begin{cases} \operatorname{div}(u_0\nabla\pi - 2u_0\,\mathbf{D}(v)\cdot\nabla\pi_0) = a(p-1)u_0^p\Delta v + H \quad \text{in } Q, \\ \pi = 0 \text{ on the parabolic boundary of } Q. \end{cases} \tag{4.8}$$

Once these problems are solved, P is restored from the second equation in (4.6).

The ensure the convergence of the sequence $\{x_n\}$ to the solution x of the nonlinear problem $\mathcal{F}(x) = 0$ we have to perform the following three steps [6]:

1. To solve the linear problem $\mathcal{G}\langle x\rangle = \mathcal{F}\langle x_0\rangle$. The degenerate parabolic-elliptic problem (4.7)–(4.8) reduces to problem (4.7) with the right-hand side P_0, while $\pi \equiv 0$. We define then P from the second equation in (4.6).
2. To show that the operators $\mathcal{F}$ and $\mathcal{G}$ are defined on the same pair of the function spaces $\mathcal{X} = \Lambda_{2k+3}\times(\Lambda_{2k+1})^2$ and $\mathcal{Y} = (\Lambda_{2k+1})^2\times\Lambda_{2k-1}$ with $k \geq 1$:

$$\mathcal{F} : \mathcal{X} \mapsto \mathcal{Y}, \qquad \mathcal{G}^{-1} : \mathcal{Y} \mapsto \mathcal{X}.$$

This is done by deriving appropriate a priori estimates for the solutions of the linear problem $\mathcal{G}\langle x\rangle = g$, (alias (4.7)–(4.8)). The existence of a unique solution of this system (for small T) is proved by means of the contraction mapping principle.

The operator $\mathcal{F} = (\mathcal{F}_1, \mathcal{F}_2, \mathcal{F}_3)$ is understood in the following way: for every element $(v,P,\pi) \in \Lambda_{2k+3}\times(\Lambda_{2k+1})^2$ with $k \geq 1$

$$\mathcal{F}_1(v,P,\pi) \equiv \mathcal{P}\langle(I+\mathbf{D}(v))\nabla v_t\rangle + \nabla(P-\pi) \equiv \nabla(f+P-\pi), \tag{4.9}$$

where $f \in \Lambda_{2k+1}$ is a solution of problem (3.1) with the right-hand side

$$\phi = \operatorname{div}\left(u_0(I+\mathbf{D}(v))^{-1}\nabla v_t\right), \tag{4.10}$$

and

$$\mathcal{F}_2(v,P,\pi) = P\det|I+\mathbf{D}(v)|^{m-1} - P_0. \tag{4.11}$$

The norm of $\mathcal{F}(x)$ is defined by

$$\|\mathcal{F}(x)\|_{\mathcal{Y}} = \|f\|_{\Lambda_{2k+1}} + \|P\|_{\Lambda_{2k+1}} + \|\mathcal{F}_2(v,P,\pi)\|_{\Lambda_{2k+1}}. \tag{4.12}$$

3. Denote by $\mathcal{H}(x)$ the Frechet derivative of $\mathcal{F}$ at the element $x = (v, P, \pi)$. The last step is to check that the $\mathcal{H}(x)$ is Lipschitz-continuous:
$$\|\mathcal{H}(x) - \mathcal{H}(y)\|_{\mathcal{Y}} \leq L\|x - y\|_{\mathcal{X}}.$$

5. Solution of the free-boundary problem (1.1)

Let us show first that formula (2.17) indeed defines a weak continuous solution of problem (1.1) in the sense of Definition 1.1. By definition
$$\Sigma = \{x : x = \xi + \nabla v(\xi, t) \text{ with } v \in \Lambda_{2k+3},\ \xi \in \sigma(0)\} \in C^{2+\alpha}.$$
According to (2.14) and (2.10)
$$\phi \equiv u(x,t)|_{\Sigma} = \left.\frac{u_0}{|I + \mathbf{D}(v)|}\right|_{\Sigma_0}.$$
It follows that $\phi > 0$ and $\phi = u_0$ on $\overline{\Sigma} \cap \{t = 0\}$. Thus, the data of problem (2.16) satisfy the zero-order compatibility condition and there exists a unique solution $v \in C(\overline{\mathcal{E} \setminus \mathcal{D}}) \cap C^{2,1}(\mathcal{E} \setminus \mathcal{D})$ [7, Chapter XIV].

According to the definition of $u(x,t)$, for all $\xi \in \overline{Q}$
$$\frac{dU(\xi,t)}{dt} = u_t(x,t) + \nabla_x u \cdot X_t(\xi,t) = u_t(x,t) + \nabla_x u \cdot \mathbf{v}.$$
On the other hand, we may calculate $U_t(\xi,t)$ differentiating in t the mass conservation law (2.10) and applying the Cauchy identity. We have:
$$|J|\,(U_t + U\,\mathrm{div}_x \mathbf{v}) = 0 \qquad \text{in } \overline{Q}\,,$$
whence
$$u_t(x,t) = U_t - \nabla_x u \cdot \mathbf{v} = -U \mathrm{div}_x \mathbf{v} - \nabla_x u \cdot \mathbf{v} = -\mathrm{div}_x (u\,\mathbf{v}) \quad \text{in } \overline{Q}.$$
Plugging the definition of $\mathbf{v}$ (2.5) and letting $t \to 0$ we conclude that the data of problem (2.16) satisfy the first-order compatibility conditions on $\overline{\Sigma} \cap \{t = 0\}$. It follows from the theory of nondegenerate parabolic equations that the Dirichlet problem (2.16) has a unique solution $v \in C^{2+\alpha,(2+\alpha)/2}(\overline{\mathcal{E} \setminus D})$ [7, Chapter XIV]. Thus, the solutions of problems (2.16) and (1.8) satisfy $\nabla v = \nabla u$ on Σ. For an arbitrary test-function η satisfying the conditions of Definition 1.1 we have then:
$$\begin{aligned}
0 &= -\int_{\mathcal{D}} \eta\,(u_t - \Delta\,u^m + a\,\eta\,u^p)\,dxdt - \int_{\mathcal{E}\setminus\mathcal{D}} \eta\,(v_t - \Delta\,v^m + a\,\eta\,v^p)\,dxdt \\
&= \int_{E(0)} u_0\eta(x,0)\,dx + \int_{\mathcal{E}} (\eta_t w - \nabla_x \eta \cdot \nabla\,w^m - a\,\eta\,w^p)\,dx\,dt \\
&\quad - \int_{\Sigma} \eta\,\left[u\,\mathbf{n}_\tau^+ + v\mathbf{n}_\tau^-\right]\,dS + \int_{\Sigma} \eta\,\left[(\mathbf{n}_x^+, \nabla\,u^m) + (\mathbf{n}_x^-, \nabla\,v^m\right]\,dS \\
&= \int_{E(0)} u_0\eta(x,0)\,dx + \int_{\mathcal{E}} (\eta_t w - \nabla_x \eta \cdot \nabla\,w^m - a\,\eta\,w^p)\,dx\,dt,
\end{aligned}$$

where $\mathbf{n}^+ = -\mathbf{n}^-$ denote the unit normal vectors to Σ exterior and interior with respect to $\mathcal{D}$.

The proof of Theorem 1.1 repeats, with some obvious changes, the proofs in [8, Subsection 8.2]. To show that the regularity of the interface velocity is better than it was at the initial instant we make use of the double representation for $\mathbf{v} = X_t(\xi, t)$: on one hand,

$$\mathbf{v}(x,t) = -\nabla p + \nabla \Pi,$$

on the other hand

$$\mathbf{v}(x,t) = \nabla_\xi v_t(\xi,t) \quad \text{with } v \in \Lambda_{2k+3}.$$

References

[1] S. Antontsev, J. Díaz, and S. Shmarev, *Energy Methods for Free Boundary Problems. Applications to Nonlinear PDEs and Fluid Mechanics*, vol. 48 of Progress in Nonlinear Differential Equations and Their Applications, Birkhäuser, 2002.

[2] V.A. Galaktionov, S.I. Shmarev, and J.L. Vazquez, *Regularity of interfaces in diffusion processes under the influence of strong absorption*, Arch. Ration. Mech. Anal., 149 (1999), pp. 183–212.

[3] ——, *Regularity of solutions and interfaces to degenerate parabolic equations. The intersection comparison method*, in Free boundary problems: theory and applications (Crete, 1997), Chapman & Hall/CRC, Boca Raton, FL, 1999, pp. 115–130.

[4] ——, *Second-order interface equations for nonlinear diffusion with very strong absorption*, Commun. Contemp. Math., 1 (1999), pp. 51–64.

[5] ——, *Behaviour of interfaces in a diffusion-absorption equation with critical exponents*, Interfaces Free Bound., 2 (2000), pp. 425–448.

[6] A.N. Kolmogorov and S.V. Fomīn, *Introductory real analysis*, Dover Publications Inc., New York, 1975. Translated from the second Russian edition and edited by Richard A. Silverman, Corrected reprinting.

[7] G.M. Lieberman, *Second order parabolic differential equations*, World Scientific Publishing Co. Inc., River Edge, NJ, 1996.

[8] S. Shmarev, *Interfaces in multidimensional diffusion equations with absorption terms*, Nonlinear Analysis, 53 (2003), pp. 791–828.

Sergei Shmarev
Departamento de Matemáticas
University of Oviedo
33007 Oviedo, Spain
e-mail: shmarev@orion.ciencias.uniovi.es

Progress in Nonlinear Differential Equations
and Their Applications, Vol. 61, 275–282

Estimates for Solutions of Fully Nonlinear Discrete Schemes

Hung-Ju Kuo and Neil S. Trudinger

Dedicated to Vsevolod Solonnikov on the occasion of his 70th *birthday*

Abstract. We describe some estimates for solutions of nonlinear discrete schemes, which are analogues of fundamental estimates of Krylov and Safonov for linear elliptic partial differential equations and the resultant Schauder estimates for nonlinear elliptic equations of Evans, Krylov and Safonov.

Mathematics Subject Classification (2000). Primary 65N12, 35J15, 39A70; Secondary 65N40, 35B05, 39A10.

1. Introduction

In this article we describe recent research of the authors and others concerning estimates for solutions of nonlinear difference schemes which are discrete analogues of nonlinear elliptic partial differential equations of second order. A general difference equation may be written in the form,

$$F[u] := F(\cdot, u, Tu) = 0, \tag{1.1}$$

where $u : E \to \mathbb{R}$ is a mesh function defined on a mesh E, which is a discrete subset of n-dimensional Euclidean space, $\mathbb{R}^n$, $Tu(x) = \{u(z) \mid z \neq x\}$ and F is a given real-valued function on $E \times \mathbb{R}^E$. We always assume that $F[u](x)$ is independent of the values $u(z)$ for $|z - x|$ sufficiently large, that is $F[u](x)$ only depends on finitely many $u(z)$ for each $x \in E$. The equation (1.1) is linear if it can be written in the form,

$$L[u](x) := \sum_{z \in E} a(x, z)\, u(z) = f(x), \tag{1.2}$$

The first author was supported by the Taiwan National Science Council; the second author was supported by an Australian Research Council Grant.

with coefficients $a(x, z)$ having finite support in z for each $x \in E$. The operator F is called *monotone* if

$$F(x,\, u,\, q+\eta) \geq F(x,\, u,\, q)\,, \tag{1.3}$$

for all $x \in E$, $u, \in \mathbb{R}$, q, $\eta \in \mathbb{R}^{E-\{x\}}$, $\eta \geq 0$, and *positive* if, in addition,

$$F(x,\, u+\tau,\, q+\eta) \leq F(x,\, u,\, q), \tag{1.4}$$

for all $x \in E$, z, $\tau \in \mathbb{R}$, q, $\eta \in \mathbb{R}^{E-\{x\}}$, with $0 \leq \eta_z \leq \tau$ for each $z \in E - \{x\}$. The linear operator (1.2) is monotone if $a(x, z) \geq 0$ for all $x \neq z \in E$ and positive if also $\sum_{z\in E} a(x, z) \leq 0$ for all $x \in E$. If F is differentiable with respect to z, q, then F is monotone if $\frac{\partial F}{\partial q_z}(x, u, q) \geq 0$ for all $z \in E - \{x\}$, $x \in E$ and positive if in addition, $\sum_{z\in E} \frac{\partial F}{\partial q_z}(x, u, q) \leq 0$ for all $x \in E$. We also call F *balanced* if

$$F(x,\, u,\, \widetilde{q}\,) = F(x,\, u,\, q\,), \tag{1.5}$$

whenever $\widetilde{q}_z = q_z + p \cdot (z - x)$ for some $p \in \mathbb{R}^n$. The linear operator (1.2) is balanced if $\sum_z a(x, z)(z - x) = 0$ and if F is differentiable with respect to q, then the operator (1.1) is balanced if $\sum_z \frac{\partial F}{\partial q_z}(z - x) = 0$. If we also assume that F is positive with equality holding in (1.4) for $\eta_z = \tau$, we may write equation (1.1) in the form,

$$F[u](x) = F(x,\, \delta u(x)\,) = 0\,, \tag{1.6}$$

where

$$\delta u(x) = \{\, u(z) - u(x) \mid z \neq x \in E\,\}.$$

By using the Taylor expansion for $u \in C^2(\mathbb{R}^n)$,

$$u(z) - u(x) = Du(x)\cdot(z-x) + \frac{1}{2}\big[D^2u(x)(z-x),\, (z-x)\big] + o(|z-x|^2)\,, \tag{1.7}$$

we see that a balanced difference scheme, of the form (1.6), is consistent with a partial differential operator of the form,

$$\mathcal{F}[u] := \mathcal{F}(x,\, D^2 u\,)\,, \tag{1.8}$$

with linearized principal coefficient matrix,

$$a^{ij} := \frac{\partial \mathcal{F}}{\partial u_{ij}} = \frac{1}{2}\sum_z \frac{\partial F}{\partial q_z}(z-x)_i(z-x)_j\,, \tag{1.9}$$

if $\frac{\partial F}{\partial q_z} = O(|x - z|^{-2})$. In (1.7), (1.8), $D^2u = [u_{ij}]$ denotes the Hessian matrix of second derivatives of the function u. The above definitions are taken from our papers [12] and [14].

In the next sections of this article we describe two types of estimates for solutions of (1.1), which are discrete analogues of fundamental estimates for elliptic equations. In Section 2, we present local pointwise estimates for solutions of linear equations, which are the analogues of the Krylov-Safonov estimates [1, 3, 9, 22] for partial differential operators, while in Section 3, we consider discrete analogues of the nonlinear Schauder estimates of Evans, Krylov, Safonov and Caffarelli [1,

2, 3, 8, 19, 20]. For the latter, we shall restrict our meshes E to be lattice meshes, that is there exists a linearly independent set of vectors $\xi_1, \xi_2, \dots, \xi_n$ such that

$$E = \{ (m_1\xi_1, m_2\xi_2, \dots, m_n\xi_n) \mid m_i \in \mathbb{Z}, \ i = 1, 2, \dots, n \}. \tag{1.10}$$

By appropriate coordinate transformation, the lattice mesh case can be reduced to the special case of a cubic mesh

$$\mathbb{Z}^n_h = \{ (m_1, m_2, \dots, m_n)h \mid m_i \in \mathbb{Z}, \ i = 1, 2, \dots, n \} \tag{1.11}$$

with mesh length h.

2. Linear equations

We consider here linear equations of the form (1.6), that is

$$L[u](x) = \sum a(x,z)\big(u(z) - u(x)\big) = f(x), \tag{2.1}$$

which are monotone and balanced. In accordance with (1.7) , (1.8) , (1.9), these correspond to degenerate elliptic partial differential equations of the form

$$\mathcal{L}[u] := a^{ij} D_{ij}\, u = f \tag{2.2}$$

with coefficient matrix

$$a^{ij} = \frac{1}{2} \sum_z a(x,z)(z - x)_i (z - x)_j \,. \tag{2.3}$$

It is thus natural to call (2.1) *elliptic* if the matrix $\mathcal{A} = [a^{ij}]$ is positive, that is

$$\frac{1}{2} \sum_z a(x,z)[(z-x)\cdot\xi]^2 \geq \lambda\, |\xi|^2 , \tag{2.4}$$

for all $\xi \in \mathbb{R}^n$ and some positive constant λ.

For local estimates, this notion is still inadequate as seen by the simple one-dimensional example, $E = \mathbb{Z}_h$,

$$L[u] = u(x+2h) + u(x-2h) - 2u(x) = 0 , \tag{2.5}$$

which has solutions

$$u(mh) = a\,(-1)^m + b$$

for constants a and b. It is thus necessary to assume that all points in E are effectively linked by L, that is for any two points $x, z \in E$, there exist points $x_0, x_1, \dots, x_k \in E$ such that $x_0 = x$, $x_k = z$ and

$$a(x_{i-1}, x_i) \geq \frac{\lambda}{(\overline{h})^2} , \tag{2.6}$$

$$k = k(x,z) \leq k_0 \frac{|x - z|}{h} ,$$

where k_0 is a positive constant,

$$h = \inf_{x \neq z \in E} |x - z| \tag{2.7}$$

denotes the minimum mesh width of E and

$$\overline{h} = \sup_{a(x,z)\neq 0} |x-z| \tag{2.8}$$

denotes the maximum mesh width with respect to L. We also define an ellipticity constant,

$$a_0 = \frac{1}{2\lambda} \sup_{x\in E} \sum a(x,z)|x-z|^2 . \tag{2.9}$$

Note that when dealing with local estimates, we can assume (by replacing E by a bounded subset) that h is positive and $\overline{h}$, a_0 are finite.

Under these assumptions we have Hölder and Harnack estimates for solutions of the scheme (2.1) which are discrete analogues of the fundamental estimates of Krylov and Safonov. In their formulation, we let for any $R > 0$ and $y \in \mathbb{R}^n$, $E_R(y)$ denote the mesh ball of radius R, and centre y, that is

$$E_R(y) = \{\, x \in E \,|\; |x-y| < R \,\}$$

and define the L_p norm, $(1 \leq p < \infty)$, of a mesh function f over a set $S \subset E$, by

$$\| f \|_{L^p(S)} = \Big\{ \sum_{x\in E} h^n |f(x)|^p \Big\}^{1/p}$$

Theorem 2.1. *Let u be a solution of (2.1) in a mesh ball $E_R = E_R(y) \subset E$. Then for any concentric ball $E_{\sigma R}(R), 0 < \sigma < 1$, we have the Hölder estimate*

$$\operatorname*{osc}_{E_{\sigma R}} u \leq C\sigma^\alpha \Big\{ \operatorname*{osc}_{E_R} u + \frac{R}{\lambda} \| f \|_{L^n(E_R)} \Big\}, \tag{2.10}$$

where α and C are positive constants depending on $n, \overline{h}/h$, k_0 and a_0. If u is non-negative in E_R and $h/(1-\sigma)R$ sufficiently small, we have the Harnack inequality,

$$\max_{E_{\sigma R}} u \leq C \Big\{ \min_{E_{\sigma R}} u + \frac{R}{\lambda} \| f \|_{L^n(E_R)} \Big\}, \tag{2.11}$$

where C depends on $n, \overline{h}/h$, k_0, a_0 and σ.

Theorem 2.1 extends earlier estimates, under slightly stronger non-degeneracy conditions than (2.4), in our papers [11, 13, 14]. To obtain the full strength of Theorem 2.1, we use our discrete Aleksandrov maximum principle [16] in place of the versions we used previously. As in our previous works, the balance condition can be relaxed to allow "lower-order" dependence and we have separate complementary estimates for sub- and supersolutions. These results may also be extended to boundary neighborhoods and parabolic equations as in [15].

3. Schauder estimates for nonlinear schemes

Now we return to the general equation (1.1), or rather the special form (1.6) where, as in the linear case of the previous section, F is assumed to be monotone and balanced. The ellipticity conditions on F are imposed on the matrix $\mathcal{A} = [a^{ij}]$, given by (1.9). However for higher-order difference estimates, we need to assume that E is a lattice mesh, which we can subsequently reduce to the case $E = \mathbb{Z}_h^n$ by a linear transformation. It is then convenient to write $z = x + y$ in (1.6) so that

$$\delta u(x) = \{ u(x+y) - u(x) \,|\, y \in Y_N \},$$

where

$$Y_N = \{ y \in E \,|\, \|y\|_\infty \le Nh \}$$

for some $N \in \mathbb{N}$.

Our Schauder estimates are discrete versions of the general Schauder estimates of Safonov for [19, 20] uniformly elliptic partial differential equations of the form (1.7), which extended the earlier estimates of Evans [2] and Krylov [8], under stronger conditions on the dependence of $\mathcal{F}$ on x. Analogously to the continuous case, we also assume that F is *concave* with respect to $q \in \mathbb{R}^{Y_N}$ and *Hölder continuous* with respect to $x \in \mathbb{R}$, in the sense that

$$|f(x,q) - f(z,q)| \le \mu(1+|q|)\,|x-z|^\gamma$$

for all $z, x \in \mathbb{R}^n$, $q \in \mathbb{R}^{Y_N}$ where μ and $\gamma\,(\le 1)$ are positive constants. Our ellipticity conditions are applied to the linearized frozen operator, in which case the condition (2.4) becomes a consequence of (2.6). Indeed fixing $z_0 \in E$, we define the *frozen* operator

$$F_0[u] = F(z_0, \delta u), \tag{3.1}$$

with *linearized* operator

$$L_0[u] = \sum_{y \in Y_N} a_0(y)\big(u(x+y) - u(x)\big), \tag{3.2}$$

where

$$a_0(y) = a(z_0, y) = \frac{\partial F}{\partial q_y}(z_0, \delta u(x)), \tag{3.3}$$

Accordingly we assume for any three points x, z, $z_0 \in E$, there exist points $x_0 = x, x_1, \ldots, x_k = z$ in E such that

$$\frac{\partial F}{\partial q_y}(z_0, q) \ge \lambda h^{-2}, \tag{3.4}$$

for all $q \in \mathbb{R}^{Y_N}$, $y = y_i = x_i - x_{i-1}$, $i = 1, \ldots, k$, where λ is a positive constant and the number $k = k(x, z; z_0)$ satisfies

$$k(x, z; z_0) \le \frac{k_0\,|x-z|}{h}, \tag{3.5}$$

for some positive constant k_0. Clearly it is enough that x and z are neighbors, that is $|x - z| = h$ and moreover the constant k_0 can be chosen to depend only on N.

A simple way of ensuring (3.4) is to assume that F_0 connects neighboring points directly, that is (3.4) holds for $y = \pm he_i$, $i = 1, \dots, n$.

We believe that these conditions should suffice for the interior Schauder estimates. So far we have obtained these estimate for operators F of the form

$$F[u](x) = G\big(x, L^1[u](x), \dots, L^K[u](x)\big), \tag{3.6}$$

where $G : E \times \mathbb{R}^K \to \mathbb{R}$ satisfies

$$\lambda_0 \le \frac{\partial G}{\partial p_i}(x, p) \le \Lambda_0, \tag{3.7}$$

for positive constants λ_0, Λ_0 and for all $i = 1, \dots, K$, $(x, p) \in E \times \mathbb{R}^K$, and the operators $L^1, \dots, L^K$ are monotone, balanced linear operators.

In order to formulate our main result, we need to define appropriate discrete seminorms and norms. As usual, we define first- and second-order difference operators by

$$\begin{aligned} \delta_i u(x) &= \frac{u(x + he_i) - u(x)}{h}, \\ \delta_{ij} u(x) &= \delta_i\big(\delta_j u(x)\big), \\ \delta^1 u &= \big(\delta_1 u, \dots, \delta_n u\big), \\ \delta^2 u &= \big[\delta_{ij} u\big]_{i,j=1,\dots,n}. \end{aligned} \tag{3.8}$$

For $\gamma \in (0,1]$ and Ω a bounded open set in $\mathbb{R}^n$ we define the Hölder semi-norm,

$$[u]_{0,\gamma;\Omega} = \max_{x \neq y, \in \Omega_h} \frac{|u(x) - u(y)|}{|x - y|^\gamma} \tag{3.9}$$

where $\Omega_h = \Omega \cap E = \Omega \cap \mathbb{Z}^n_h$. We can then define interior semi-norms, similarly to the continuous case [3], by

$$\begin{aligned} [u]^*_{0,\gamma;\Omega} &= \sup_{\Omega' \subset \Omega} (d')^\gamma [u]_{0,\gamma;\Omega'}, \\ [u]_{k;\Omega} &= \max_{\Omega_h} |\delta^k u|, \\ [u]^*_{k;\Omega} &= \sup_{\Omega' \subset \Omega} (d')^k [u]_{k;\Omega'}, \\ [u]^*_{k,\gamma;\Omega} &= \sup_{\Omega' \subset \Omega} (d')^{k+\gamma} [\delta^k u]_{0,\gamma;\Omega'}, \end{aligned} \tag{3.10}$$

where $k = 0, 1, 2, \dots,$ and $d' = \operatorname{dist}(\Omega', \partial\Omega) > kh$. We also need to define the quantity

$$a_0 = \frac{1}{2\lambda} \max_{\Omega_h \times \mathbb{R}^{Y_N}} \sum_y \frac{\partial F}{\partial q_y}, \tag{3.11}$$

corresponding to (2.9).

Theorem 3.1. *Let Ω be a bounded domain in $\mathbb{R}^n$ and u a solution of the difference equation* (1.6) *in Ω_h. Then we have the estimate*

$$[u]^*_{2,\gamma;\Omega} \leq C$$

for any $\gamma < \alpha$, where $\alpha > 0$ is a constant depending on n, a_0, N, Λ/λ_0 and C is a constant depending additionally on μ/λ, γ, $|u|_{0;\Omega}$ and $\operatorname{diam}\Omega$.

The special case when $L_1, \ldots, L_K$ are independent of x is proved in [17]. It extends earlier work of Holtby [4, 5] where the operators L_i are pure second-order differences. The proof of Theorem 3.1 is through a perturbation from the frozen case (3.1), (as in the continuous case [19, 20, 24]), which is treated in [17] using the results for linear equations [14] and an idea from [23]. The discrete Schauder estimates for linear equations (1.2)[21] will also be a special case of Theorem 3.1, which also extends to embrace lower-order dependence.

References

[1] L.A. Caffarelli and X. Cabré, **Fully Nonlinear Elliptic Equations**, Amer. Math. Soc., 1995.

[2] L.C.Evans, *Classical solutions of fully nonlinear, convex, second-order elliptic equations*, Comm. Pure Appl. Math., **35** (1982), 333–363.

[3] D. Gilbarg and N.S. Trudinger, **Elliptic partial differential equations of second order**, 2nd edition, Springer Verlag, 1983.

[4] D. Holtby, *Higher-order estimates for fully nonlinear difference equations I*, Proc. Edinb. Math. Soc. **43** (2000), 485–510.

[5] ——, *Higher-order estimates for fully nonlinear difference equations II*, Proc. Edinb. Math. Soc. **44** (2001), 87–102.

[6] J.S. Hwang, *Rational approximation to orthogonal bases and of the solutions of elliptic equaitons*, Numer. Math. **81** (1999), 561–575.

[7] M. Kocan, *Approximation of viscosity solutions of elliptic partial differential equations on minimal grids* Numer. Math. **72** (1995), 73–92.

[8] N.V. Krylov, *Boundedly inhomogeneous elliptic and parabolic equations*, English translation: Math. USSR Izv., **20** (1983), 459–492.

[9] N.V. Krylov and M. Safonov, *A property of the solutions of parabolic equations with measurable coefficients*, Izv. Akad. Nauk SSSR Ser. mat., **44** (1980), 161–175 (in Russian); English translation in Math. USSR Izv. **16** (1981), 151–164.

[10] T. Kunkle, *Lagrange interpolation on a lattice: bounding derivatives by divided differences* , J. Approx. Theory, **71** (1992), 94–103.

[11] H.J. Kuo and N.S. Trudinger, *Linear elliptic difference inequalities with random coefficients*, Math. Comp. **55** (1990), 37–53.

[12] ——, *Discrete methods for fully nonlinear elliptic equations*, SIAM J. Numer. Anal. **29** (1992), 123–135.

[13] ——, *Maximum principles for difference operators*, Topics in Partial Differential Equations & Applications: Collected papers in Honor of Carlo Pucci, Lecture Notes in Pure and Applied Mathematics Series/**177**, 209–219, Marcel Dekker, Inc. 1996.

[14] ——, *Positive difference operators on general meshes*, Duke Math. J. **83** (1996), 415–433.

[15] ——, *Evolving monotone difference operators on general space-time meshes*, Duke Mathematical Journal, Vol. 91, (1998), 587–607.

[16] ——, *A note on the discrete Aleksandrov-Bakelman maximum principle*, Taiwanese Journal of Mathematics, **4** (2000), 55–67.

[17] ——, *Schauder estimates for fully nonlinear elliptic difference operators*, Proc. Roy. Soc. Edin. 132A, 1395–1406 (2002).

[18] T.S. Motzkin and W. Wasow, *On the approximation of linear elliptic differential equations by difference equations with positive coefficients*, J. Math. Phys. **31** (1953), 253–259.

[19] M.V. Safonov, *On the classical solution of nonlinear elliptic equations of second order*, Izv. Akad. Nauk SSSR, Ser. Mat. **52** (1988), 1272–1287. (Engish transl.: Math. USSR Izvestiya) **33** (1989), 597–612.

[20] M.V. Safonov, **Nonlinear elliptic equations of second order**, Lecture Notes Univ. di Firenze, 1991.

[21] V. Thomée, *Discrete interior Schauder estimates for elliptic difference operators*, SIAM J. Numer. Anal. **5** (1968), 626–645.

[22] N.S. Trudinger, *Local estimates for subsolutions and supersolutions of general second order elliptic quasilinear equations*, Inventiones Mathematicae, **61** (1980), 67–79.

[23] N.S. Trudinger, *Regularity of solutions of fully nonlinear elliptic equations*, Boll. UMI, **3** (1984), 421–430.

[24] N.S. Trudinger, **Lectures on nonlinear elliptic equations of second order**, Lectures in Mathematical Sciences, **9**, Univ. Tokyo, 1995.

Hung-Ju Kuo
Department of Applied Mathematics
National Chung-Hsing University
Taichung 402, Taiwan
e-mail: `kuohj@nchu.edu.tw`

Neil S. Trudinger
Centre for Mathematics and its Applications
Australian National University
Canberra, ACT 0200, Australia
e-mail: `neil.trudinger@anu.edu.au`

GPSR Compliance
The European Union's (EU) General Product Safety Regulation (GPSR) is a set of rules that requires consumer products to be safe and our obligations to ensure this.

If you have any concerns about our products, you can contact us on

ProductSafety@springernature.com

In case Publisher is established outside the EU, the EU authorized representative is:

Springer Nature Customer Service Center GmbH
Europaplatz 3
69115 Heidelberg, Germany

www.ingramcontent.com/pod-product-compliance
Lightning Source LLC
Chambersburg PA
CBHW071241230926
51849CB00001B/185
* 9 7 8 3 7 6 4 3 7 1 6 5 4 *